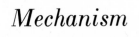

Mechanism

Mechanism

Virgil Moring Faires
PROFESSOR OF MECHANICAL ENGINEERING
UNITED STATES NAVAL POSTGRADUATE SCHOOL

Robert McArdle Keown
LATE ENGINEER, INDUSTRIAL COMMISSION OF WISCONSIN
AND ASSOCIATE PROFESSOR OF MACHINE DESIGN
UNIVERSITY OF WISCONSIN

FIFTH EDITION

McGRAW-HILL BOOK COMPANY, INC.

New York   Toronto   London

1960

MECHANISM

THE MAPLE PRESS COMPANY, YORK, PA.

19899

Preface

This revision has been taken largely from a somewhat more complete work by the author entitled "Kinematics." It is designed to fit into the same situations as the previous editions of "Mechanism," but in view of the many years since the fourth edition was published, it is necessarily a significant departure from the old. Some ideas new to this edition as compared with the old include the concept of jerk, especially as applied to cams; cycloidal motion; an improved treatment of cam design that encourages the student to think through the layouts; methods of computing the pressure angle in cams; more on the design of Geneva mechanisms; roller-gear drives; more extensive treatment of relative velocities and relative accelerations; equivalent mechanisms for acceleration analysis; more complete analysis of gear-tooth action, including methods of computing the contact ratio, the amount of addendum on the gear for no interference, and the number of pinion teeth for no interference; additional design information for helical gears on nonparallel axes; a method of finding tooth numbers for a close approach to an exact velocity ratio; planetary trains by relative velocity, by formula, and by tabulation; and more on universal joints. Since the book is completely rewritten, there are other changes and modernizations too numerous to be mentioned.

It should be noted that most of the problems are new, or at least different. A noteworthy feature is that most of those intended for graphical solution have the location and the details of the mechanism so defined that they can be laid out by inexperienced student labor. (I'll ask forgiveness now for those that are poorly defined.) With a minimum of supervision by the teacher, a student or other draftsman can prepare master copy for lithoprint, mimeograph, or ditto, making it possible to

v

present the class with prepared layouts, thus saving valuable time for learning more kinematics.   Having such layouts, the instructor can use the same ones for many different problems of his own design.   In any event, a set of prepared layouts, and there are several on the market, is well worth the cost in terms of learning.   I have referred in this text to the set by Professor R. M. Wingren ("Kinematic Problems," Prentice-Hall, Inc.).   The advantage of problem sheets that are $8\frac{1}{2}$ by 11 in. in size is that they are entirely suited for homework.   Saving time on homework is as advantageous as saving time during a class drafting period.

Acknowledgments belong mostly to the numerous authors of technical papers; boldface numbers in parentheses identify the paper, and the list is in the back of the book.

I shall appreciate reports on errors found by the reader and suggestions for future changes.

Virgil Moring Faires

Monterey, California
January, 1960

Contents

vii

Contents

Symbols and Abbreviations

The symbols agree in general with the recommendations of the American Standards Association (ASA). The insignificant differences are ones that the author feels are justified by virtue of the overlapping or conflict of various standards, or the limited number of letters in the English and Greek alphabets.

a	linear acceleration; addendum in gearing; a_Q or a_q, acceleration of point Q; a_h, a_e, accelerations during harmonic and cycloidal motions respectively; etc.
a^n, a^t	normal aceleration, tangential acceleration; a^n_q, normal acceleration of point Q; etc.
B	a constant.
C	a constant; center distance.
c	clearance.
D	diameter.
d	sometimes distance or diameter.
e	train value.
F, F_1	forces.
f	coefficient of friction; frequency.
g	acceleration of gravity.
h	height; a dimension or distance.
I	moment of inertia.
j	jerk or pulse (da/dt).
K	scale, as of a drawing or spring; K_s, space scale; K_v, speed or velocity scale; K_a, acceleration scale; K_j, jerk scale.
L	length; stroke of piston or slider; total movement of follower.

M_a	mechanical advantage.
m	mass (in slugs).
m_c	contact ratio (gearing).
m_ω	angular velocity ratio.
N	number of anything; number of teeth in a gear; N_p, number of teeth in pinion; etc.
n	angular velocity, usually rpm.
P	name of point; pitch.
P_b	base pitch (gearing).
P_c	circular pitch, inches.
P_d	diametral pitch.
Q	name of point.
r	radius.
s	displacement; distance; $\dot{s} = ds/dt$; $\ddot{s} = d^2s/dt^2$.
T	period of harmonic motion; torque; total time.
t	time.
v	speed; velocity; v_P or v_p is speed of point P, v_Q or v_q is speed of Q, etc.
W	load; force of gravity.
x, y, z	coordinates of a point, displacements in the x, y, and z directions.
Z	length of line of action (gearing).
α (alpha)	angular acceleration; angle of approach.
β (beta)	an angle; cam angle for total throw of follower; angle of recess.
γ (gamma)	pitch angle (bevel gears).
δ (delta)	an angle; pitch angle, spur gears; dedendum angle, bevel gears.
Δ (delta)	indicating a change of something, as speed or velocity (Δv).
ϵ (epsilon)	an angle.
θ (theta)	angular displacement; angle of action, gearing.
λ (lambda)	lead angle, in gearing and screws.
π (pi)	$3.1416 \cdots$.
ρ (rho)	variable radius.
Σ (sigma)	summation sign; shaft angle.
ϕ (phi)	an angle; pressure angle, cams and gearing.
ω (omega)	angular velocity, usually radians per unit time; ω_B or ω_b, is the angular velocity of link B. (Use upper case when lower case would be confusing, as in the study of velocities by centro method.)
AGMA	American Gear Manufacturers Association.
ASA	American Standards Association.
ASME	American Society of Mechanical Engineers.

CC	counterclockwise.
CL	clockwise.
cpm	cycles per minute.
cps	cycles per second.
DRD	dwell-rise-dwell.
DRRD	dwell-rise-return-dwell.
fpm	feet per minute.
fps	feet per second.
fps²	feet per second-second
ips	inches per second.
ips²	inches per second-second.
LDH	long dimension horizontal.
LH	left hand.
ln	logarithm to base e.
LOC	line of centers.
log	logarithm to base 10.
mph	miles per hour.
rev.	revolution.
RH	right hand.
rpm	revolutions per minute.
rps	revolutions per second.
RRR	rise-return-rise
SAE	Society of Automotive Engineers.
SHM	simple harmonic motion.
UARM	uniformly accelerated and retarded motion.

1

Equations of Plane Motion

1. Introduction. ***Kinematics,*** a branch of physics, is the science of motion. The study of the motion and related attributes of elements of machines is a phase of kinematics often called ***mechanism.*** This book should have the latter classification, namely, kinematics applied to situations of interest to the engineer, especially the graphical solution of problems on velocity and acceleration. Traditionally, the layout of cams (which is a motion problem) and the study of gear-tooth action are included in this text. The importance of kinematics applied to machines has been and is being more clearly appreciated. Indeed, there is such an active interest that one must go to current literature to learn of new applications. See the list of references at the end of the book, which is intended as an aid to further study. Boldface numbers (**1**) in parentheses designate a particular reference.

2. Historical Note. The laws of motion were investigated scientifically first by Galileo Galilei (1564–1642), who was born at Pisa to an impoverished noble family. Although he was educated for medicine, his interest shifted to the physical sciences. Most of his experimentation on motion was conducted in later life, although the famous one of two masses released from the Tower of Pisa was reputedly performed when he was twenty-six.

Aristotle had reasoned that heavy bodies fall faster than light bodies, a logical and plausible conclusion, and nearly everyone of Galileo's day believed that this was true. No one thought to question Aristotle's idea until, some 2000 years later, Galileo decided to drop simultaneously a half-pound stone and a 100-lb. cannonball from the Leaning Tower of Pisa. When the spectators observed that both objects struck the ground at the same time, they roundly hissed Galileo, thinking that perhaps he was practicing witchcraft. We all "know" so many things which are not so, and such a result was contrary to what the people "knew."

Galileo asked himself how bodies fell. His first assumption, that the velocity of a falling body was proportional to the distance traversed ($v = Ch$), was in error. However, he reasoned himself out of this notion, and decided that the velocity would be proportional to the time of descent ($v = Ct$, where C turns out to be the constant acceleration a; $v = at$). Finding no fallacy in this assumption, he proceeded to experimental proof. First, he decided to reduce the gravitational acceleration (and velocities within a particular time) by using an inclined plane.

Another problem was the matter of timing. The pendulum clock had not been invented; sand and water clocks were traditionally used. Galileo used a large vessel of water with a small orifice at the bottom over which he could hold his finger. The amount of water to run out upon the removal of his finger was practically proportional to the elapsed time because the drop in the water level was negligible during his experiments.

From his assumption ($v = at$) and using the definition of v ($v = s/t$), he derived the relation $s = at^2/2$. From this latter equation, he learned that in 1, 2, 3, 4, . . . units of time, the spaces traversed were proportional to 1, 4, 9, 16, . . . , etc. Marking these spaces on the inclined plane and using a rolling ball, Galileo soon had experimental verification of his law, thus introducing the conception of acceleration a. The discrepancies between his experimental results and the algebraic relations he attributed to friction and air resistance, although air pressure had not been established. The ramifications of the deductions that Galileo made from his experiments on the inclined plane are too numerous to give here (see E. Mach, "Science of Mechanics").

Galileo evolved the concept of inertia (Newton's first law), established the forms of beams of uniform strength, conceived the pendulum clock, rediscovered Archimedes' law of fluid buoyancy (which had been lost during the centuries), and made many astronomical discoveries.

3. Relation of Kinematics to Engineering. One has only to read the news of rockets to know that many kinds of scientists and engineers are concerned with problems of motion (kinematics). The machine designer (in the broad sense, a rocket is a machine) nearly always gets involved in kinematics, although often this part of his problem is simple. However, the kinematic aspects are frequently highly important in automatic machines, computers, instruments, controls, and in general in high-speed machinery, especially when combined with kinetics. On the way to placing kinematics in the scheme of things, we need a few definitions.

A **kinematic chain** is any connected group of elements, such as gears, links, and cams, whose parts have motions which may or may not be constrained (§59); but our dealings will be only with constrained chains.

A chain is constrained when the motion of one element relative to another induces certain definable *relative* motions of all the elements. A kinematic chain does not necessarily have any links stationary; nor is it necessary that the chain be intended for a machine, witness the popular folding chairs, some of which are quite ingenious.

A **mechanism** is a constrained kinematic chain in which one link is stationary or considered so during a motion analysis. For example, if an automobile is moving over a rough, curving, and hilly road, the motion with respect to the earth of a point on the crankshaft-and-piston mechanism of the engine would be quite complex, and it would take a tremendous effort to define it even approximately. However, with respect to the frame of the engine, it is easily defined in an ideal manner and it is with this definition that the motion of the mechanism would be studied. The motion of one element (a gear, link, or cam) in a mechanism compels a predictable and definable motion of each of the other elements in the mechanism—at least ideally. Actual machines are composed of elastic parts which deform under load. Moreover, there must be clearances here and there for one reason or another. These departures from the ideal rigid connections and parts often cause disconcerting differences between a motion as planned and as it actually occurs.

Reuleaux (**4**) defines a **machine** as *a combination of resistant bodies so arranged that by their means the mechanical forces of nature can be compelled to produce some effect or work accompanied by certain determinate motions.* Many machines consist of one or several mechanisms so arranged that their motions are coordinated to produce a desired result.

In the design of a machine, it often happens that a major effort is required to solve the kinematic problems involved. On the other hand in some machines, as in a simple vise, the motion problem is almost negligible once we have the machine's function in mind. If the machine consists of a group of mechanisms, the links, gears, etc., may be represented by lines and circles which can be manipulated to find an arrangement and proportions which will accomplish the desired result. A subsequent study of displacements, velocities, and accelerations may or may not be required. In either case, this phase of the design of a machine is an application of the principles of mechanisms and might be considered as the first phase of the design.

Since the elements of the actual machine must have substance, the second phase of the design might be to determine the forces acting on the various elements, so that each element may be given form and dimensions to resist these forces. This phase is an application of the principles of *analytic mechanics* (**2**) and *strength of materials*, together with the judgment of the experienced designer [sometimes called *machine design* (**5**) in college courses].

A third phase or aspect of the design of a machine has to do with dynamic effects, that is, with the effective forces which exist because of accelerations of bodies having mass (Newton's law, $F = ma$). This phase is an application of the science of *dynamics*. Of course, some of the dimensions found in the second phase may be adjusted as a result of the findings of the dynamic analysis. Indeed, these various phases are not likely to be separate and independent, but must be carried on more or less simultaneously, often with a trial-and-error approach.

There are many other phases of design (**5**), but this discussion will suffice to locate the application of kinematics in the general scheme. For pedagogical reasons, it usually seems necessary for the student to concentrate his energies on one subject at a time, but it will be rewarding for him always to be alert to the relations between subjects. Because of the modern trend toward higher and ever higher speeds of machines and toward automatic machines (to increase productivity and reduce overhead and labor cost per unit), the kinematic study is becoming more and more important. And we herewith get on with it.

4. Plane Motion. A body which is in the process of changing its position is said to be in motion. The change of position can be noted only with respect to the position of some other body which may or may not be at rest; that is, *motion is purely relative.* Two bodies may be at rest relative to each other but in motion relative to a third body. For example, the two front wheels of an automobile on a smooth, straight road have no motion relative to each other, but they have one motion relative to the *frame* of the car (the frame itself is a moving body), and another motion relative to the *road*.

In problems dealing with machinery, the motions of the various parts are usually taken with reference to the frame of the machine. This is not always the case, however, for sometimes it is easier to compare the motion of one part of a machine directly with that of some other moving part, as, for example, the number of revolutions that the cylinder of a printing press makes to each stroke of the knife for cutting off the sheets. Thus, we might logically presume that, in order for the machines to be useful, the motions of the various parts must be controlled or constrained; they must be related in a definable manner.

Although some machine parts have motions which are not planar, most engineering problems in kinematics are concerned with plane motion only, which shall be our principal concern. A *particle* in plane motion is moving at all times in a single plane and may move in any path, prescribed or unfettered. If a *body* is in plane motion, each point in the body is moving at all times in the same plane. Plane motion of a body in a machine is generally either **rotation** or **translation,** or a motion, such as that of a rolling wheel, that can be reduced to a combination of rotation and translation. *Rectilinear translation* occurs when each point in a body is moving in a straight line; it is the limiting case of rotation about a center at infinity. As we shall learn later, general plane motion can be reduced to a series of instantaneous rotations.

5. Vector Quantities. We shall usually be dealing with vector quantities, as opposed to scalar quantities. Scalar quantities, like energy, have magnitude only. If energy is flowing into a system, it does not matter from what direction the flow comes, the accumulation of energy within the system will be the sum of the various energy quantities. If forces act on a system, it is important to know (1) the **magnitude,** (2) the **sense** or direction, and (3) the **location** of its line of action, because force is a vector quantity and these are the three attributes of a vector. Forces are mentioned only incidentally in this book, but we shall deal extensively with displacement, velocity, and acceleration, all vector quantities.

6. Adding and Subtracting Vectors. Since the matter of direction is involved, vector quantities have **components** in directions other than that of the given vector. Thus, in Fig. 1(a), v_x and v_y are the horizontal and vertical components of the vector v. However, the x and y axes can be taken in any desired directions. When the components are at right angles, Fig. 1(a), usually the most convenient kind, they are called

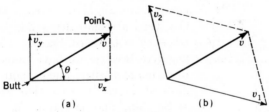

Fig. 1. Components of a Vector.

rectangular components. Two other components v_1 and v_2, chosen at random, are shown in Fig. 1(b). In any case, the components of a vector add vectorially to give a vector that is the **resultant** of the vectors being added. Any series of vectors may be said to be components of that vector which is the sum of the series. In Fig. 1, v is the resultant of v_x and v_y and also of v_1 and v_2.

The method of adding vectors is shown in Fig. 2. Given the vectors a_1, a_2, and a_3, as seen in Fig. 2(a), lay out these vectors to some convenient scale* as shown in Fig. 2(b), starting from any convenient point *O*, called a *pole*. To add vectors, place butt to point; the same sum is obtained no matter in what sequence the vectors are added. The angle θ

---

* An engineers' scale is too convenient in this work to be without.

that the resultant makes with the x direction in Fig. 2 is defined by

$$(a) \qquad\qquad \tan\theta = \frac{\Sigma a_y}{\Sigma a_x},$$

where the symbol Σa_x means the sum of the rectangular x components of all the individual a's (for example, a_{3x} is the x component of a_3) and Σa_y means the sum of the rectangular y components of all the individual

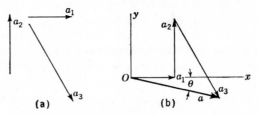

Fig. 2. Adding Vectors.

vectors (for example, a_{3y} is the y component of a_3). This sentence is interpreted in Fig. 3 where the vectors a_1, a_2, and a_3 have all been arbitrarily drawn through a common origin. Since a_1 is horizontal, it has no vertical component and $a_{1x} = a_1$. Since a_2 is vertical, it has no horizontal component and $a_{2y} = a_2$. Thus, the sum of the horizontal components Σa_x is $a_{1x} + a_{3x} = Oc$; the sum of the vertical components Σa_y is $a_{2y} - a_{3y} = Oh - Of = Ob = cd$, Fig. 3. The sign of a_{3y} is negative because we have taken the upward direction as positive and a_{3y} points downward. Now it is easily seen that $\tan\theta = cd/Oc = \Sigma a_y/\Sigma a_x$, and equation (a) is confirmed. For the simple case of rectangular components only as in Fig. 1(a), $\theta = \tan^{-1} v_y/v_x$, where θ is the angle the resultant makes with the x component. The foregoing vectors may be imagined to represent forces or displacements or any other vector quantities.

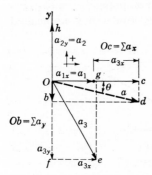

Fig. 3. Sense of Resultant.

There are various ways of indicating vector sums and differences. We shall use the signs $\nrightarrow$ and $\rightarrow$; thus

$$v = v_x \nrightarrow v_y \qquad \text{and} \qquad a = a_1 \nrightarrow a_2 \nrightarrow a_3.$$
$$\text{[\scriptsize FIG. 1]} \qquad\qquad\qquad \text{[\scriptsize FIG. 2]}$$

When these signs appear, it is to be understood that vector quantities are involved.

To subtract two vectors, they are placed point to point or butt to butt. In Fig. 4(a), the heavy vector Ae represents v_A; the vector Bf represents

v_B. Lay off from A a vector equal to v_B and find the vector $he = v_{A/B}$, which is the velocity of A minus the velocity of B; $v_{A/B} = v_A \rightarrow v_B$. If the difference of the vectors is taken in the opposite sense, Fig. 4(b), it means that the vector in the sense $dk = v_{B/A} = v_B \rightarrow v_A$. If the vectors represent velocities, the rule for deciding in which sense the difference vector points comes from the principle of relative velocities: *The velocity of any point, say P, is equal to the velocity of any other point, say Q, vectorially plus the velocity of P with respect to Q;*

(b) $v_p = v_q \nleftrightarrow v_{p/q}$ or $v_{p/q} = v_p \rightarrow v_q.$

The difference of two velocity vectors is by definition the velocity of one point (or particle) P relative to another point (or particle) Q. Thus we

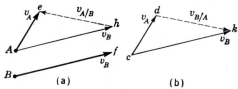

Fig. 4. Subtracting Vectors.

read the symbol $v_{p/q}$ as the velocity of P relative to Q and symbol $v_{q/p}$ is the velocity of Q relative to P.

In dealing with vectors, it is important to decide on the positive sense so that in algebraic sums, the terms can be given appropriate signs. If the rightward sense is positive, a vector which points toward the left is negative; if the upward sense is positive, a vector (or component) pointing downward is negative.

7. Speed and Velocity. The **speed** of a particle or point at any instant is the time rate with which it is traversing distance. Although the words *speed* and *velocity* are often used as though their meanings were identical, **velocity** is technically a vector quantity, possessing magnitude and **sense,** whereas speed is a scalar quantity. In other words, the speed is the magnitude of the velocity. This distinction is important because changes of velocity are important; and a change in the *sense (direction)* of a velocity is as truly a *change of velocity* as is a change in the *magnitude* of a velocity. Consider an airplane propeller as the engine warms up at constant angular speed. A point on the tip of the propeller is moving at constant *speed* in the path of a circle, but its velocity is continuously varying, since the direction of the velocity vector changes through 360° with every revolution of the propeller.

Let a point traverse a distance Δs during some time interval Δt, where the Greek delta (Δ) signifies a "change of." (For instance, Δx is always the second value of x minus the first value; $\Delta x = x_2 - x_1$.) Then the *average* speed of the point is

(1) $$v = \frac{\Delta s}{\Delta t} \qquad \text{or} \qquad s = vt$$
$$[s_1 = 0,\ t_1 = 0]$$

This equation gives the actual speed when the speed is constant. If the speed is not constant, one must know in detail how it varies in order to be able to find its instantaneous value. See § 10. Speed may be expressed in any convenient units of distance and time. In engineering work, the most common units are feet per second (fps), feet per minute (fpm), miles per hour (mph), inches per minute (ipm), and inches per second (ips).

8. Example—Converting Units. (a) Convert 960 yards per day to feet per per minute.
Solution.

$$v = \left(960\,\frac{\text{yd.}}{\text{day}}\right)\left(3\,\frac{\text{ft.}}{\text{yd.}}\right)\left(\frac{\text{day}}{24\,\text{hr.}} \times \frac{\text{hr.}}{60\,\text{sec.}}\right) = \frac{(960)(3)}{(24)(60)} = 2\ \text{fpm.}$$

(b) Convert 60 mph to feet per second.
Solution.

$$v = \left(60\,\frac{\text{mi.}}{\text{hr.}}\right)\left(5280\,\frac{\text{ft.}}{\text{mi.}}\right)\left(\frac{\text{hr.}}{3600\,\text{sec.}}\right) = \frac{(60)(5280)}{3600} = 88\ \text{fps.}$$

See the conversion numbers given on p. 23.

9. Example—Uniform Motion. Motion at constant speed is often called uniform motion. If an automobile driver takes 2 sec. to react and apply the brakes after an emergency arises, how far does he go if his speed is 60 mph during this reaction time?

Solution. From the previous example, we know that 60 mph is equivalent to 88 fps. Therefore the distance traveled, by equation (1), is $(2)(88) = 176$ ft. or 58.7 yd.

10. Acceleration. *Acceleration* a is the time rate of change of velocity. The velocity may change in either magnitude or direction (or both); hence the acceleration of a point (or particle) may be due to a change in the magnitude or in the direction, or in both the magnitude and direction, of the velocity. The acceleration a may be constant or variable.

A point in plane motion may move in a straight path, which is *rectilinear motion,* or in a curved path, which is *curvilinear motion.* If it moves rectilinearly, there is no acceleration due to change of direction since there is no change of direction. If a point moves in a curved path, we generally determine separately that part of the acceleration due to the change of direction (see § 21) and that part due to the change in magnitude of the velocity. The component arising from a change in magnitude of velocity is always in a direction tangent to the path and is called the *tangential acceleration.* Considering only the magnitude now, let Δv be the change in speed during a time interval Δt.   Then the *average* acceleration is

$$(2) \qquad a = \frac{\Delta v}{\Delta t} = \frac{v_2 - v_1}{t} \qquad \text{or} \qquad v_2 - v_1 = at,$$

in which $t_1 = 0$ and $\Delta t = t_2 = t$.   This equation gives the actual "speed-acceleration" when the acceleration is constant.   If the acceleration is not constant, one must know in detail how it varies in order to be able to find its instantaneous value.   If a point is moving in a curved path, equation (2) gives the tangential acceleration.

11. Equations and Graphs for Constant Acceleration.   *If the acceleration is constant*, its plot with coordinates a and t is a straight line of zero slope, $a = C$, Fig. 5(a).   Moreover, the curve of v against t is a straight line, as seen from equation (2) and as illustrated in Fig. 5(b) where the

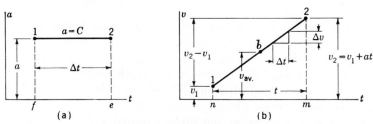

Fig. 5. Constant Acceleration.

constant slope is seen to be $\Delta v/\Delta t$, where Δv is the change of speed during the time Δt.   When the acceleration vector has the same sense as the velocity vector, the point is speeding up; this is the situation pictured in Fig. 5, where $v_2 > v_1$.   When the acceleration vector is in the opposite sense to the velocity vector, the point is slowing down; in this case, the acceleration is often called the **deceleration** or **retardation** (but not necessarily).

Observe in Fig. 5 that the average (mean) speed is $(v_2 + v_1)/2$ and substitute this value for the *average* velocity v in equation (1):

$$(c) \qquad v = \frac{s}{t} = \frac{v_2 + v_1}{2} \qquad \text{or} \qquad v_2 + v_1 = \frac{2s}{t}.$$

Change the signs of all terms in equation (2) and add the result to the foregoing equation as follows:

$$v_2 + v_1 = \frac{2s}{t}$$
$$-v_2 + v_1 = -at$$
$$2v_1 = \frac{2s}{t} - at,$$

from which

(3)
$$s = v_1 t + \frac{at^2}{2} \quad \text{and} \quad s = \frac{at^2}{2}$$
$$[v_1 = 0]$$

This equation gives the distance traversed by a point during a time interval t when it has a *constant* acceleration a and an initial speed v_1.

To get another useful relation for motion with constant acceleration, eliminate t from equations (2) and (c):

$$t = \frac{v_2 - v_1}{a} = \frac{2s}{v_2 + v_1},$$
$$\text{[FROM (2)]} \quad \text{[FROM (c)]}$$

from which

(4)
$$v_2{}^2 = v_1{}^2 + 2as \quad \text{and} \quad v_2{}^2 = 2as.$$
$$[v_1 = 0]$$

The most common unit of acceleration in engineering is feet per second-second (fps²); other units include inches per second-second (ips²), feet per minute-minute (fpm²), miles per hour-second. For a freely falling body in a vacuum and near the surface of the earth, equation (4) becomes $v^2 = 2gh$, where $g = 32.174$ fps² (use 32.2 on the slide rule) and h ft. is the vertical distance through which the body falls.

There are some interesting and useful properties of diagrams such as those of Fig. 5. (1) We have already noted that the slope of the vt

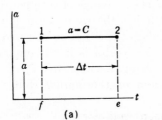

(a)

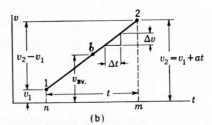

(b)

Fig. 5. Repeated.

curve is equal to the acceleration; $a = \Delta v / \Delta t$. This is true even when the slope (and acceleration) is not constant. If the slope varies, its value at a particular point is to scale the instantaneous acceleration at that point. (2) Observe that the area 12*ef*, Fig. 5(a), is $a \Delta t$, which from equation (2) is seen to be the change of speed, $\Delta v = a \Delta t$. Thus, the area "under" the at curve represents to some scale the change of speed and this is true whether or not the acceleration is constant. If the scales to which time and acceleration have been plotted are, respectively, K_t sec./in. (as 6 sec. to the inch) and K_a ft./sec²-in. (as 20 fps² to the inch), the scale of the area is $(K_t \text{ sec./in.})(K_a \text{ fps}^2/\text{in.}) = K_v$ ft./sec-in.², the scale of the speed.   (3) From equation (1), $v = \Delta s/\Delta t$, the definition of average speed, we see that if the st diagram for constant speed were drawn, its slope would be to scale this constant speed.   It is true that even if the speed varies, the slope of the st curve at a particular point is to scale the instantaneous speed at that point.   (4) Note that the area $12mn$ "under" the vt curve is the average height $v_{av} = (v_1 + v_2)/2$ times $\Delta t = mn$; or $t(v_1 + v_2)/2$, in which t is the duration Δt of time.   Comparing this expression with equation (c), we see that it represents the distance traversed; $s = t(v_1 + v_2)/2$.   Thus, areas "under" the vt curve represent to some scale the distance moved by the point under consideration and this is true whether or not the vt curve is a straight line.   If the speed is plotted to a scale of K_v ft./sec-in. and time to a scale of K_t sec./in., the scale of the area is $(K_v$ ft./sec-in.$)$ $(K_t$ sec./in.$)$ = K_s ft./in.^{2}.   To summarize: the area under the at curve represents a change of speed; the area under the vt curve represents a change of s; the slope of the st curve represents speed; the slope of the vt curve represents acceleration.

12. Example—Constant Acceleration.   A body is projected vertically upward with an initial velocity of $v_1 = 70$ fps.   (a) What is its velocity and displacement after 4.22 sec.?   (b) What is its velocity when its displacement is 60 ft. above the starting point?   What is the elapsed time?   Neglect air resistance.

Solution.   (a) Choose the upward direction as positive; then the acceleration is the acceleration of gravity, $a = -g = -32.2$ fps^2 if the air resistance is negligible.   From equation (2), the definition of acceleration, we have

$$v_2 - v_1 = at \qquad \text{or} \qquad v_2 - 70 = -(32.2)(4.22),$$

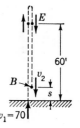

Fig. 6.

from which $v_2 = -65.9$ fps.   The minus sign indicates that the velocity is *downward;* that is, the body has gone to the top of its motion and is now moving down, as at B, Fig. 6.   The displacement is obtained from equation (3);

$$s = v_1 t + \frac{at^2}{2} = (70)(4.22) + \frac{-(32.2)(4.22)^2}{2} = 8.7 \text{ ft.,}$$

or the body is almost back to its starting point.   Since the sign of s is positive, it is measured upward.

   (b) To get the velocity directly from the displacement, use equation (4).

$$v_2{}^2 = v_1{}^2 + 2as$$
$$= 70^2 + (2)(-32.2)(60) = 1036,$$

from which $v_2 = \pm 32.2$ fps, the signs meaning that the velocity may be in either direction, on the way up or on the way down, as at E, Fig. 6.   To get the time at $s = 60$ ft., use the st relation, equation (3).

$$s = v_1 t + \frac{at^2}{2} = 70t + \frac{(-32.2)t^2}{2} = 60,$$

from which $t = 1.174$ sec. (going up) and $t = 3.174$ sec. (going down).

13. Example. A cam follower is moving in a manner that the acceleration changes with time as shown in Fig. 7(a). At the instant that $t = 0$ in Fig. 7(a), the velocity of the follower is 1 ips vertically upward. Determine the velocity of the follower at station 10 where $t = 1$ sec. and also its displacement from its position at station 0.

Solution. From $\Delta v = a\Delta t$, we know that the change of velocity is the area between the *at* curve and the $t = 0$ axis. Thus from station 0 to 4, $\Delta v = (15)(0.4)$ $= 6$ ips, which, added to $v_0 = 1$ gives $v_4 = 7$ ips. The construction of the *vt* curve, Fig. 7(b), can now be started. Lay off $0k = 1 = v_0$ at station 0 to some convenient scale, locating the beginning at the *vt* curve. Inasmuch as the acceleration is constant, the slope of the *vt* curve is constant (the *vt* curve is a straight line). Hence, lay off $4m$, Fig. 7(b), equal to $v_4 = 7$, and draw line km. During the next phase of the motion, the acceleration is negative $a_4 = -5$ ips² (that is, its vector is pointing down) and the velocity is decreasing. The change of velocity is area $46ed = -(0.2)(5) = -1$ ips; hence, $v_6 = v_4 - 1 = 7 - 1 = 6$ ips. Plot $6n = 6$ in Fig. 7(b) and draw the straight line $(a = C)$ mn.

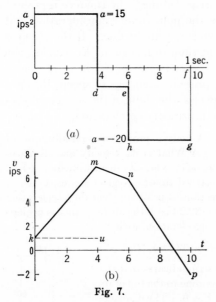

(a)

(b)

Fig. 7.

Since the acceleration continues negative, the area $6hgf = -(0.4)(20) = -8$ ips and $v_{10} = v_6 - 8 = 6 - 8 = -2$ ips. The negative sign means that the follower has reached the top of its movement (when $v = 0$) and is now moving downward. Plot $tp = -2$ and draw np, which completes the *vt* diagram as far as the problem goes.

To get the displacement of the follower as measured from station 0, we recall that areas on the *vt* diagram represent displacement, that the area of a triangle is *half the base times the altitude,* and that the area of a trapezoid is *half the sum of the bases times the distance between bases.* Thus,

$$\Delta s_{0-4} = \text{area } 0ku4 + \text{area } kum = (0.4)(1) + (0.2)(6) = 1.6 \text{ in.}$$

Similarly,

$$\Delta s_{4-6} = \left(\frac{7+6}{2}\right)0.2 = 1.3 \text{ in.,}$$

$$\Delta s_{6-9} = \frac{(6)(0.3)}{2} = 0.9 \text{ in.,}$$

$$\Delta s_{9-10} = -\frac{(2)(0.1)}{2} = -0.1 \text{ in.,}$$

from which $\Delta s_{0-10} = 1.6 + 1.3 + 0.9 - 0.1 = 3.7$ in. You might note that the maximum displacement from station 0 is 3.8 in.

14. Jerk.   Relatively recently, we have become aware of the significance of the time rate of change of acceleration, the *average* value of which
is $\Delta a/\Delta t$, especially with respect to cams (**7, 8**) and from the standpoint
of human comfort.   A descriptive name for this characteristic is **jerk**
(also called *pulse* and other names), because when its value is large, there
is occurring a rapid change of acceleration (and force, because $F = ma$).
A human being subjected to a high value of $\Delta a/\Delta t$ has a sensation of being
jerked.*   When the displacement of a point can be expressed mathematically as a function of time, the jerk is usually easily found by the
calculus, a step beyond the scope of this book.   In any event, the slope
of the *at* diagram is to scale the jerk ($\Delta a/\Delta t$), and if the slope varies, the
slope at a particular point represents the instantaneous value of the jerk
at that point.

15. Angular Motion.   In thinking of linear velocity and acceleration,
concentrate on a certain *point* in a body (if the body is moving with
rectilinear motion only, all points in it have the same velocity).   However, a body with finite dimensions may undergo angular displacement θ,
which should be thought of as the angle through which a *line* attached to
the body turns.   By definition and logical deductions, equations analogous to those above for linear motion may be obtained for angular motion
when the angular acceleration α is constant; to wit,

$$(5) \qquad \omega = \frac{\Delta\theta}{\Delta t} \qquad\qquad \text{or} \qquad\qquad \theta = \omega t;$$
$$[\theta = 0, t_1 = 0]$$

$$(6) \qquad \alpha = \frac{\Delta\omega}{\Delta t} = \frac{\omega_2 - \omega_1}{t} \qquad \text{or} \qquad \omega_2 - \omega_1 = \alpha t;$$

$$(7) \qquad \theta = \omega_1 t + \frac{\alpha t^2}{2} \qquad \text{and} \qquad \theta = \frac{\alpha t^2}{2};$$
$$[\omega_1 = 0]$$

$$(8) \qquad \omega_2{}^2 = \omega_1{}^2 + 2\alpha\theta \qquad \text{and} \qquad \omega_2{}^2 = 2\alpha\theta;$$
$$[\omega_1 = 0]$$

in which the angular velocity ω is the time rate of change of angular displacement θ, the angular acceleration α is the time rate of change of
angular velocity ω.   Compare these equations with the similar ones for
linear units.   The same limitations apply.   When the words speed,
acceleration, etc., are used without an adjective, the context should make
clear whether linear or angular is meant; usually when these words stand
alone, linear values are intended.

---

* When a driver "guns" a car from low speed or stops suddenly, the passengers sway
suddenly in the opposite sense of the acceleration.   When a driver begins to slow
down, the jerk may be kept low by a gradual application of the brake and the pas-
senger's muscles have time to counteract the inertia forces, with the result that they
experience small jerk and sway little.

The units in equations (5) to (8) may be any combination of angular measure and time. The angle θ may be measured in degrees, radians, or revolutions and the time t in seconds, minutes, hours, etc.; common units for angular velocity are radians per minute (rad./min.), radians per second (rad./sec.), revolutions per minute (rpm) and per second (rps). A **radian** is the angle subtended by an arc of a circle which is equal in length to the radius of the circle. The entire circumference subtends an angle of 2π rad.; there are $360/(2\pi) = 57.29°$ per radian.

16. Relations between Linear and Angular Motion Equations. Recalling that the arc length s is equal to the radius r times the subtended angle θ in radians ($s = r\theta$), we may easily find other relations between linear and angular measures. Using $\Delta s = r\,\Delta\theta$, Fig. 8, applied to the special case of a constant radius r, we see that

(d) $$v = \frac{\Delta s}{\Delta t} = r\frac{\Delta\theta}{\Delta t} = r\omega \qquad \text{or} \qquad v = r\omega.$$

Then

(e) $$a = \frac{\Delta v}{\Delta t} = r\frac{\Delta\omega}{\Delta t} = r\alpha \qquad \text{or} \qquad a = r\alpha.$$

While the foregoing conclusions were obtained from equations of average speed and acceleration, they nevertheless are valid for instantaneous situations, as can be proved.

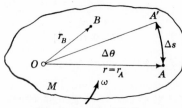

It is common practice to use ω for angular velocities in radian units and n for angular velocities in revolutions. Since there are 2π rad./rev., we have

Fig. 8. Angular Motion.

(9) $$\omega = 2\pi n,$$

in which the time unit is usually either seconds or minutes. Note that since all lines in a rigid rotating body move through the same angle during the same time, they all have the same angular velocity and angular acceleration. The following equation is very useful in finding the linear speed of a point or radius r in a body that is rotating:

(10) $$v = r\omega = 2\pi r n.$$

Note that the linear speeds of points in a rotating body are proportional to their radii (distances from the center of rotation); as in Fig. 8, $v_A/v_B = r_A/r_B$.

17. Velocity Ratio. Whenever two bodies have angular motion relative to each other, we often find it convenient to use a term called the *velocity ratio* m_ω, defined as the angular velocity of the driver divided by the angular velocity of the driven member; that is,

$$(f) \qquad m_\omega = \frac{\omega \text{ (driver)}}{\omega \text{ (driven)}} = \frac{n \text{ (driver)}}{n \text{ (driven)}}.$$

This term may be applied whenever it is convenient; however, it is very commonly applied to gear trains and to other situations where the relation between angular velocities is significant. For example, if we speak of the velocity ratio between an automobile engine and the rear wheels, we mean the turns of the engine per turn of the rear wheels. Notice that if $m_\omega > 1$, the angular velocity of the driver is greater than that of the driven; and if $m_\omega < 1$, then ω (driver) $< \omega$ (driven).

Let the two wheels of Fig. 9 be rolling on each other without slipping, as in the case of two gears, with A driving B. Since there is no slipping at the point of contact E, we know that the tangential speed $v_P = v_Q$, where P and Q are points on the circumferences of A and B, respectively. But $v_P = r_P\omega_A$ $= (D_A/2)\omega_A$ and $v_Q = r_Q\omega_B = (D_B/2)\omega_B$. Equating these two speeds, we find the ratio of ω_A/ω_B which is m_ω, or

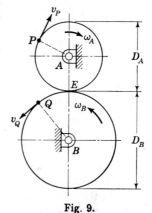

Fig. 9.

$$(g) \qquad m_\omega = \frac{\omega_A}{\omega_B} = \frac{n_A}{n_B} = \frac{D_B}{D_A},$$

the velocity ratio of two wheels running in contact with *no slipping*. In other words, the angular velocities of two wheels rolling on each other are inversely proportional to their diameters (or radii). It is customary in engineering to state the size of wheels and other cylindrical bodies as the diameter in inches; a 10-in. pulley is 10 in. in diameter. The reason for this custom is that diameters somehow or other can be measured, whereas radii are often difficult if not impossible to measure satisfactorily.

18. Slider-crank Mechanism. One of the most useful of linkage arrangements is the four-link system called the slider-crank mechanism, Fig. 10(a). It represents kinematically any type of reciprocating engine, such as the automotive engine or Diesel engine, Fig. 10(b), and also a host of similar linkages on other machines. In the Diesel engine, link D represents the piston (just a slider in general) and the pin cd is called the **wrist pin;** link C is the connecting rod, pin bc is the **crankpin,** link B is the crank, pin ab represents the crankshaft; the short marks, as at A, indicate the frame of the machine, or in general the reference link.

Observe that when the crank makes a complete revolution, the slider makes two strokes, to and fro. If the stroke of the slider is S ft, and the

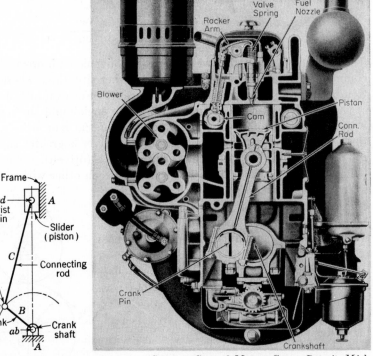

Fig. 10. Slider-crank Mechanism and Diesel Engines.   In a 6-cylinder engine, for example, there are six slider-crank mechanisms as in (a), representing the six pistons, etc.   In studying motions, it is much easier to work with a line sketch than with the drawing of the machine.   All the piston-crank mechanisms in a particular engine have the same motion—but not at the same time.   The engine shown in (b) is a supercharged (see blower), two-stroke-cycle Diesel engine.   The intake valve itself is not in sight, being behind the fuel nozzle and under the spring.

engine turns n rpm, the slider therefore moves at the *average* speed of $2Sn$ fpm.

19. Example.   An automobile engine is 3×3.2 in. (always bore by stroke) in size.   The car, with 30-in. tires (always diameter), is traveling 60 mph.   The velocity ratio, engine to rear wheels, is 3.8 in high gear.   Compute (a) the angular speeds of the wheels and engine, (b) the crankpin speed relative to the frame, (c) the average piston speed relative to the frame, (d) the number of revolutions the engine would make in 50,000 mi. of high-gear driving.

Solution.   (a) If the wheels are not slipping, the speed of a point on the surface about the center of the wheel is the same as the car speed, to wit, 60 mph or $(60)(5280)/(60) = 5280$ fpm; also it is $\pi D n_w$; therefore,

$$v = \pi D n_w = \pi \tfrac{30}{12} n_w = 5280,$$

from which the angular speed of the wheels is $n_w = 673$ rpm. With $m_\omega = 3.8$, the speed of the engine is

$$n_c = (3.8)(673) = 2557 \text{ rpm}.$$

(b) The absolute velocity of the crankpin, that is, relative to the ground, changes direction throughout an engine revolution, but its speed relative to the frame of the car is constant when the speed of the car is constant.

$$v_p = \pi D n_e = \pi \left(\frac{3.2}{12}\right) (2557) = 2140 \text{ fpm}.$$

The diameter D of the crankpin circle is equal to the stroke.

(c) The piston speed relative to the frame varies from zero at the ends of the stroke to a maximum value near the middle of the stroke. Means of finding this speed at any position of the piston is explained in detail later. At this stage, the average speed can be computed as

$$v_p = 2Sn_e = (2) \left(\frac{3.2}{12}\right) (2557) = 1364 \text{ fpm}.$$

The average piston speed of automotive engines is often used as an index of performance.

(d) The revolutions of the wheel per mile are $5280/(\pi 30/12) = 673$, and of the engine, $(3.8)(673) = 2557$ rev./mi. In 50,000 mi., the revolutions of the engine are

$$(25.57)(5 \times 10^6) = 128 \times 10^6 \text{ rev}.$$

20. Curvilinear Motion. A point moving in the path of a plane curve which is not a straight line is said to have curvilinear motion. Since the velocity of the point at any instant is in the direction of the motion at that instant, the velocity of a point in a curved path is in the direction of a tangent to the path at the instantaneous position of the point. In short, the velocity vector is always tangent to the path of the point.

A point may move in a curved path with a constant *speed*, but inevitably its *velocity* varies because the direction in which the point travels varies. This variation of the *direction* of the velocity results in an acceleration, even with constant speed—an acceleration which we term the **normal acceleration,** because its direction is normal to the path.

21. Tangential and Normal Accelerations. Consider a point moving in the curved path shown in Fig. 11. Let the two positions A and B, defined by radius vectors ρ_A and ρ_B and the angles θ and $\theta + \Delta\theta$, be a distance Δs apart. Let the speed of the point at A be v, and at B, $v + \Delta v$. And let the time taken by the point to move from A to B be Δt. Remembering that the *change* of velocity is that vector which when added to the first velocity gives the second velocity, we may construct the vector diagram in Fig. 11(b). Thus, with a change of velocity equal to the vector EF, the average acceleration is $EF/\Delta t$, since this change EF occurs

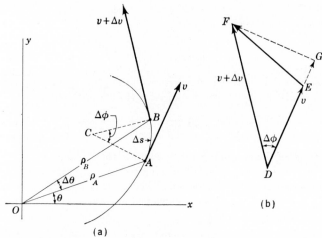

(a)

(b)

Fig. 11. Normal Acceleration. Point C is the center of curvature of an arc Δs. In the limit, as Δt approaches zero, point C is the center of curvature at point A.

in the time Δt; thus

$$a = \frac{\text{vector } EF}{\Delta t}.$$

We find it convenient to use components EG and GF of the vector EF; substitute $EG \nrightarrow GF$ for EF and get

$$(\boldsymbol{h}) \qquad a = \frac{EG}{\Delta t} \nrightarrow \frac{GF}{\Delta t}.$$

The vector EG, when angle $\Delta\phi$ is very small is the change in speed, $v + \Delta v - v = \Delta v$, Fig. 11(b); hence, the first term of $(\boldsymbol{h})$ is

$$(11) \qquad a^t = \frac{\Delta v}{\Delta t}$$

and is called the tangential acceleration (indicated by the superscript ^{t}), because it is always directed tangent to the path of motion at a particular instant. The other component GF is

$$GF = (v + \Delta v) \sin \Delta\phi$$

from triangle DGF, and the corresponding acceleration is the normal component

$$a^n = \frac{(v + \Delta v) \sin \Delta\phi}{\Delta t}.$$

If the time Δt is taken very small, then $\Delta\phi$ is quite small, so that

$$\sin \Delta\phi = \Delta\phi,$$

and the product of two very small numbers $\Delta v \, \Delta\phi$ becomes negligible (or Δv becomes negligible compared with v); thus the foregoing expression becomes

(12) $$a^n = v \frac{\Delta \phi}{\Delta t} = v\omega \qquad \text{and} \qquad a^n = r\omega^2 = \frac{v^2}{r},$$

in which the last two forms are obtained by using $v = r\omega$; v is tangent to
the path and perpendicular to the radius of curvature of the path $r = AC$,
Fig. 11. The normal acceleration is *always* directed toward the center
of curvature of the path being followed by the point. Thinking of the
time Δt as being an instant, we see that points A and B coincide and that
all the foregoing expressions can be considered as instantaneous values.
Equation (11) becomes

(11a) $$a^t = \frac{\Delta v}{\Delta t} = r \frac{\Delta \omega}{\Delta t} = r\alpha$$

where α is the instantaneous angular acceleration of the radius vector
$r = AC$ to A, the point whose tangential acceleration is a^t.

The tangential and normal accelerations, always at right angles,
are rectangular components of the resultant acceleration a; that is,

(13) $$a = a^t + a^n = [(a^t)^2 + (a^n)^2]^{1/2}.$$

22. Example. A 30-in. wheel, Fig. 12, is turning counterclockwise (CC) with a
clockwise (CL) acceleration of 2 rad./sec.².
At a certain instant,

$$\omega = 11.4 \text{ rad./sec.}$$

When a point A on the rim has the position
A_2 as shown, what is its total acceleration?

Solution. The tangential acceleration is,
equation (11a),

$$a^t = r\alpha = (1.25)(2) = 2.5 \text{ fps}^2. \searrow$$

The normal acceleration is, equation (12),

$$a^n = r\omega^2 = (1.25)(11.4)^2 = 162.5 \text{ fps}^2. \nearrow$$

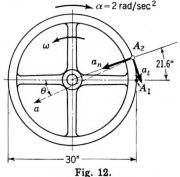

Fig. 12.

The approximate directions are indicated by small arrows. The total acceleration is

$$a = [(a^t)^2 + (a^n)^2]^{1/2} = [(2.5)^2 + (162.5)^2]^{1/2} = 162.52 \text{ fps}^2,$$

a value which is very nearly equal to a^n since a^n is very large compared with a^t
in this example. The slope of the resultant vector must be found before it is
completely defined. The angle that it makes with the vector a^n is

$$\phi = \tan^{-1} \frac{a^t}{a^n} = \tan^{-1} \frac{2.5}{162.5} = 0.88°.$$

The total acceleration is therefore directed downward toward the left at an angle
of $\theta = 21.6 + 0.88 = 22.48°$ with the horizontal, as shown slightly exaggerated
in Fig. 12. Observe that the normal acceleration may be quite significant with
respect to the forces on a body ($F = ma$).

23. Harmonic Motion. There are many definable motions during which the acceleration is not constant. Perhaps the most important of these is called simple harmonic motion (abbreviated SHM) and is defined mathematically by $a = -Cs$, which is to say that the acceleration a is proportional to the displacement s from some origin but oppositely directed. It turns out that if a point Q, Fig. 13, moves on the circumference of a circle at constant speed, its projection P on a diameter moves with harmonic motion (SHM). From Fig. 13, we see that

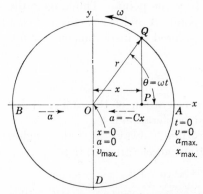

Fig. 13. Harmonic Motion. Since $v = -\omega r \sin \omega t$, which says that v is proportional to $\sin \omega t$, we note that PQ represents the speed of P to some scale; the speed varies as the ordinate to the circle.

(**i**) $\qquad x = r \cos \theta = r \cos \omega t;$

in which the angular velocity ω of the radius vector OQ is $\omega = \theta/t$ or $\theta = \omega t$ and the displacement x is measured from the center of the circle. The component of the velocity of Q in the x direction is always

(**j**) $\qquad v = -r\omega \sin \omega t,$

in which $v_Q = r\omega$. Because Q is moving with constant speed, Q's acceleration is entirely normal, $r\omega^2$ directed toward O. Its component in the x direction is the acceleration of P and is always given by

(**k**) $\qquad\qquad a = -r\omega^2 \cos \omega t.$

Since $x = r \cos \omega t$, this equation becomes $a = -\omega^2 x$, which is seen to accord with the definition of harmonic motion given above, $a = -Cs$, in which $C = \omega^2$.

Applied to vibrations, the motion is said to have **amplitude,** which is the maximum displacement of the moving point from the point of zero acceleration (the mid-point of the stroke); in Fig. 13 the amplitude is $r = OA$. The **period** T of a harmonic motion is the time required for the point to complete a cycle; that is, Fig. 13, the time for P to move from A to B to A; $T = 2\pi/\omega$. The **frequency** f is the number of cycles per unit time or the reciprocal of the period; $f = \omega/(2\pi)$. In this book, our principal interest in harmonic motion is as it is applied to cams, in which case we shall move the origin to the end of a stroke (§38).

The equations (**i**) and (**k**) show that the displacement and acceleration diagrams are cosine curves, and equation (**j**) shows that the velocity diagram is a sine curve. These equations may be plotted, but they are easily constructed as sine and cosine curves. This has been done in Fig. 14, where enough construction lines are drawn to show the method.

This illustration and its caption should be studied until you have a good idea of what is happening in harmonic motion.   The Scotch yoke, Fig. 15, which is used in various kinds of machines, is a mechanism which produces harmonic motion in the reciprocating member when the crank C rotates

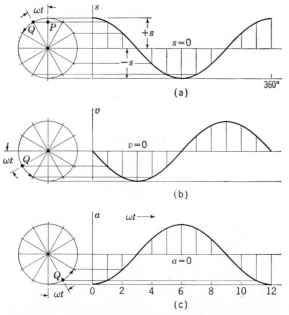

Fig. 14. Motion Curves for Harmonic Motion.   In the displacement diagram (a), note that the point P is moving on a vertical diameter.   If the upward direction is taken as positive, the velocity starts at zero and immediately becomes negative (the velocity vector points downward), and the acceleration is initially a maximum negative value (the acceleration vector points downward).   With the velocity and acceleration vectors both pointing in the same sense, the velocity at first increases, until the acceleration crosses zero at station 3; then the velocity begins to decrease in magnitude, becoming zero when the acceleration has its maximum positive value at station 6. Observe the phase relations of s, v, and a.   The v curve is 90° ahead of the s curve; the a curve is 90° ahead of the v curve.

with constant speed.   The velocity of the projected point P is always the projected length of the constant vector v_Q, as shown in Fig. 15.

24. Example—Harmonic Motion.   The rectilinear motion of a point is defined by $a = -16x$; the amplitude is 3 in.   (a) Find the period and the frequency. (b) Determine the displacement, velocity, and acceleration after 10 sec.

Solution.   (a) From the equation, we recognize that the motion is harmonic. Accordingly, $\omega^2 = 16$, whence $\omega = 4$ rad./sec.   Thus the period and frequency are,

$$T = \frac{2\pi}{4} = 1.57 \text{ sec.,} \qquad f = \frac{1}{1.57} = 0.637 \text{ cps.}$$

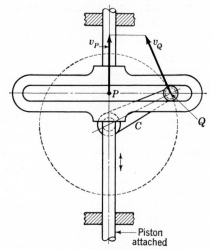

Fig. 15. Scotch Yoke.

(b) After 10 sec., the number of cycles is $(10)(0.637) = 6.37$ cycles. Hence,

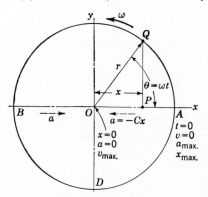

Fig. 13. Repeated.

the point has completed 0.37 part of a cycle from point A, the origin of time, Fig. 13. The corresponding time is $(0.37)(1.57) = 0.581$ sec. In order that the angle ωt be less than $360°$, we thus use $t = 0.581$ sec., instead of 10 sec., in the equations of harmonic motion. The angle ωt is then

$$(4)(0.581) = 2.32 \text{ rad.} = 133°.$$

The displacement after 10 sec. is

$$x = r \cos \omega t = 3 \cos 133° = -2.045 \text{ in.},$$

where the negative sign shows that the point is on the left side of the origin of displacement. The velocity is

$$v = -r\omega \sin \omega t = -(3)(4) \sin 133° = -8.78 \text{ ips},$$

the negative sign showing leftward motion. The acceleration is

$$a = -r\omega^2 \cos \omega t = -(3)(16) \cos 133° = +32.75 \text{ ips}^2,$$

the positive sign indicating that the acceleration is toward the right.

25. Nonplanar Motion. Luckily most motion problems are of plane motion or can be considered so for many practical purposes. Of those which are not planar, helical motion is relatively common. The path traced by a point moving at a constant distance from a straight axis and with uniform speed in the direction of the axis is a *helix*, and a point moving in such a path is said to have **helical motion.** Perhaps the most common example of helical motion is that of a point on a nut being turned onto a screw, or vice versa; each point on the nut has helical motion. Both limits of helical motion are plane motion; that is, if the pitch of the helix (screw) is zero, the resulting motion is rotation; if the pitch is infinite, the motion will be translation.

Another nonplanar motion sometimes found in practice is **spherical motion,** in which a point (or all points in a body) moves in such a manner that it remains at a constant distance from another point (the center of a sphere that would be the locus of all positions). Illustrative of this motion is the fly-ball governor.

Table 1. Conversion Constants

(Abbreviations: cm. = centimeter, km. = kilometer, m. = meter; others as usual.)

LINEAR

$12 \dfrac{\text{in.}}{\text{ft.}}$	$0.394 \dfrac{\text{in.}}{\text{cm.}}$	$30.48 \dfrac{\text{cm.}}{\text{ft.}}$	$5280 \dfrac{\text{ft.}}{\text{mi.}}$
$3 \dfrac{\text{ft.}}{\text{yd.}}$	$2.54 \dfrac{\text{cm.}}{\text{in.}}$	$3.28 \dfrac{\text{ft.}}{\text{m.}}$	$1.609 \dfrac{\text{km.}}{\text{mi.}}$

TIME

$60 \dfrac{\text{sec.}}{\text{min.}}$	$3600 \dfrac{\text{sec.}}{\text{hr.}}$	$60 \dfrac{\text{min.}}{\text{hr.}}$	$24 \dfrac{\text{hr.}}{\text{day}}$

ANGULAR

$6.2832 \dfrac{\text{rad.}}{\text{rev.}}$	$57.3 \dfrac{\text{deg.}}{\text{rad.}}$	$360 \dfrac{\text{deg.}}{\text{rev.}}$	$9.549 \dfrac{\text{rpm}}{\text{rad./sec.}}$

SPEED

$1.152 \dfrac{\text{mph}}{\text{knot}}$	$88 \dfrac{\text{fpm}}{\text{mph}}$	$0.6818 \dfrac{\text{mph}}{\text{fps}}$	$1.467 \dfrac{\text{fps}}{\text{mph}}$

26. Closure. Table 1 may be useful in converting units.

This chapter is perhaps a review for some readers, and it covers some elementary notions which will be repeatedly useful as one proceeds with the study. In passing, it is interesting and instructive to note that a point may have zero velocity (or speed) but not zero acceleration or zero acceleration but not zero velocity.

In applications, the engineer has to do with transformations of motion. Quite commonly, the input motion is a constant rotary motion that may be converted into another rotary (including oscillating) motion or into reciprocating motion. In other instances, as in automotive and Diesel engines, a reciprocating motion is converted into rotary motion. Linkages that perform these functions are seen on other pages.

PROBLEMS

Constant Speed and Velocity Ratio

1. (a) A certain depth and width of cut may be taken on soft cast iron in a lathe with a high-speed cutting tool at a cutting speed of 139 fpm. At what rpm should a 2-in.-diameter piece turn? (b) The same as (a) except that the material is forged steel and the cutting speed is 125 fpm. (c) The same as (a) except that the material is titanium and the cutting speed is 55 fpm.

2. (a) A 16-in. friction gear, turning at 80 rpm, drives another gear without slipping (surface speeds the same) with a velocity ratio of 3. Compute the size and speed of the driven gear and the center distance. (b) The same as (a) except that $m_\omega = 0.8$.

3. It is desired to connect two shafts by spur (cylindrical) friction wheels. The center distance is 30 in., the driver turns 200 rpm, and the velocity ratio is 0.4. (a) What should be the diameters of the wheels? (b) What is the linear speed of a point on the surface of the large wheel; (c) of the small wheel?

Ans. (a) $D_1 = 42.85$ in., (b) 2250 fpm.

4. The same as **3** except that $m_\omega = 2.8$.

5. A tractor moves a distance of 6 mi. in 1 hr. and 3.5 min. The driving wheels are 7 ft. in diameter and the velocity ratio between the engine and wheels is 20. Determine the average rpm of the engine.

Ans. 454 rpm.

6. An automobile goes 156 mi. in 4 hr. 13 min. The engine stroke is $4\frac{1}{16}$ in., the outside diameter of the tires is 26 in., and the velocity ratio, engine to rear wheels, is 4.4. Determine (a) the average speed of the car, (b) the average rpm of the engine, (c) the average speed of the crankpin relative to the frame, and (d) the distance moved by the piston relative to the frame.

Ans. (a) 54.2 fps, (b) 2105, (c) 43 fps, (d) 78.7 mi.

7. If the surface speed of the grindstone of Fig. P1 remains constant at 300 fpm, what is the average speed of point F on the treadle? At the end positions, of F, the connecting rod C is vertical.

Ans. 198 fpm.

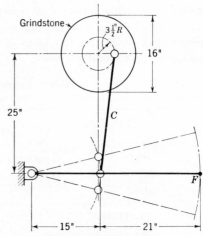

Fig. P1. Problems 7, 8.

8. The same as **7** except that the peripheral speed is 400 fpm.

9. A $2 \times 2\frac{1}{2}$-in. (always bore by stroke) automobile engine drives the rear wheels with $m_\omega = 4.3$; outside diameter of the wheels is 29 in. If the car is traveling 60 mph, determine (a) the rpm of the rear wheels, (b) the rpm of the engine, (c) the speed of the crankpin relative to the frame, (d) the average piston speed, and (e) the number of revolutions the engine will make in 50,000 mi. of high-gear driving.

Ans. (b) 2990 rpm, (c) 1960 fpm, (d) 1245 fpm, (e) 14.95×10^7.

10. A compressor piston makes 350 strokes per minute. The crank radius is 8 in. and the connecting-rod length is 36 in. (a) Compute the linear speed of the crankpin and the average speed of the piston.

11. A reciprocating compressor mechanism, with a 1-ft. crank and a 4-ft. connecting rod, is represented in Fig. P2.

The crank turns at a uniform speed of 50 rpm. Thus, each of the divisions 0-1, 1-2, etc., on the crank circle represents a certain time unit. Let 1 in. = 0.2 sec. of time and 1 in. = 1 ft. of displacement. Construct a displacement *st* diagram for the piston, using the positions of the crankpin numbered in Fig. P2.

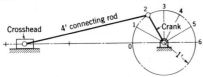

Fig. P2. Problem 11.

12. An automotive engine with a crank radius of 2 in. and a connecting-rod length of 10 in. turns at 2000 rpm. Determine the average piston speed and the crankpin speed.

Constant Acceleration

13. An automobile made a test stop in 300 ft. from 60 mph on a certain type of road. If the driver sees a roadblock 250 ft. ahead and if it takes him 0.6 sec. to apply the brakes, find the speed at which he would hit the roadblock, assuming uniform acceleration at the test rate.

Ans. 35.1 mph.

14. Two automobiles *A* and *B* are traveling in a line on a straight highway at the same speed of 100 fps. At time $t = 0$, *A*'s brakes are applied to give a constant deceleration of -15 fps². After 1 sec. reaction time, *B*'s brakes are applied to decelerate at -20 fps². If the distance between the cars is reduced to zero without collision, what is the least distance at which *B* could have been following *A* at the time *B* applied his brakes? Find the time *t* and speeds v_A and v_B when this bumper-to-bumper condition occurs. The initial position of *A* is suggested as the origin of displacement.

Ans. 30 ft.

15. A body is projected upward with an initial velocity of v_o at $t = 0$ and permitted to fall back to the origin. Neglecting air resistance, sketch the *st*, *vt*, *at*, and *jt* curves. What does the area under each curve represent? What does the slope of each curve represent?

Angular Motion, Constant Acceleration

16. A wheel which is rotating 300 rpm is slowing down at the rate of 2 rad./sec.². (a) What time will elapse before the wheel stops? (b) At what rate in rpm is the wheel revolving after 10 sec.? (c) Through how many revolutions has it turned during the first 10 sec.? (d) What is the total angular displacement? (e) Compute the number of revolutions from the time $t = 10$ sec. until the wheel stops.

Ans. (a) 15.7 sec., (b) 109, (c) 34.06, (d) 247 rad., (e) 5.25.

17. (a) A turbine turning at 1200 rpm is brought to rest at the uniform rate of 100 rpm². What are the number of revolutions turned and the elapsed time? (b) If this turbine slows to 300 rpm from 1200 rpm during 7500 rev., what is the corresponding uniform acceleration in radians per second-second?

Ans. (a) 7200 rev., 12 min., (b) 0.157 rad./sec.².

Motion Diagrams

18. The acceleration of an automobile is to be increased gradually in a straight-line function from 0 to 20 fps² during 10 sec. (a) Sketch the *at* diagram and compute the maximum speed. (b) Compute the speed for 2-sec. intervals and sketch a *vt* curve. (c) Estimate the distance traveled.

Ans. (a) 68.2 mph.

19. The *vt* diagram of a point is shown in Fig. P3. Let the area of this diagram be 0.58 sq. in., the time scale 1 in. = 0.12 sec., and the velocity scale 1 in. = 30 fps. Find the acceleration of the point at (a) position *B*, (b) position *A*, (c) position *C*. (d) Determine the displacement during the time *OD*. (e) Sketch freehand the *st* and *at* diagrams for this *vt* curve.

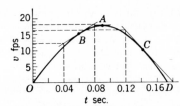

Fig. P3. Problem 19.

20. The vt diagram for the rectilinear motion of an automobile between two points is shown in Fig. P4. (a) What is the acceleration at the end of the 5th sec.; during the first 10 sec.; during the next 10 sec.; during the last 3 sec.? (b) What distance is traveled by the car after 10 sec., after 20 sec., and after 23 sec.?

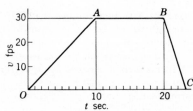

Fig. P4. Problem 20.

21. An automobile undergoes a constant acceleration of 6 fps² for 10 sec., moves at constant velocity for 20 sec., comes to rest with constant acceleration in 15 sec. Determine the maximum velocity, the distance traveled, and the average velocity for the total displacement. Sketch the at, vt, and st curves in a qualitative manner.

22. An automobile speeds up from rest to 60 mph in 17 sec. (a) Assuming that the acceleration is constant at 3.53 mi./hr-sec., compute the speed and the distance traveled after 8 sec. How far did the car go in 17 sec.? (b) Figure P5 shows a smooth curve through actual test data obtained with a calibrated speedometer and a stop watch. Using the test curve, answer part (a) and estimate the percentage error of the answer in (a) for the displacement of the car after 17 sec.

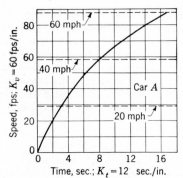

Fig. P5. Problem 22.

23. The same as **22** except that another car takes 21 sec. to attain 60 mph as shown in Fig. P6.

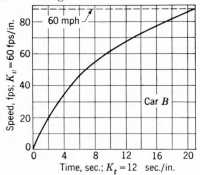

Fig. P6. Problem 23.

Normal Acceleration

24. If a man has demonstrated that he can safely stand an acceleration of $8g$, where g is the standard acceleration of gravity, what is the corresponding sharpest radius of curvature in a plane going 1200 mph?

25. A motorcycle moves around a circular path of radius 200 ft. If its normal acceleration is 68.9 fps², what is its tangential speed in mph?
Ans. 80.

26. A point on a rotating body at a radius of 6 ft. changes speed uniformly from 10 fps to 20 fps while it moves a distance of 120 ft. If the tangential acceleration continues constant, what is the absolute acceleration of the point at the instant its speed is 20 fps?
Ans. 66.7 fps².

27. A point at a radius of 2 ft. in a rotating body has an initial speed of 160 fps. During a period of 6 sec., the angular deceleration of the body is 5 rpm each 1.5 sec. Compute the tangential and normal components of the acceleration (a) after 3 sec., (b) after 6 sec.

Ans. (a) 0.698 fps², 12,450 fps², (b) 0.698 fps², 12,120 fps².

28. A body A, Fig. P7, is suspended from a cable wound around a 5-ft. drum and is moving down with a constant acceleration of 2 fps². At $t = 0$, we find $v_o = 10$ fps. When $t = 3$ sec., (a) deter-

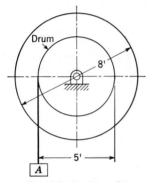

Fig. P7. Problem 28.

mine the angular and linear velocities of a point P, which is on the flywheel attached to the drum, and (b) the normal and tangential accelerations of a point P.

Ans. (a) 6.4 rad./sec., 25.6 fps, (b) 164 fps², 3.2 fps².

Harmonic Motion

29. The 3-in. crank in a Scotch yoke, Fig. 15, rotates at 4 rad./sec. Set up the equations for the motion of the slider and determine (a) the period of the motion, its frequency, and its amplitude. (b) After 10 sec. from the time the slider is at the end of its stroke, what are its velocity and acceleration? What is its displacement from the mid-point of the stroke?

Ans. (a) 1.57 sec., 0.637 cps, 3 in., (b) −8.78 ips, +32.75 ips², −2.045 in.

30. In Fig. 13, p. 20, the point Q moves in the circle with a constant

tangential speed of 10 fps. The radius of the circle is 4 ft. Five seconds after Q, going counterclockwise, passes the point D, what are the displacement, velocity, and acceleration of P?

Ans. −3.18 in., 9.98 fps, 1.58 fps².

31. The cam shown in Fig. P8, turning 60 rpm, raises and lowers the follower a distance $s = 3$ in. with harmonic motion. (a) Determine the maximum velocity and the maximum acceleration of the follower. The follower makes a stroke in each half revolution of the cam. (b) What are the velocity and acceleration at $s = 2$ in. from the lowest position, moving down?

Ans. (a) 9.42 ips, 59.2 ips², (b) −4.44 ips, −1.57 ips².

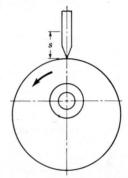

Fig. P8. Problems **31, 32.**

32. The cam shown in Fig. P8 raises and lowers the follower a distance $s = 5$ in. with harmonic motion. (a) If the period is 1 sec., find the maximum velocity and acceleration of the follower. (b) What are the velocity and acceleration at $s = 2$ in. from the lowest position, moving down?

Ans. (a) −15.7 ips, 98.5 ips², (b) −15.38 ips, +19.7 ips².

33. Thompson Indicator Mechanism. —Fig. P9. (a) Plot the path of point T for a length of over 1 in. (= 3 in. distance on 11 × 17 paper) on *each* side of the horizontal line through point A. Locate Q at $X = 13\frac{1}{2}$ in. and $Y = 3$ in. on 11 × 17-in. paper, LDH. Use 6 to 12 positions of the linkage. Indicate a

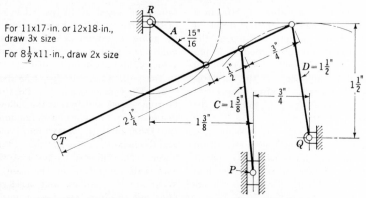

For 11x17·in. or 12x18·in., draw 3x size

For $8\frac{1}{2}$x11·in., draw 2x size

Fig. P9. Problem 33. Thompson Indicator Mechanism.

length on this path that would be best for drawing the indicator card. (b) What is the ratio of the distance moved by *T* divided by that moved by the actuating piston *P*? See Fig. 79.

34. Watt's Straight-line Mechanism.—Fig. P10. (a) Plot the complete path of the mid-point of link *C*. Compare with Fig. 75. (b) How long is that part which is nearly a straight line? Locate *Q* ($10\frac{1}{2}$, 3), LDH. (See Note 1 below.)

35. Use plate No. 1, Wingren. Solve problem 1A as stated in Wingren. (d) Write the equation for the ellipse generated by point *X*. (See Note 2 below.)

36–70. These numbers may be used for other problems as convenient to the instructor.

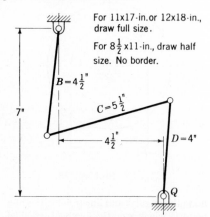

For 11x17·in. or 12x18·in., draw full size.

For $8\frac{1}{2}$ x11·in., draw half size. No border.

Fig. P10. Problem 34. Watt's Straight-line Mechanism.

Important Note 1. On occasion, a mechanism will be located on a drawing by such a designation as *Q* ($10\frac{1}{2}$, 3), as in **34**. In every instance, this is interpreted in the following manner:

Point *Q* is located at *X* = $10\frac{1}{2}$ in. and *Y* = 3 in., full scale, as measured from the left and bottom edges, respectively, of the paper; LDH means long dimension horizontal.

Important Note 2. If a prepared set of problem plates is used, there will result a considerable saving in time which might be devoted to a deeper or broader coverage of kinematics. For this reason, there will be a number of problem statements for which plates from Wingren's "Kinematic Problems" (Prentice-Hall) are to be used. So that the reader will be immediately alerted, these problems will open with the sentence "Use plate No. 00, Wingren." In general, our problems will be different from Wingren's. Also, the same or a similar problem can usually be used with other sets. There will often be more than one problem for a particular plate. Even though it is not practicable to work two such problems on the same plate, the problems can be used in alternate years. The $8\frac{1}{2}$ × 11-in. size of plate is suited for homework assignments. Perhaps the instructor would care to prepare some of the problems in this text on mimeographed or dittoed sheets for student use. See **35** for an example of the statements of this paragraph.

2

Cams

27. Introduction. A cam mechanism consists of a minimum of three essential links—the **cam** (Fig. 16), the **follower,** and the **frame** of the machine which supports the bearing surfaces for the cam and follower. Additional links may be and are often used for a reason, as in the operation of overhead valves in internal combustion engines. The cam mechanism is a versatile one. It can be designed to produce almost any type of motion of the *follower* (while the *cam* ordinarily turns at constant speed), and it is easily adapted as a substitute for manual operations. Thus it is nearly indispensable in automatic machines, certainly in many shop machines as automatic lathes, in linotype and shoemaking machinery, in phonograph record changers, servomechanisms, and in countless other machines. Cams could easily be voted as the mechanical (as distinguished from electrical or hydraulic) mechanism which has contributed most to modern civilization. Decide now which is the cam and which the follower, and be sure to recall the correct mental image as you read.

28. Types of Cams. There are two principal types of cams: (1) **disk** or **plate cams,** in which the follower moves in a plane perpendicular to the axis of rotation of the camshaft, and (2) **cylindrical cams** (Figs. 44 to 47), in which the follower moves in a plane parallel (sometimes approximately parallel) to the axis of rotation of the cam. There are variations of cam forms from these basic types too numerous to detail here (see reference **3**). The form of the follower or its manner of movement, translation or oscillation, often serves to provide a name for certain cam mechanisms. Also, the shape of the cam may provide a name. For example, the cam of Fig. 16 is a plate cam called a *tangential cam*, because the cam surface is outlined by two circular arcs connected by tangent straight lines. We might also refer to it as a *disk cam with roller follower*

29

Moreover, it is a **radial cam** because the follower moves along a radial line. Tangential cams were formerly regularly used for automotive valve-lift cams and now are used in some of the slower-speed engines. A cam roller with needle bearing is shown in Fig. 17, an example of available commercial components. You may see what one actual cam looks like in Fig. 18.

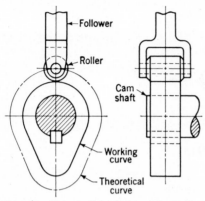

Fig. 16. **Tangential Cam with Roller Follower.**

In a general way, a cam should be designed and manufactured so as (1) to cause the driven mass to move in accordance with a desired plan or program, (2) to be smooth and quiet in operation (little vibration), (3) to withstand the applied loads, including the dynamic (inertia) loads without rapid wear or early failure.

29. Definitions. We shall give a few definitions here to which the reader may refer as the need arises. In laying out a cam surface, one chooses a convenient reference point on the follower, which has been called the **trace point** (3); it is the center of the roller in case of a roller

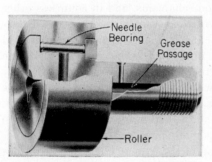

Fig. 17. **Cam Follower Roller.** Available roller diameters from 0.5 to 4 in.; roller widths from $\frac{3}{8}$ to $2\frac{1}{4}$ in.

Fig. 18. **Plate Cam.** This cam on a form grinder is designed to advance the grinding wheel rapidly, to withdraw it slowly, to return it rapidly to the cycle stop. Such a cam is designed for each particular job on the machine.

follower, the knife-edge for a knife-edge follower, Fig. 19(a), and on a flat-face follower, it is usually but not necessarily the point where the trace of the face cuts the vertical center line of the camshaft, Fig. 19(b).

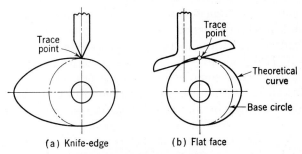

(a) Knife-edge   (b) Flat face

Fig. 19. Trace Point.

The *theoretical curve* or *pitch curve,* Fig. 16, is the path which would be followed by the trace point if the follower moved about the cam (instead of the cam rotating); or it is the curve which always passes through the trace point as the cam rotates.

The *working curve* or *cam profile* is the projection of the actual cam surface (for a plate cam); it is the outline of the plate cam. For the knife-edge follower, the theoretical and working curves coincide.

The *base circle* is one with its center at the center of the camshaft and a radius equal to the distance *from* this center *to* the nearest point on the theoretical curve (Figs. 20 et seq.).

The *stroke* or *throw* of a follower is the maximum movement away from the camshaft from its "lowest" position. Perhaps because on the printed page the follower is nearly always shown as moving toward the top of the page, we not only speak of the stroke, but we also call this *outward* movement the *rise* or *lift L*, although of course the follower is likely to be moving in some other direction in the actual machine. The opposite movement is the *return.* A *dwell* occurs when the follower remains stationary during *some finite rotation* of the cam.

The *cam angle* θ is an angle the cam turns through while the follower is displaced s. The cam turns through the *total cam angle* β when the follower undergoes its entire stroke L. It is possible for the follower to make several strokes during a cam rotation, in which case there would be several total cam angles in a revolution of the cam.

30. Motions Used for Cam Curves. When machines move slowly, the kind of motion curve on a cam is often not important; french curves and circular arcs have been used freely. As speeds increased, engineers became conscious of the significance of the kinds of motions used, at least to the extent of adopting motions which, as far as the analyses at the time went, seemed to promise the best results. The traditional motions are constant-velocity, constant-acceleration, and harmonic motions, all of which have been defined in a general way in Chapter 1. Let the direction of the outward (up) movement of the follower be the positive direction for all vectors unless otherwise specified.

As machine speeds continued to increase, there came a point where engineers realized that these motions were inadequate in some manner. As it was eventually discovered (**7, 8**), the manner in which they were deficient was that when combined with dwell periods, the *jerk* at one or more points during the movement was theoretically infinite. We say "theoretically" because infinite jerk $j = da/dt$ corresponds to an instantaneous change of acceleration; but because of elasticities of materials and clearances at joints, actual instantaneous changes of acceleration are impossible. As the discussion proceeds, you will see the significance of providing finite jerk at high speeds.

Curves which will provide finite values of jerk throughout the motion include the cycloidal motion and certain of the polynomials.

All of the motions to be discussed may be classified as polynomial motions (constant acceleration, $s = Ct^2$, is an example) or trigonometric motions [harmonic, $s = f(\cos \theta)$, is an example]. The events of a follower's motion are simply designated in abbreviated form as follows:

DRD, which we translate as dwell-rise-dwell. We could as well think of DRD as meaning dwell-return-dwell; the significance is the same— this symbol means that a movement of the follower is preceded *and* followed by a dwell. After DRD, a return is normal but not essential; some sort of additional movement is essential.

DRRD means that the follower is dwelling, rises, and returns to a dwell. This classification also includes dwell-return-rise-dwell, that is, with the dwell at the "top" of the stroke. The difference between DRD and DRRD is that there are at least two dwell periods in a DRD, but only one in DRRD.

RRR means rise-return-rise, without dwells. If this sort of movement is involved, it is likely that some linkage other than cams (Chapter 3) would do the job—in most cases, better.

31. Methods of Drawing Cam Profiles. Certain of the motion curves used for cams are easily laid out by graphical procedures, as explained below. However, when a high degree of accuracy becomes important, as at high speed, the ordinates of the curve must be computed by machine, and the accuracy of manufacture must be correspondingly adequate. The general plan of graphical solutions for plate cams is outlined as follows:

1. Draw the base circle. In practice, the engineer decides upon its size in accordance with the machine-design requirements (perhaps space

available)—see § 48. In a text, its diameter is usually given because of the difficulty of defining its complete environment; or its diameter may be chosen to provide some maximum pressure angle, § 35. The theoretical curve touches the base circle when the follower is closest to the camshaft.

2. The path of the trace point on the follower is drawn in its proper position with respect to the camshaft, either a straight line for reciprocating followers (06', Fig. 20) or a circular arc for oscillating followers (09', Fig. 32).

3. Now the follower is to rise a certain distance L while the cam turns through some cam angle β. Both of these quantities are usually dictated by the requirements of the design. With L and β known, *divide the cam angle into a suitable number of equal parts to represent equal time units* (equal angles are proportional to equal times because the cam is presumed to turn at constant speed). Then

4. *Divide the path of the trace point on the follower into the same number of divisions as there are time units (equal angles)* in β, but the magnitudes of the divisions are such that if the follower moves over one division in one unit angle, the next division in the next unit angle, etc., it will be moving with the desired kind of motion.

5. Find points on the theoretical curve. In making this step, we *think* of the cam as remaining stationary and imagine *the path of the follower as moving about the camshaft center.* See the subsequent examples for detail.

6. Draw the contact face of the follower at each of the points found on the theoretical curve. (For a roller follower, the contact face is the surface of a cylinder, represented on the drawing by a circle whose center is at a theoretical point.) In the design of actual cams, a large number of theoretical points should be found so that the position of the follower will appear many times; in school, we shall be satisfied with fewer points in order to save learning time.

7. Draw a smooth curve tangent to *all* positions of the face. This curve is the *working curve.*

The details of the procedures will be shown in examples following the definitions of certain motions. After you have solved a few problems, reread this article to be sure that you have in mind the basic ideas; that is, do not memorize each procedure and work by rote.

32. Constant Acceleration. As you know from your previous experience (Chapter 1), motions are often defined in terms of displacement and time $[s = f(t)]$. However, when applied to cams, these motions are much more conveniently expressed in terms of angles on the cam. For example, for constant acceleration (parabolic motion) starting from rest, we have $s = at^2/2$, equation (3), § 11, where $a/2$ is some constant. For cams, we prefer $s = C\theta^2$, where θ is any cam angle corresponding to the displacement s of the follower and C is a constant. The value of C is determined from conditions of the motion.

If a follower is started with constant acceleration, its speed will continue to increase until the acceleration changes sign. Therefore, somewhere along its rise, the follower must be slowed in order that it come to rest at the top of its stroke. If, at the mid-point of the stroke, the acceleration is reversed, but kept the same magnitude, this deceleration will bring the follower to rest in the nick of time. Such motion is named **uniformly accelerated and retarded motion,** to be abbreviated here by UARM; the magnitude of a is constant but its sense is reversed at mid-stroke.*

To lay out the motion graphically, we note that the displacement s is proportional to the square of the time, $s = Ct^2$, a second-order polynomial and the equation of a parabola. For consecutive equal time units, the displacements of the follower are therefore:

t	s	Δs	t	s	Δs
0	0	0	3	$9C$	$5C$
1	C	C	4	$16C$	$7C$
2	$4C$	$3C$	5	$25C$	$9C$

in which the values of Δs are the displacements in successive units of time. Thus, to move with constant acceleration, the follower moves $1C$ during the first time interval 0–1, $3C$ during the second time interval 1–2, $5C$ during the third time interval 2–3, etc.; that is, in successive and equal units of time, the follower moves distances proportional to the numbers $1:3:5:7:9$, etc. It follows that if the movement is to be accomplished with constant acceleration, it is only necessary to divide the angle into a certain number of equal parts and to divide the corresponding rise into the same number of spaces proportional to $1:3:5:7$ . . . . It should seem logical that if the body is slowing down, the spaces traversed in equal units of time would be reversed, proportional say to $7:5:3:1$. Observe that mathematically the st curve for UARM consists of two equal parabolas that are tangent at the mid-point of the stroke. However, if it is desired to have different magnitudes of the acceleration during each part of the stroke, this may be done, but with unequal tangent parabolas. More detail is best given by an example.

33. Example—Radial Follower, Constant Acceleration. A reciprocating radial follower is to move outward 1.8 in. from a dwell with UARM while the cam

---

* Constantly accelerated motion as used in cams is often mistakenly called "gravity motion," or a similar alternative. However, it is only rarely and usually coincidental that the acceleration of a follower is the same as that of gravity.

turns 150° CC; dwell for 30°; return with UARM in 180°. The cam speed is 420 rpm. (a) Find a working curve by graphical solution. (b) Compute the acceleration and the maximum velocity on the outward stroke. How does the jerk vary?

Solution. (a) Since it is not given, a size of base circle, say 4 in. in diameter (half size), must be chosen. The first choice may or may not be a good one—see §§ 34 and 48. The angle 0A6, Fig. 20, which is turned through by the cam while the follower moves outward, should be divided into a number of equal parts to represent equal time units. (If this cam were to be manufactured, many more divisions than shown would be desirable.) Since the follower accelerates during half of the angle 0A6 (= 0A3) and decelerates during the remaining half (this is UARM by definition), the first half of the rise is divided into 3 spaces proportional to 1:3:5, because there are 3 time units. This is accomplished by drawing any line 0d through point 0, laying a scale along it and counting off equal divisions to get spaces, 1 unit, 3 units, and 5 units (any convenient unit) long. Lines drawn parallel to b3′ will intercept proportional spaces on 03′. Therefore, 01′:1′2′:2′3′ = 1:3:5. [See also the method of laying out $a = C$ motion in Fig. 22(b).]

With equal deceleration from 3′ to 6′, the spaces should be proportional to 5:3:1 = 3′4′:4′5′:5′6′. A new line could have been drawn for this motion, but it is quicker to use the original line 0bd and divide it proportionally to 1:3:5:5:3:1 at the outset. (Notice that 1:3:5:3:1 does *not* produce UARM when all time angles are equal.) We notice that UARM is such that the total cam angle for the movement should be divided into an *even* number of parts. To get the points 1, 2, 3, . . . on the theoretical curve, we imagine the follower to be moving about the cam and outward at the same time. If the cam turns counterclockwise (CC), the equivalent motion of the follower with the cam stationary is clockwise (CL). It is a good idea to number (or letter) the points 1′, 2′, etc., on the path of the follower and the corresponding positions of the follower 1, 2, etc., with matching numbers in order to avoid error. As the follower moves through the angle 0A1, it also moves outward so that it ends up at point 1, corresponding to 1′ on the actual follower path. It may come in handy to note that in this type of cam, the rise may be laid out on *any* radial line, as convenient. Now study the caption to Fig. 20 for further detail.

It is entirely feasible to have accelerations of different magnitudes during the speeding-up and slowing-down periods; that is, the follower may decelerate at a smaller or greater rate than it accelerated. In this case, the cam angle for acceleration is not the same as for deceleration; and one must be careful that the maximum speed for each motion is the same. There should not be any instantaneous change of speed (infinite acceleration).

(b) The magnitude of the acceleration is easily computed from the basic data. Since the usual units in cam work are inches and seconds, we note that

$$\frac{420}{60} = 7 \text{ rps} \qquad \text{and} \qquad \frac{1}{7} = 0.1429 \text{ sec./rev.}$$

Now since the curve is discontinuous where the acceleration changes from positive to negative, we should apply $s = at^2/2$ to no more than half the rise. (The *point of inflection* is 3, Fig. 20, on the way up.) Thus, the angle for half the

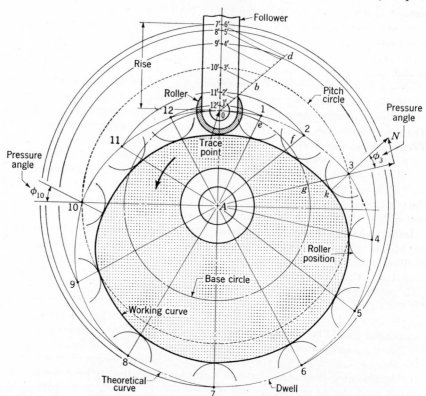

Fig. 20. Plate Cam, Radial Follower, Constant Acceleration. *Procedure.* Draw the base circle. Lay off path of the trace point 06′ equal to 1.8 in. The rise occurs through angle 0*A*6, which is divided into 6 equal parts. Therefore, divide rise into 6 parts, proportional to 1:3:5:5:3:1 for UARM, obtaining points 1′, 2′, . . . , 6′ from the convenient divisions on any line 0*bd*, by drawing parallels to *d*6′ as shown. Name the points. With center of compass at *A*, and radius *A*1′, strike an arc (CL because the cam turns CC) to find 1 on the center line of the follower *A*1; with radius *A*2′, draw arc to get point 2; etc. This procedure carried to point 6 defines the theoretical curve to that point. Since from 6 to 7 is a dwell, this part of the curve is a circular arc of radius *A*6. Inasmuch as the return is also UARM, we may use the same divisions on 06′ as used on the outward stroke, *provided* the return angle, 180° in this example, is divided into the same number of equal time units, 6 in this example. Swinging arcs as before locates points 8, 9, . . . , 12. A smooth curve through all these points is the theoretical curve. Let the radius of the roller be ½ in., say, and with centers at the points on the theoretical curve, strike arcs representing the roller positions. *A smooth curve tangent to all positions of the roller is the working curve.* A different diameter of roller would give a different working curve.

rise is 75° or $\frac{75}{360} = \frac{5}{24}$ rev.; hence the time for half the rise is

$$t = \left(\frac{5}{24} \text{ rev.}\right)\left(\frac{1}{7} \text{ sec./rev.}\right) = \frac{5}{168} \text{ sec.},$$

so that for $s = 0.9$ in.,

$$a = \frac{2s}{t^2} = (2)(0.9)\left(\frac{168}{5}\right)^2 = 2030 \text{ ips}^2.$$

The acceleration during the last half of the upward stroke is $a = -2030$ ips^2. The speed may be found from $v = at$:

$$v_3 = at = 2030\left(\frac{5}{168}\right) = 60.4 \text{ ips},$$

the maximum speed during the rise. The values of a and v_{max} during the return stroke, if different, are found in the same manner.

During a DRDR motion, the acceleration theoretically changes instantaneously at the beginning, in the middle, and at the end of the stroke; thus the time rate of change of a, which is the jerk, must theoretically be infinite at these points.

34. Pressure Angle. As you can see in Fig. 20, the point of contact between roller and cam is not always on or even near the follower's

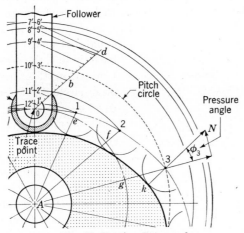

Fig. 20. Portion Repeated.

center line. See point k at position 3, for instance. In the absence of friction (and deflection), which is relatively small, the force at the point of contact is normal to the surfaces. The normal at a point on the circumference of a circle goes through the circle's center; hence, $k3$ is the line of action on the follower of force N. The angle which $k3N$ makes with the center line $g3$ of the follower is called the pressure angle ϕ at point 3. Thus, we may define the **pressure angle** for roller followers on plate cams as *the angle between the line of travel of the follower and the normal to the surfaces in contact.* The component of force N that is perpendicular to the center line, namely, $N \sin \phi$, produces a bending moment on the follower; the larger the component, the greater the bending moment and the reactions on the guides (bearings). Such side thrust does no useful work and can easily become excessive, especially at high speeds. It is therefore considered desirable to limit the maximum value of ϕ. Naturally such a limiting value must be based on experience, since we would expect it to be affected markedly by the details of the design of the cam mechanism. A maximum permissible value often seen in print is 30°, and this value or perhaps one somewhat smaller would be a good starting point for cam designs where there is no background experience. Among the factors which affect the magnitude of the total force are: the spring force, if any, the force of gravity, the mass of the follower system, the maximum acceleration ($F = ma$) that in turn is affected by the jerk. The actual use of the pressure angle is in the force analysis of the cam mechanism. The maximum force on the follower may occur at some point in the cycle at which the pressure angle is not a maximum; if so, the dynamical effects should be investigated for this position as well as for the one with maximum ϕ (and perhaps for others).

The size of the base circle affects the value of the pressure angle ϕ. Suppose a roller follower is to rise $\frac{1}{2}$ in. as shown in Fig. 21 (rise b to c is equal to rise d to e). For the smaller base circle 1, the pressure angle is ϕ_1. Comparing this value with that for base circle 2, we easily see that $\phi_2 < \phi_1$.

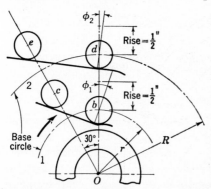

Fig. 21. Size of Base Circle and Pressure Angle.

To obtain the value of the pressure angle at any point of the motion, the cam profile may be found as in Fig. 20, and the desired angles measured. For example, we may compare the pressure angles at the points of inflection of the st curve, points 3 and 10, Fig. 20 (where the acceleration changes from $+a$ to $-a$, or vice versa). Since the angle of rise in Fig. 20, $OA6$, is less than the angle of return, $7A0$, we expect the pressure angle at the inflection point on the rise in this example to be greater than that during the return. It is, and we find $\phi_3 \approx 25°$. Since certain inaccuracies are inherent in such graphical solutions, it will usually be satisfactory to assume ϕ to be a maximum at the point of maximum speed. See § 35. If space should happen to be at a premium where the cam in

Fig. 20 is to be used, the designer might decide on the basis of the 25° pressure angle to redesign the cam with a smaller base circle, letting the value of ϕ_3 approach (or exceed) 30°.

35. Computing the Pressure Angle. On an *st* diagram, and on a reciprocating (translating) cam, Fig. 36(a), the maximum slope occurs at the point of inflection, which is the point of maximum speed.

There are mathematical techniques for finding the actual maximum pressure angle, in magnitude and location, on any roller-follower cam (**3, 23**). Confining our attention now to *plate cams with radial roller followers*, we can note that on an *st* diagram the slope of the curve is to some scale the speed, Fig. 22; under certain conditions, it is also the pressure angle. The difference between the speed and the slope is that speed has units of length and time, whereas slope is dimensionless (length/length). Thus, the actual slope at any point of a displacement *st* diagram can be obtained by multiplying the actual follower speed v_f by a time scale K'_t; thus

$$(a) \qquad\qquad \tan \phi = \frac{dy}{dx} = \left(v_f\ \frac{\text{in.}}{\text{sec.}} \right)\left(K'_t\ \frac{\text{sec.}}{\text{in.}} \right).$$

However, a certain value of the time scale, distinguished by a prime mark (K'_t), must be used and this scale corresponds to the length of the circumference of the *"reference" circle*. To understand what is meant by the reference circle, consider point 3, Fig. 22(a). The line $3k$ is normal to the cam profile at k (also to the roller at k) and to the theoretical profile at 3; the pressure angle ϕ_3 is $A3k$. Now the line mn is tangent to the theoretical curve (perpendicular to the normal $3k$) and $3p$ is tangent to the circle of radius $A3$ (perpendicular to the follower center line $A3$ in the position 3). Hence angle $n3p$ is also ϕ_3. In other words, if the circle of radius r_3 is rectified into a displacement diagram, the slope of the displacement curve at point 3 will be the true pressure angle at that position. Similarly the true pressure angle at position 2 would be shown by an *st* diagram whose length is equal to the circumference of the circle of radius $A2$, because the tangent $2u$ to the circle $A2$ and the tangent to the curve $2w$ define the pressure angle ϕ_2 at 2. Thus for point 3, the 360° of cam rotation would correspond to an *st* diagram of $2\pi r_3$ in./rev., Fig. 22(b); for point 2, the 360° corresponds to $2\pi r_2$ in./rev. If we know or can find the cam speed n_s rps, then $2\pi r n_s$ is in inches per second and $K'_t = 1/(2\pi r n_s)$ sec./in. for a diagram drawn to full scale $2\pi r$ in. long. Consider any point X on the theoretical curve, Fig. 22(a), at a radius of $r = r_b + s$, where r_b is the radius of the base circle and s is the displacement of the follower from its initial position 0. The length of the corresponding rectified circle is $2\pi r = 2\pi(r_b + s)$ in., Fig. 22. This is one revolution and if the cam makes n_s rps, we have $2\pi(r_b + s)n_s$ ips or

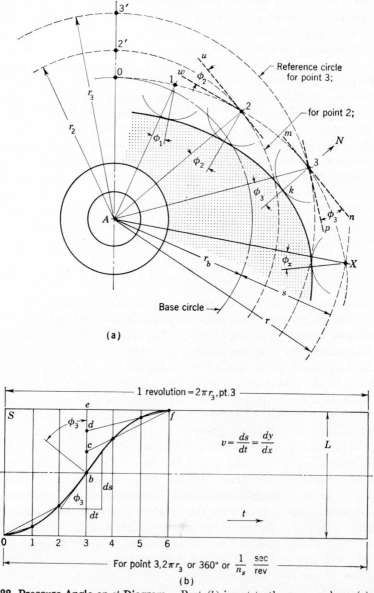

(a)

(b)

Fig. 22. Pressure Angle on *st* **Diagram.** Part (*b*) is not to the same scale as (*a*); also note that the slope of the curve in (*b*) is equal to the pressure angle only at position 3. An alternate construction for $a = C$ motion is shown at *fbe*. Divide the displacement *be* into a number of equal parts, *bc*, *cd*, *de*, equal to the number of time units in the corresponding cam angle. Draw lines *fc*, *fd*, etc. The intersections of these lines with the vertical lines are points on the curve. For the curve 0*b*, divide 3*b* into equal parts and draw lines from 0.

$K'_{tx} = 1/[2\pi(r_b + s)n_s]$ sec./in.; thus

$$(14) \qquad \tan \phi_x = v_x K'_{tx} = \frac{v_x}{2\pi(r_b + s)n_s} = \frac{v_f}{(r_b + s)\omega} = \frac{v_f}{v_c},$$

[CAM WITH RADIAL ROLLER FOLLOWER]

in which ϕ_x is the pressure angle at any displacement s of the follower whose speed is $v_x = v_f$ ips; v_c is the speed of the point on the cam at the radius $r_b + s$. (Of course, feet could be consistently used for the length unit just as well as inches.) Equation (14) is valid no matter what kind of motion the follower has (see §§ 37, 40, 43, 44, 47), but only when the *roller follower is radial.* It may be helpful to note that $2\pi(r_b + s)n_s$ is the linear speed of a point X on the cam. Equation (14) may be used to find the pressure angle ϕ at any point, or it can be used to compute the size of the base circle r_b for a given pressure angle. See problem **232,** Chapter 3, for a method of computing the pressure angle for an offset follower.

36. Example—Pressure Angles. For the cam of § 33, compute (a) the pressure angle at 3 and (b) the size of base circle that would result in a pressure angle of 30° at **3.**

Solution. From § 33, we have the following data:

$$r_b = 2 \text{ in.}, \qquad n_s = 7 \text{ rps}, \qquad a = 2030 \text{ ips}^2, \qquad v_3 = 60.4 \text{ ips.}$$

(a) At point 3, $r_3 = r_b + s = 2 + 0.9 = 2.9$ in. Thus, from (14),

$$\tan \phi_3 = \frac{v_3}{2\pi(r_b + s_3)n_s} = \frac{60.4}{2\pi(2.9)(7)} = 0.473,$$

from which $\phi_3 = 25.3°$. (The close agreement with the graphical answer of 25° is fortuitous.)

(b) For $\tan 30° = 0.577$ and $s = 0.9$ in., we may solve for r_b from equation (14);

$$r_b = \frac{v_s}{2\pi n_s \tan \phi_3} - s = \frac{60.4}{2\pi(7)(0.577)} - 0.9 = 1.48 \text{ in.}$$

That is, a base circle of diameter $2 \times 1.48 = 2.96$ in. will result in a maximum pressure angle of approximately 30° when the motion is as defined in § 33.

37. Harmonic Motion. As we know, § 23, harmonic motion may be defined in terms of a point moving at constant speed in the circumference of a circle; the motion of the projection of this point on any diameter is *simple harmonic motion* (SHM). This is the key to the graphical layout of a cam curve for harmonic motion. The details are best given in the following example. We shall also take advantage of this example to explain the method of finding the working curve for a *flat-face follower,* including some important discussion which should not be missed.

Harmonic motion for the entire 360° of the cam is produced by a con-

tinuous curve (see Fig. 14); there are no instantaneous changes of acceleration and the jerk is finite at all times. However, if a harmonic movement
is preceded or followed by a dwell, there is a theoretically instantaneous
change of acceleration ($j = \infty$) at the point of change of motion.

38. Example—Disk Cam, Flat-face Follower, Harmonic Motion. Find the
working curve for a disk cam with flat-face follower. The plane of the face is
perpendicular to the line of motion. The follower is to move outward 1.5 in.
with harmonic motion in 150° of cam rotation, return to the starting point with
harmonic motion in 150°, dwell 60° (DRRD). Rotation is clockwise at 700 rpm.
Let the size of the base circle be $3\frac{1}{4}$ in.

Solution. Draw the base circle, Fig. 23, and use a trace point T at the intersection of a radial line and the circumference of the circle. This is the point
which follows the theoretical curve. The motion may be laid out from the base
circle along any radial line, taken as $0'5'$ for convenience here. (The position of
the follower may be drawn at any point desired, here at the middle of the dwell
period, just to keep it in the clear. But draw this last.) Decide upon the number of divisions for the cam angle $0A5$, say 5 (not enough, but it keeps the construction clearer). Then divide the rise into 5 spaces such that if the follower
moves through the distance $0'1'$ during one division of the cam angle, through $1'2'$
during the next division, etc., on through $4'5'$, it will be moving with harmonic
motion. This division is done by drawing a semicircle whose diameter is equal
to the rise and dividing the circumference of the semicircle into the same number
of equal arcs as there are equal time units in the cam angle for the rise. The arcs
are chosen equal because the point in the circumference whose projection is
desired is moving with constant speed. Project the points marking the divisions
of the arcs onto the diameter (rise) to obtain harmonic displacements $0'1'$, $1'2'$,
etc., for the follower. See detail in Fig. 23.

The harmonic curve as used here is continuous from dwell to top and back to
dwell. Hence, the return cam angle may be divided into the same number of
parts as the rise angle and the same divisions along $0'5'$ may be used again, as
shown, Fig. 23. If the angle for the return had been less than that for the rise,
then the maximum acceleration on the return would have been greater than on
the rise; and the curve would have been discontinuous at the peak of the rise.
However, the same divisions along $0'5'$ could be used with a different return angle;
for example, if the return angle is 100° (instead of 150°), divide the 100° into 5
equal parts and proceed as before.

Knowing the location of the theoretical points, draw through these points lines
which represent the face of the follower. When the face is in position 2, it
touches the cam curve at point F, Fig. 23. Thus the length of face needed to
the left of the center line BK is the maximum of such distances as $2F$. This
maximum occurs at that position where the speed of the follower is greatest,
which in this cam is midway between positions 2 and 3. It happens that this
distance is easily computed (**3, 23**); the minimum necessary

$$\text{Face length, one side} = \frac{\text{max. vel. while that side in contact}}{\text{angular vel. of cam (rad./sec.)}},$$

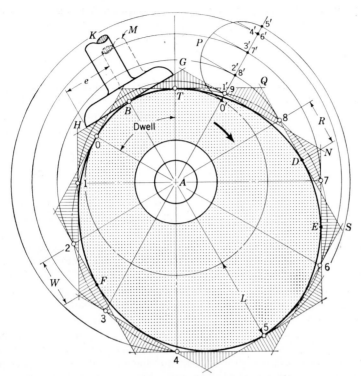

Fig. 23. Plate Cam, Flat-face Follower, Harmonic Motion. *Procedure.* After drawing the base circle, divide the cam angle for the rise into say 5 equal parts. Lay off the required rise of 1.5 in. along some radial line $0'5'$. With a diameter equal to the rise, draw semicircle $0'P5'$ and divide it into 5 equal arcs. Drop perpendiculars from these points and obtain $1', 2', \ldots, 5'$. With radii $A1', A2', \ldots, A5'$ and with the center at the center of the camshaft, draw the arcs intersecting the radial lines to locate points $1, 2, \ldots, 5$ on the theoretical curve. Since the return angle is equal to the rise angle, the same points, numbered $6', 7', 8', 9'$ (9 and $9'$ are coincident), may be used to locate $6, 7, 8, 9$ as shown. Draw this curve if you desire (or are so instructed). Through each theoretical point, draw a line of indefinite extent representing the face of the follower, as HG, GN, QS, etc. Crosshatch the triangles formed by these lines and draw the working curve tangent to the mid-points of the inside bases. If the working curve touches all face positions, it will provide the motion as designed (except for the error in the graphical solution). The necessary length of face to the *left* of the trace point $T = B$ is determined by measuring from the theoretical point (small circles) to the farthest tangent point in the counterclockwise direction. Upon checking points $1, 2$, etc., it is found that the maximum distance is about $2F = W \approx 0.8$ in.; use $W = 1$ in. to allow for the inaccuracy of the graphical method and to cover the contingency of not having the face drawn at the maximum position. The necessary face length to the *right* of T is determined by measuring from the theoretical points to the farthest tangent point in a clockwise direction. By checking points $9, 8, 7$, etc., the maximum distance is found to be in the vicinity of points 7 and 8 (actually midway between these points—see text). We find $8D = R \approx 0.83$ in.; use $R = 1$ in. Total length of face $HG = 1 + 1 = 2$ in. The stem may be in any convenient position M, other than the radial position BK.

or in brief equation form, the minimum necessary face length on one side of follower center line, Fig. 23, is

(15)
$$e = \frac{v_{max}}{\omega},$$

wherein e in. and v_{max} ips are corresponding values.   (See problem 233, Chapter 3.)
If the face is a plane disk (see below), its radius should be the largest value from

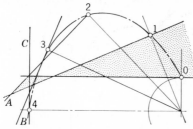

Fig. 24. False Curve, Flat-face Follower.

equation (15), plus a small allowance, considering both the outward and return strokes.

This face HG, Fig. 23, is not necessarily perpendicular to a radial line BK and often is not [Fig. 19(b)]. If it is inclined, it should be in such a direction as to cause the line of action of the contact force on the working stroke to pass closer to the center of the shaft, thus reducing its moment arm and the torque on the shaft.

The face is not always flat; sometimes it is curved or spherical (more or less). Whatever its configuration, the face is drawn in many positions, enough to define accurately a curve *tangent to all positions*. If an infinite number of positions is used, the resulting envelope is the working curve. If the theoretical points are close together, a working curve drawn *tangent to the mid-points* of the inside bases of the triangles will closely approach the correct curve. When there are as few points as in Fig. 23, an ill-defined working curve should be expected.

The purpose of crosshatching the triangles, Fig. 23, is to be certain that no face position is missed and to make it easy to spot the approximate mid-points. If every thing is all right, it will be found that there is a string of triangles running from one to the other, including all face positions, and returning to the original triangle. *If there is any position of the face of the follower to which the working curve cannot be tangent without crossing some other face position, the correct curve to give the desired motion has not been found.* This situation is illustrated in Fig. 24, where it is seen that the curve cannot be drawn tangent to the face position $2A$ without crossing positions $3B$ and $4C$, and it is most likely to occur when the deceleration is large (**3**). The remedy for this dilemma is to use a larger base circle (§ 48) when it is not practicable to reduce the

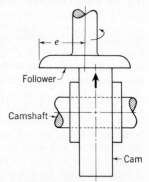

Fig. 25. Mushroom Follower.

deceleration. This condition may exist and not be revealed if the points on the theoretical curve are too far apart (time angles too large).

Often the face of a flat-face follower is made circular (looking at its "bottom"), Fig. 25, in which case the radius e of the face should be somewhat greater than the maximum measurement $8D$ found in Fig. 23. It is common practice in this type of cam to offset the center plane of the cam from the axis of the follower so

that the consequent rotation of the follower distributes the wear. See Fig. 25,
where the amount of the shift shown is greater than normal.

The follower faces for automotive engines are flat because such faces perform
better at high speeds and sharp lifts, and also it is easier to compensate for wear
with flat faces than with rollers on bearings. The only *side thrust* on the follower
(when the plane of the face is perpendicular to the line of motion) is that due to
the frictional force between cam and face. On Diesel engines, roller followers are
used, Fig. 10, because at the slower speeds of operation, the accelerations are not
excessively high. Also rollers are used on radial aircraft engines. These cams,
which turn say at one-eighth engine speed, are made with a series of lobes in the
plane of one circumference, and they are designed for positive return (Fig. 40).

39. The Case against Infinite Jerk.

39. The Case against Infinite Jerk. If the speed of the cam is not too
great, there is no case against infinite jerk, and the parabolic and harmonic motions (and many others) are completely satisfactory. Assume
that a cam laid out with such a motion is operating successfully at a
certain speed—the follower essentially performs the desired motion, the
forces and vibrations imparted by the action of the cam mechanism are
reasonable and untroublesome, and the cam has a long life (that is, the
cam and roller surfaces remain virtually undamaged). Now if the speed
is increased in successive steps, stages will be reached where vibration
becomes excessive and the cam surfaces rapidly wear out. Vibrations
result from periodic changes of acceleration (and force; $F = ma$) and
they increase as the magnitude of a_{max} (and F_{max}) increases. These
magnitudes naturally increase as the speed of the machine increases and
the cyclic displacement of the follower remains the same. As the forces
increase, a point is reached where the contact surfaces are damaged and
the cam wears rapidly. Also, it may in the meantime become evident
that the vibrations originating in the cam mechanism are about to destroy
some other part of the machine. These effects are accentuated if
the change of acceleration is sudden (an instantaneous change of acceleration results theoretically in infinite jerk), because then the effect of the
force is as though it were suddenly applied—not an impact such as a
right to the jaw,* but more analogous to the results of placing your
hands on someone and suddenly thrusting strongly forward (as con-
trasted to a gradual increase in the thrust). More specifically to the
case at hand, we have experimental [and theoretical (7)] proof of the
effects. Let us call the values of the accelerations obtained from our
simple motion equations the *computed* accelerations. Now look closely
at Fig. 26. For each motion, the wavy line represents the actual variation of acceleration, and in each case, the acceleration period is preceded
and followed by a dwell. In Fig. 26(a), notice the computed value, represented by a short horizontal line which changes to a negative value at

---

* With excessive clearances in the system, impacts are also possible.

about the third time interval from the left end of the chart. Then observe that the actual peak of acceleration is more than twice this computed value (and the actual maximum force is more than twice that indicated by the simple theory). The follower system is elastic, and under the action of the impulsive force, which is suddenly applied with the instantaneous acceleration, the system vibrates. The vibration gradually damps out, or tends to until the next sudden change of acceleration. Notice especially the effects of the sudden change of sign, from $+a$ to $-a$ (no change in magnitude), in Fig. 26(a). This is at the point of inflection, the mid-point of the rise (point 3, Fig. 20), where the jerk is theoretically infinite.

In Fig. 26(b), we note the same general phenomenon, a peak acceleration some twice the computed value at a point of instantaneous

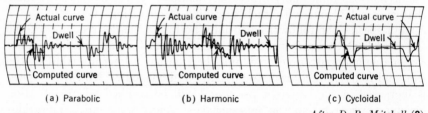

(a) Parabolic   (b) Harmonic   (c) Cycloidal

After D. B. Mitchell (**8**)

Fig. 26. Actual vs. Computed Accelerations. The solid wavy curve is the actual record of variation of acceleration as obtained by test.

change of acceleration, with a gradual damping until the next sudden change.

As speeds of machines were increased and these varying and high accelerations began to do their damage, someone began to wonder what would happen if the sudden changes of acceleration were eliminated. Why not start the motion with zero acceleration? This was done and there are now a number of mathematically definable motions which provide zero accelerations at the ends of the stroke (adjacent to dwells). One of the most popular is the so-called *cycloidal motion.* In this motion, the acceleration varies as a sine function, thus starting at zero. Now study Fig. 26(c) and notice that in the absence of sudden changes, the actual acceleration follows fairly closely the computed acceleration (dotted curve). The more or less violent oscillations are now absent. Let the *computed* maximum acceleration for parabolic motion be represented by the index number 1 ($a_a = 1$); then the *computed* maximum accelerations of harmonic a_h and cycloidal a_c motions are $a_h = 1.23$ and $a_c = 1.57$. While the computed maximum acceleration in cycloidal motion is 57% greater than the maximum a_a, the *actual* maximum acceleration in cycloidal motion is less than the actual maximum acceleration in parabolic motion. The maximum forces in a damped system (there is naturally some damping) are less than those in an undamped system. For an undamped system, Mitchell (**8**) found the *actual* forces to be in the ratio $1:0.834:0.584$ for parabolic:harmonic:cycloidal. With damped systems, the differences in the forces are not so great. We may safely conclude that in high-speed applications, it is generally desirable to keep the *actual* maximum accelerations as low as possible.

As usual with good things, there is a catch—price. In order to approach a true cycloidal curve, it is necessary for the manufacture to be very accurate, especially at the beginning and end of the curve, and precise manufacture is costly. The needed precision is such that the total tolerance on the radial dimension needs to be of the order of half a thousandth of an inch (a visible tool mark may introduce a large sudden change of acceleration). On this account, one would not use the new curves with initial $a = 0$ without a reason for doing so. At slow speeds, the kind of motion used is not significant.

40. Laying Out Cycloidal Motion. Cycloidal motion is a trigonometric type and therefore periodic (as is harmonic), defined by the equation

$$(16) \qquad\qquad s = \frac{\theta}{\beta} L - \frac{L}{2\pi} \sin \frac{2\pi}{\beta} \theta.$$

For a given L and β, values of s may be computed at various cam angles θ; then the computed displacements s may be laid out on the cam or on a displacement diagram. Since, as pointed out in the previous article, the accelerations are sensitive to small errors in s, the various values of s *should* be computed by machine for dimensioning the drawing for manufacturing purposes. Tables giving the cycloidal displacements in terms of angles are in print (**12**).

For our purposes, the motion can be laid out graphically, as shown in Fig. 27, whose caption should be studied now. The proof of the construction is not difficult. Consider any point P on the curve, located by θ and s. From the dimensions on Fig. 27, $s = n - m$. By similar triangles ($0BQ$ and $08A$), $n/\theta = L/\beta$ or $n = (\theta/\beta)L$. The distance m is shown on the vertical diameter of the reference circle, where it is seen to be $m = r \sin \delta = (L/2\pi) \sin \delta$. The angle δ is the angle on the reference circle, Fig. 27, corresponding to the cam angle θ; that is, while θ changes from 0 to β, δ changes from 0 to 2π, both at a uniform rate. Hence, $\theta/\beta = \delta/2\pi$, or $\delta = 2\pi\theta/\beta$; thus $m = (L/2\pi) \sin (2\pi\theta/\beta)$. Combining all of these findings, we have

$$s = n - m = \frac{\theta}{\beta} L - \frac{L}{2\pi} \sin \frac{2\pi}{\beta} \theta,$$

which is the defining equation (16). Notice that the first term is the equation of the diagonal $0A$ and the second term subtracts (adds when

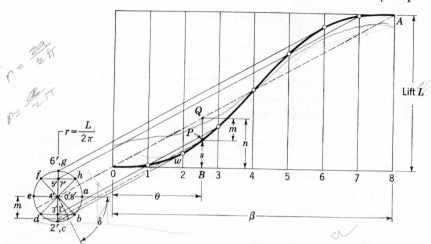

Fig. 27. Laying Out Cycloidal Motion. *Procedure.* Lay out angle β and lift L to suitable scales. Draw the diagonal $A0$. Anywhere along this diagonal, as convenient, draw a circle of radius $r = L/(2\pi)$, here drawn at $0'$, to the same scale as L is drawn. The reference point on this circle moves clockwise in the circumference at constant speed and the start of the motion is on the horizontal center line at a; that is, a corresponds to 0 on the displacement diagram. Next, divide angle β into a number of equal parts (conveniently an even number); and divide the circumference of circle $0'$ into the same number of equal parts (8 herewith), locating points a, b, . . . , h. Project these points a, b, . . . , h onto the vertical diameter locating $0'$, $1'$, $2'$, $3'$, . . . , $8'$. Draw a line through $1'$ parallel to the diagonal $0A$; where it intersects the line for station 1 is a point on the diagram. A line through $2'$ parallel to $0A$ intercepts the station line 2 at point w for another point on the diagram; etc. A smooth curve through the points found is a displacement (st) diagram for cycloidal motion.

$\theta > \beta/2$) from the diagonal to locate the point on the curve. This curve is continuous beyond $\theta = \beta$.

41. Offset Roller Follower. It is not always possible or desirable to locate the center line of the follower on a radial line. Perhaps the configuration of the machine is such that the follower must be moved to another position. If the follower is offset to the *left* of a parallel radial line, as in Fig. 28, the cam generally rotates *clockwise* (CL); if the offset is to the *right*, the cam generally rotates *counterclockwise* (CC). The reason for this relationship and perhaps the reason for offsetting is that the pressure angle at the point of maximum velocity during the rise, and therefore the side thrust on the follower, is decreased; while the pressure angle ϕ on the return will be increased as compared with the same motion of a radial roller follower. See problem 232, Chapter 3. Since the maximum forces normally occur during the rise, this varying of the pressure angle may be advantageous, either in keeping the side thrust at a satis-

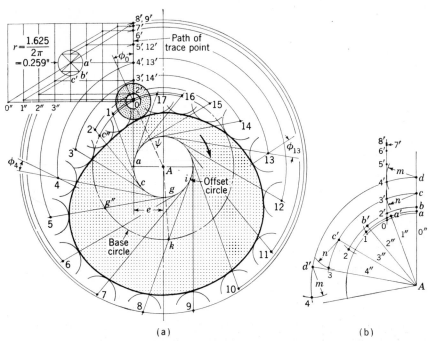

(a)   (b)

Fig. 28. Cam with Offset Follower. *Procedure.* Draw base circle and offset circle ($e = \frac{5}{8}$ in.) and the position of the center line of the follower motion ($a08'$). The trace point is 0. Measure angle $0Ak = 160°$, the angle of rise, and divide it into 8 equal angles (20° each) from 0 (CC since cam rotates CL). Measure angle $aAi = 160°$ and divide it into 8 equal parts, obtaining $a, \ldots, c, \ldots, i$. Now draw the various center-line positions of the follower, using these points, as line cc'' for position 2. Divide the displacement $L = 08' = 1\frac{5}{8}$ in. (half scale) into cycloidal motion, according to the method given in Fig. 27, using 8 divisions to match the 8 time angles for the 160°. The reference circle is divided into 8 equal angles starting at a'. The method of locating points $1', 2', \ldots, 8'$ should be evident from the illustration. Swing arcs as before intersecting the center-line positions of the follower to find points $1, 2, \ldots, 8$, which define the theoretical curve. With the roller radius, strike arcs with centers at $1, 2, \ldots, 8$ and the working curve is tangent to these arcs. Since 8 to 9 is a dwell, this part of the curve is a circular arc. The return occurs with the same motion; hence the return angle is divided into 8 parts in order to use the positions $1', 2',$ etc., already found. The dwell from 17 to 0 is a circular arc. The return points are then $9, 10, \ldots, 17$.

You may prefer the alternative construction shown in (*b*), where points $1', 2', \ldots, 8'$ are located as in (*a*). Measure 160° from Aa and divide into 8 parts, obtaining the 20° angles $aAa', a'Ab',$ etc. Point 3 on the theoretical curve is obtained as follows: Swing the arc of radius $A3'$ from c to 3; in position 3, the radial line $A0''$ has swung to $A3''$ and point c has moved to position c'; point 3 has the same relation to c' as point $3'$ has to c, namely, on the arc $c3'c'3$ and at a radius n from c; with center at c', swing arc of n radius to locate 3. The other points on the theoretical curve are found in the same manner. Go through the motions mentally in locating point 4.

factory value or in permitting a smaller cam (if the thrust is no great problem). The greater the amount of the offset, the more the pressure angle is affected. The method of laying out the cam profile is given in the following example.

42. Example—Cam with Offset Follower; Cycloidal Motion. A follower is to move $1\frac{5}{8}$ in. with cycloidal motion during 160° of cam rotation, dwell for 20°, return with cycloidal motion in 160°, and dwell for 20°, whence the cycle starts over. The cam turns at 1200 rpm CL, and it has been decided to offset the follower $\frac{5}{8}$ in. toward the left. Roller diameter is $\frac{3}{4}$ in.; base-circle diameter is 3 in.

Solution. The details of the procedure are given in the caption to Fig. 28. There are a number of graphical constructions which may be used, an alternate being shown in Fig. 28(b). What one does is to imagine the follower moving about a stationary cam in a direction opposite to the direction of cam rotation, and to follow mentally the positions of the trace point 0. You may invent your own way of locating points on the theoretical curve. Observe, for example, that the center line of the follower is always tangent to a circle, called the *offset circle.* Use this observation for drawing the positions of the follower.

After the lift L has been laid out on the path of the follower, 08', this displacement is divided into the desired motion, cycloidal in this case. The reference circle has been located at the midway point of the curve; $r = L/2\pi = 1.625/6.28 = 0.259$ in. Otherwise the construction is identical with that of Fig. 27. The reader should consider the construction that locates 1, 2, 3, etc., until he perceives and understands why the method is correct.

43. Uniform Motion for Cams. As defined in § 9, uniform motion is motion at constant velocity. With $v = C = ds/dt$, then $s = s_o + Ct$,

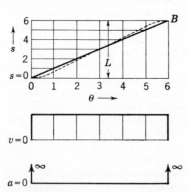

Fig. 29. Motion Diagrams—Uniform Motion. Observe the equal divisions of s for equal time units.

and $dv/dt = a = 0$. We see that the displacement is directly proportional to time (also to θ in cams) and the st ($s\theta$) curve is a straight line. See Fig. 29 for the motion diagrams. Constant-velocity motion is seldom used for cams following or preceding a dwell, because theoretically the follower achieves this constant speed instantaneously and the corresponding acceleration is infinite. Actually of course, $a \neq \infty$ (which corresponds to an infinite force) because of the elasticity and deformation of the parts. However, there is a distinct bump when motion starts or ends, which would limit the use of the unmodified uniform motion to the slowest speed applications.

There are many situations in which for one reason or another, such as delivering an object onto a conveyor, it is desired that the follower move for a while at constant speed. The start and stop can be eased by (1) using circular arcs on the $s\theta$ diagram at the beginning and end, as shown dotted in Fig. 29, (2) starting the motion with constant acceleration, § 44, (3) starting the motion with a cycloidal curve if extreme smoothness is important, or (4) giving the follower any other harmonious starting motion, as desired.

44. Combination of Constant Acceleration and Constant Velocity. This is one of the most common combinations of motions, often called *modified uniform motion.* One of the conditions to be met is that at the point where the motion changes from constant acceleration to constant

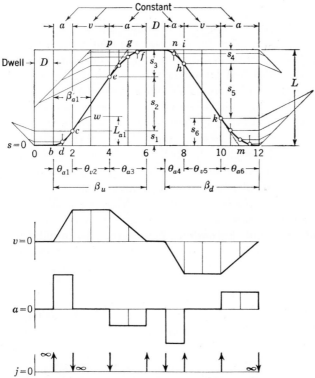

Fig. 30. Motion Diagrams—Modified Uniform Motion. To find the *st* diagram, locate the mid-point of the acceleration period at *d*; locate the mid-point of the deceleration period *g*; draw a straight line *dg*; that part of this line *ce* lying in the $v = C$ angle, station 2 to 4, is a true part of the *st* curve. The part *bc* at constant acceleration is constructed with 2 time intervals, station 1 to 1.5, 1.5 to 2; hence the corresponding rise 2*c* is divided into 2 parts proportional to 1:3 for $a = C$. The part *ef* at constant deceleration is constructed with 4 time intervals between stations 4 and 6; hence the corresponding rise *ep* is divided in 4 parts proportional to 7:5:3:1. See text for justification of method.

velocity, the velocities must be identical. Since it is helpful in dealing with combinations of motion to distinguish symbols standing for the same concept in different motions, we may, for example, let v_a = the maximum velocity in the $a = C$ motion and v_v = the velocity in the $v = C$ motion.

Suppose that a follower is to move some distance s in. with $v = C$; then $v_v = s/t_v$. If a follower moves the same distance s with $a = C$, then the final $v_a = 2s/t_a$ (from $s = at^2/2$ and $v = at$), where t_a is the time it takes for the follower to move s with $a = C$. Equating these two values of v, we get

$$(\boldsymbol{b}) \quad v_v = v_a \quad \text{or} \quad \frac{s}{t_v} = \frac{2s}{t_a} \quad \text{or} \quad t_a = 2t_v \quad \text{or} \quad \theta_a = 2\theta_v,$$

which is to say that the time t_a (cam angle) taken to move a certain distance s by constant acceleration is twice the time t_v (cam angle) it would have taken to move the same distance s with constant velocity. This is the key to the graphical layout of this motion: make the time for the $a = C$ period twice what it would have been if the motion had been $v = C$. See Fig. 30; the construction procedure is explained in the caption. (Read it now.) To demonstrate the correctness of the construction, consider the slowing-down period *ef*. If the follower continued at $v = C$, it would complete the stroke, displacement s_3, in cam angle *pg* (following dotted line, Fig. 30). However, it actually requires cam angle *pf* to complete the displacement s_3. Since cam angle *pf* is twice cam angle *pg*, the condition in (*b*) is satisfied, and the curve *ef* may be an $a = C$ curve if properly constructed, which it is.

45. Oscillating and Secondary Followers. An oscillating follower is one which is in the form of an arm or crank which pivots about a fixed point B, Fig. 31, and the motion of the arm imparted by the cam is oscillatory. The cam may be a disk cam or cylindrical (§ 54). The follower may have a roller or a flat face in contact with the cam. The general problem of finding the working curve is much the same as for other cams: select a convenient point on the follower as the trace point (the center of

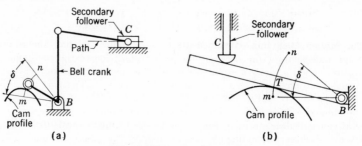

Fig. 31. Secondary Follower.

the roller for a roller follower); divide the path of the trace point into divisions which when traversed in equal times result in the desired motion of the follower; imagine the cam as stationary and imagine the follower moving about the cam and outward at the same time; devise a suitable method for locating points on the theoretical curve; draw in numerous positions of the follower's contact surface and draw a tangent to *all* these positions to obtain the working curve. There is one significant change having to do with the fact that the trace point on the follower moves in a circular arc.

If there is a reason for the trace point to move with a particular kind of motion, say harmonic, the circular arc *mn*, Fig. 31, which is the path of the trace point, needs to be rectified; then the resulting straight line is divided into harmonic motion and the points obtained are stepped off on the arc *mn*. For example, in Fig. 32, the straight line 09″ is the same length as the arc 09′; the divisions for the required motion are made on 09″; then the divisions are short enough that the chord is virtually equal to the arc, and the spaces can be laid off on the arc with dividers. (Unless done to large scale, there is considerable error inherent in this method. With a known motion and displacement of the trace point, the positions can be calculated.)

Since graphical rectifying is inaccurate, the total length of the arc should be checked by calculation, arc $= r\delta$, where δ is the angle of oscillation of the follower. Since the smaller the angle δ, the more nearly the arc approaches the chord in length, we can *for academic purposes* set an arbitrary limit on the magnitude of arc to be rectified: if the arc is 30° or less, do not rectify; if the arc is greater than 30°, do rectify. When the arc is not rectified, the motion may be laid out on the chord— with probably no more error than exists in graphical work done to small scale.

As likely as not, an oscillating follower drives another body via connecting links, say. The body so driven is called a **secondary follower**. Perhaps the oscillator drives a reciprocating follower *C*, Fig. 31, which reciprocates in a horizontal, Fig. 31(a), vertical, Fig. 31(b), or any other direction. Some such arrangement may be necessary because the configuration of the machine is such that a cam acting directly on *C* is impossible to install—a sort of remote control being necessary. It is also as likely as not that it is the motion of the secondary follower *C* which is defined for a reason. In this event, the path of *C*, Fig. 31, is divided into the required motion, and the corresponding points on arc *mn* may be found graphically by using the known relationships of the links.

If the design of a machine is such that the natural location of the cam is offset considerably from the line of travel of the final follower, some kind of oscillating follower may be an easy solution to the difficulty. Also

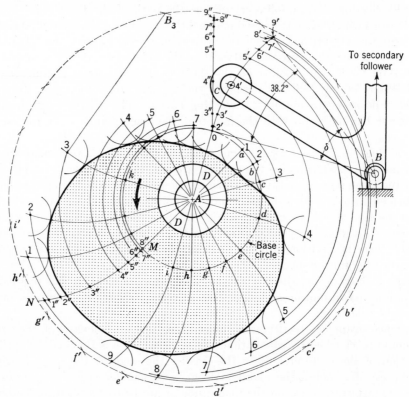

Fig. 32. Disk Cam with Oscillating Follower. *Procedure.* Rectify arc 09′ onto any convenient line 09″. Divide the cam angle into appropriate divisions, and divide the rectified arc 09″ into the desired motion, in this case modified uniform motion. Transfer points to arc 09′, locating 2′, 3′, . . . , 9′. With a radius AB, draw dotted circle which is the locus of all positions of B as follower is imagined moving around the cam. The first 30°, arc 0a, is a dwell. Divide the 30° arc ac into 2 equal parts to match the 2 spaces, 02′ and 2′3′, of $a = C$ motion. The $v = C$ is for 60°; this locates d and e. The deceleration is also for 60°; this locates f, g, h, and i for 4 spaces, 5′6′, 6′7′, 7′8′, 8′9′, on path of follower. With radius $B4′$ and centers on B's locus, circle $Bb'g'$, draw arcs from b, c, . . . , i. The center of the roller lies on the arc at each position. For example, c' is the center for the arc from c on which 3 (up) is located; B_3 (down) is the center for the arc from k. When B is at point g', the center of the roller has moved up curve $g7$ and is located at 7. With center at A, strike arcs 2′, 3′, . . . , 9′ to find 2, 3, . . . , 9, as shown. From 9, there is a 30° dwell. The points for the return are found similarly. For convenience, we started with the number 1 again at N and laid out the return motion on arc MN. Study this construction until you see why it correctly locates the theoretical points. Draw in face positions (arcs of a circle with a radius equal to the radius of the roller) and then draw the working curve tangent to all face positions. See Fig. 33 for an alternate construction.

as in Fig. 31, the configuration can be such that a large stroke of the secondary follower is provided with a relatively small cam. Not all secondary followers are driven by an oscillator.

If the oscillating follower has a flat face, choose any convenient point on the contact face as the trace point, say where the vertical center line of the camshaft intersects the lowest position of the face. It is hoped that with one more example explained in detail, the reader will be able to lay out any disk cam with any kind of follower system.

46. Example—Disk Cam with Oscillating Follower. In Fig. 32, given a base circle (2.5 in.), the location B of the fixed pin in the arm (relative to A), the size of roller ($\frac{3}{4}$ in.), the radius of the arm ($B4' = 3$ in.). With the offset to the right, it would be better for the cam to rotate counterclockwise. To be economical of what has been done, let the motion of the center of the roller be the same as that defined in the displacement diagram of Fig. 30. Find the working curve.

Solution. For this motion, the arc $09'$, Fig. 32, is 2 in., corresponding to an angle $\delta = 38.2°$; the rectified arc $09''$ is of course also 2 in. The points on the st diagram, Fig. 30, are transferred to $09''$. Since 01, Fig. 32, is a dwell, the follower begins to move at 1, constantly accelerated to 3, constant velocity 3 to 5, and $0a$ and $9N$ are circular arcs (dwells). More detail on the procedure is given in the captions to Figs. 32 and 33.

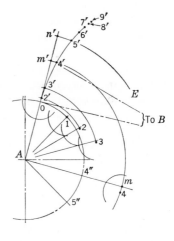

Fig. 33. Alternate Construction, Oscillating Follower. *Procedure.* Divide arc of rise $09'$ into the desired motion as before. Divide the cam angle into appropriate "time units" as before, and draw radial lines $A1$, $A2$, etc., of length to obtain intersections with arcs from $2'$, $3'$, . . . , $9'$. Use point 4 for purposes of explanation. Swing arc of radius $A4'$ until it intersects the proper radial line at m. (Numbering the radial lines, as $4''$, will aid greatly in keeping the radial lines and arcs correctly matched.) Now at the top of the figure, we note that $4'$ is a chord distance $m'4'$ *clockwise* from a radial line drawn through the point 0 on the base circle where the arc path $09'$ starts. Therefore, if point 4 is laid off by dividers from m an amount equal to $m'4'$ *in a clockwise direction*, the correct location of the trace point for this position is found. The method is the same for all other points. If point 5 were found, it would be a distance $n'5'$ *clockwise* along the arc $n'5'E$ from the point where this arc intersects the radial line $A5''$.

47. Other Motions Used in Cam Mechanisms. There are a number of other kinds of motions used for cams, many entirely arbitrary, as for example cams whose profiles are made up of a series of circular arcs with or without straight-line tangents such as shown in Fig. 16. Tangent circular arcs impose no discontinuity in velocity, but the acceleration instantaneously changes ($j = \infty$) at the junction of two arcs of different radii. Pieces outlined by circular arcs are relatively easy to manufacture.

Not only may the displacement program be arbitrarily assumed, as the circular arcs above, but a certain velocity or acceleration program may be desired. For example, Fig. 34, we may desire a very smooth start

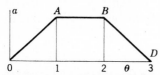

Fig. 34. Combination Motion.
This motion is called *trapezoidal acceleration* and also *modified constant acceleration*.

($a_o = 0$) with finite jerk, part $0A$, Fig. 34, a constant acceleration AB, and a gradually decreasing acceleration BD. Note that during $0ABD$, the acceleration is positive and the follower has been increasing in speed. To bring the follower to rest, there must be negative acceleration such that the "negative" area on the at (or $a\theta$) diagram is equal to the positive area shown. This would bring the follower to rest and complete one stroke.

48. Effect of Size of Base Circle. The size of the base circle has already been related to the pressure angle, § 34, but there are additional particulars that are of interest. Since the size of the cam varies with the size of the base circle, costs are involved, so that the engineer often compromises the maximum pressure angle to obtain lower cost. If a large number of cams are to be made, as in some of our mass-production industries, a small reduction in size may represent a substantial total saving in cost. Also, the maximum size of cam is often limited by the space available for it. On the other hand, the cam size must be large enough to provide plenty of stock around the shaft—unless the cams are made integral, as in automotive-engine camshafts.

It has been reiterated that the working curve must be tangent to the follower's face at all its positions, and a possible difficulty in this regard is explained with respect to a flat-face follower in Fig. 24. In a similar manner, it may happen

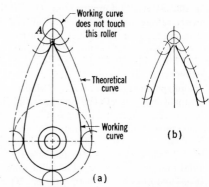

Fig. 35. Working Curve Not Tangent to All Positions. If the motion as defined by the theoretical curve is required, the remedy for the situation in (a) is to use a larger base circle. With a smaller roller in (b), the center of the roller would approach closer to its proper peak position.

that a true working curve cannot be found for a roller follower when the change of motion is abrupt. In Fig. 35(a), the working curve falls far short of lifting the center of the roller to its supposedly highest position at A. If the theoretical curve were pointed at A, only a knife-edge follower on the theoretical curve would have the defined motion. If the size of the roller is reduced, the working curve more nearly moves the follower to its top position, and it theoretically would do so if the radius of the roller is not greater than the minimum radius of curvature of the theoretical curve. Of course, there is some irreducible minimum for the diameter of the roller; there must be room enough for an adequate shaft and bearing, and it also happens that the contact stresses increase as the size of roller is decreased. Thus, a larger base circle is more likely to be the remedy for the condition in Fig. 35(a).

49. Special Forms of Cams. A *translating cam,* Fig. 36(a) and 46, is one which moves in a straight line. Evidently it must reciprocate and it may be driven by any one of a variety of linkage arrangements, or by virtue of the action of another cam.

A circular disk, Fig. 36(b) or (d), rotating about a pin A eccentric to the center of the circle is a common type of cam, called an **eccentric cam.** An *eccentric* as such is usually an arrangement as shown in Fig. 36(c), having a strap about the eccentric disk which is integral or connected with an arm that extends to a slider H. Figure 36(c) is not considered to be a cam, but the motion of the follower H is defined by the same equations which define the motion of C in Fig. 36(b); namely, those for a slider crank, § 18. In Fig. 36(b), the equivalent slider crank mechanism is crank AB, connecting rod BC, slider C; in Fig. 36(c), it is crank DG, connecting rod GH, slider H. The motion of the flat-face follower on the eccentric, Fig. 36(d), is harmonic, like that of the Scotch yoke, Fig. 15. Imagine the point K (center of circle) moving about the axis A at constant velocity. The velocity of the follower is seen to be always the projection of the velocity vector for K on the vertical diameter—which is a characteristic of harmonic motion (§ 23). The stroke of each of these followers, Fig. 36(b), (c), and (d), is $L = 2e$, where e is called the eccentricity.

The **toe-and-wiper cam,** Fig. 36(e), is used to operate valves on steam engines; the cam oscillates in this mechanism.

The **involute cam,** Fig. 36(f), is one whose theoretical curve is an involute. The method of laying out an involute is described in the appendix (§ 179). When the center line of the follower is tangent to the base circle from which the involute is generated, the follower moves with constant velocity while it is in contact with the involute. Such cams are used on ore crushers in stamp mills (low velocities). The beginning of the curve may be modified to provide a smoother start. The outline of this cam may be $ABCD$, Fig. 36(f); the follower runs off of the cam

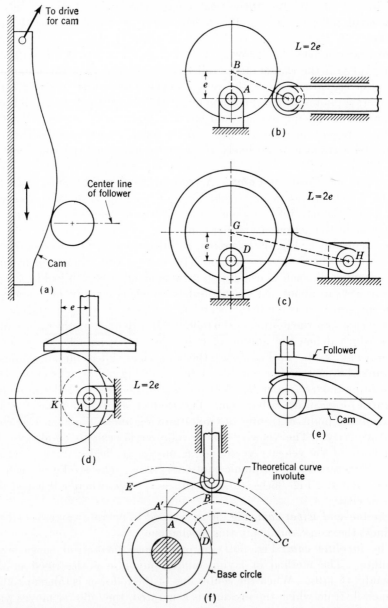

Fig. 36. Some Special Cam Forms. (a) Translation cam. (b) Eccentric cam with roller follower. (c) Eccentric. (d) Eccentric cam with flat-face follower. (e) Toe and wiper. (f) Involute cam.

curve at C, drops, and crushes ore. The curve EBC may be taken as typical of a portion of a continuous cam curve, part of which is an involute.

50. Inverse Cams. Another special form of cam is an inverse cam, Fig. 37, in which the functions of the parts are reversed; the body with

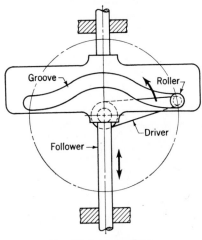

the *path* is the driven member and the roller is the driver. A Scotch yoke, Fig. 15, is an inverse cam. The driving arm that carries the roller may either oscillate or make complete revolutions. The groove for the roller may be shaped to give any desired motion for 180° of turn of a rotating arm. These mechanisms are used where the loads are light, such as in sewing machines. The procedure for layout is as usual to assume a series of positions of the driving member at unit time intervals apart, and determine the corresponding positions of the driven member for the kind of motion de-

Fig. 37. Inverse Cam.

sired; then project the points in such a way as to define the center line of the groove.

A *Geneva mechanism,* so called because it was early used as a device to prevent overwinding of watches made in Geneva, is one which might be described under the heading of inverse cams, although there is a distinctive difference between this one and other cams described in this chapter. Whereas the driven members in the others oscillate or reciprocate, the driven member in the Geneva mechanism rotates intermittently (it has dwell periods) and always in the same sense. The operation of this mechanism can be understood from Fig. 38. The member C, which might be a gear driven by another gear P, rotates at constant speed (usually) and contains the driving pin (or roller) A. Rotating clockwise, say, the pin A enters a radial slot (but these slots are not necessarily radial) on the driven wheel G and turns G while A moves to position D, where it is on the point of leaving wheel G. Slot K moves to the present position of A. While the pin is moving from position D to A (CL via DEA), the driven wheel remains stationary. Back again to position A, the pin enters the next slot and turns the wheel again through angle θ. A plate H, shown shaded, is attached to the driver C, which could be just a crank carrying the pin A. The shape of the driven wheel between slots has virtually the same radius as H. When the pin leaves at D, H is

engaged with a circular part on G and holds G in position until the pin
engages the next slot.   The arc MdN is cut from the plate H to provide
clearance for the largest radius of G, as at K.   Let the distance Ad be a
little greater than ab (you explain why), and the angle $MBA = \delta = aBD$
(also $ABa = ABN = \delta$) in order for the plate to lock wheel G.

As seen, one may use various numbers of slots (**stations**), 3 or more,
with the top limit dependent on the required size of pin or roller and
other dimensions.   In Fig. 38, the crankpin A makes one turn for $\frac{1}{5}$

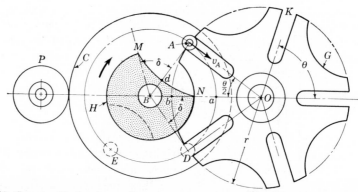

Fig. 38. Geneva Mechanism.   *Procedure.*   Having decided upon radius r (this may
be a trial solution), draw outside boundary AKD of G.   Divide this circumference
into the number of movements (slots) desired, placing 2 slots A and D equidistant
from the line of centers BO.   From A, drop a perpendicular to AO, which intercepts
point B on the line of centers; B is the best axis of rotation for C.   This defines
radius AB.   Curve MdN is such as to provide clearance for the driven member;
make angle MBN = angle ABD; make distance Ad a little greater than ab.   If the
size of parts is not large enough to accommodate shafts, etc., choose a larger radius r
and repeat.   [Greater masses, for a particular acceleration, mean larger forces and
moments—$F = ma$ and $T = I\alpha$.   See Johnson (**39**) for optimum masses as balanced
against stresses.]

turn (5 slots) of G.   The driven member naturally speeds up and decelerates with each engagement of pin A.   At quite high speeds, which correspond to high accelerations (and forces, $F = ma$), the maximum acceleration can be and has been reduced by shaping the slots other than
radial.   But this is expensive.   The maximum acceleration occurs near
the $\frac{1}{3}$ and $\frac{2}{3}$ positions, counting from beginning to end of one movement
of G.

As designed in Fig. 38, the pin A is moving radially AO at the instant
of initial engagement, so that there is no component of v_A in the direction
of the sides of the slots.   This situation gives the smoothest entry (without impact), which is the reason why angle BAO should be made 90°.
(See the caption to Fig. 38.)   However, the initial acceleration is not

zero; hence, there is infinite jerk, suddenly applied loading, which is a source of trouble if the speeds are high enough. See Fig. 47.

The Geneva mechanism is widely used as an indexing mechanism on machines: moving tools, conveyors, feeding devices, and other elements which should have intermittent motion. It was the first successful device for moving film in motion-picture projectors, intermittently moving a frame into position. One could, in Fig. 38, place another driving pin shown dotted at E, in which case C would make $2\frac{1}{2}$ rev. for 1 rev. of G. Geneva mechanisms appear in numerous variations. See references **47, 48.**

51. Positive-motion Cams. In most instances of disk cams, the follower returns to its starting point by virtue of a spring force or, occasionally in vertical positions, by the force of gravity. Since it is not

Courtesy The Rowbottom Machine Co., Waterbury, Conn.

Fig. 39. Face Cam. Shown in position in a cam-milling machine; end mill cutting groove.

always practical or desirable to depend on these forces, a more positive return is sometimes needed. A *positive-motion cam* is one which has been designed to drive the follower on both strokes (no dependence on springs or gravity).

Since a groove is a logical first thought to provide positive motion, the *face cam*, Fig. 39, is often used. The inside surface of the groove drives the follower up; the outside surface drives it down. This change of contact surface (reversal of force) is a source of pounding and vibration, which is minimized by keeping the clearances, especially the clearance between the roller and the groove, a minimum. Since zero clearance is impracticable, the effects of force reversal cannot be entirely eliminated. The alternative of the groove would be a ridge or rib with rollers on each side, Fig. 40, an arrangement in which the clearances could be very small. This latter plan is used for valve cams in aircraft radial engines. In laying out such a cam, the trace point T could be taken midway between the roller centers, then a number of roller positions drawn as indicated at A for the purpose of defining the working curves.

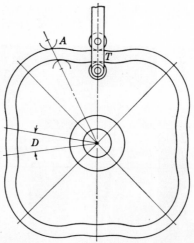

Fig. 40. Ridge Cam. This follower is raised and returned to its starting position four times during one revolution of the cam, with a short dwell between movements.

A *constant-diameter cam* gets its name because the centers of the rollers are a constant distance apart, Fig. 41. It can be laid out for any desired motion for 180°, but the remainder of the motion must satisfy the condition of the roller centers being a fixed distance apart. Thus in Fig. 41, find points 1, 2, . . . , 6 in the usual manner. Then with the dividers set with the distance between roller centers and with centers at 1, 2, . . . , 5, locate points a, b, . . . , e on the respective radial lines. The theoretical curve is now defined. Find the working curve in the usual manner.

A *constant-breadth cam,* Fig. 42, is similar to a constant-diameter cam, except that the follower has flat faces a fixed distance apart ($= D_b + L$). It can be laid out for any motion for 180°; the remaining 180° must satisfy the constant-breadth condition. The theoretical points are located as

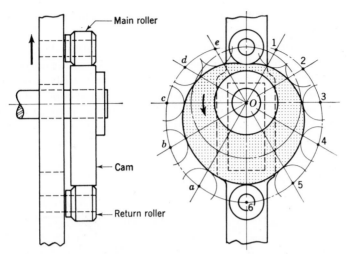

Fig. 41. Constant-diameter Cam.

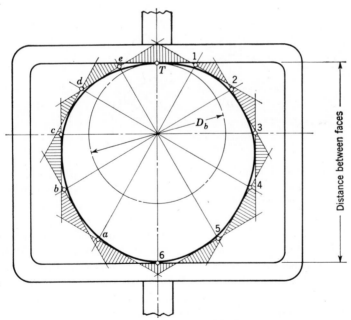

Fig. 42. Constant-breadth Cam.

before; the working curve can be determined as for any other flat-face follower.

Main-and-return cams, also called **conjugate cams,** Fig. 43, as the name implies, consist of two cams. The so-called main cam is designed for 360° of motion as desired, using methods previously described for roller followers. In Fig. 43(a), this produces the theoretical points 1, 2, . . . , 11 and the solid cam curve. With the dividers set at the desired distance between roller centers and with centers at 1, 2, . . . , 11, locate

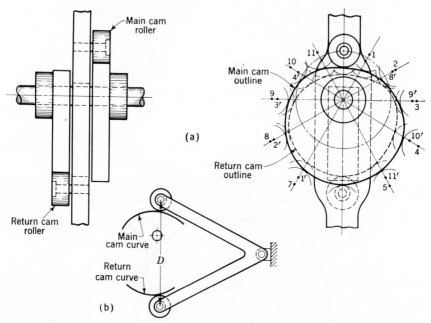

Fig. 43. Main-and-return Cams. Conjugate cams may be superior to an ordinary cam in high-speed situations where it is sometimes impracticable to provide spring loadings large enough to maintain follower contact.

points 1′, 2′, . . . , 11′ on radial lines to define the theoretical curve for the return cam. The working curve for this cam is shown dotted.

Either the constant-diameter cam or the main-and-return cams may be constructed with oscillating followers, as suggested by Fig. 43(b). In either case, the distance *D* is constant. Main-and-return cams can be designed and assembled with minimum clearances between parts, thus minimizing troubles which arise as a result of force reversals.

52. Cylindrical Cams. Cylindrical cams are generally positive-motion cams, except when the cam surface is on the end of a cylinder, as in Fig. 44. The follower for an **end cam** may either reciprocate or oscillate.   The usual variety of cylindrical cams is one with a groove or grooves providing the cam surfaces, Fig. 45, into which pins or conical rollers fit.   (If the path crosses itself, § 53, the follower must be oblong.) The follower may oscillate, as in Fig. 45, or reciprocate.   A third type of cylindrical cam is the *barrel cam,* commonly used on automatic lathes;

Courtesy The Rowbottom Machine Co., Waterbury, Conn.

Fig. 44. End Cam.   Also called *side cam.*

Courtesy The Cincinnati Milling Machine Co., Cincinnati

Fig. 45. Cylindrical Cams.   Cams in feedbox of milling machine for shifting gears and clutch, which are behind the visible cams.

see Fig. 46, whose caption briefly states some of the operations controlled by the cams.   The feature of barrel cams is that the motion and the timing of the motion may be varied to do a particular job by using different hardened-steel cam plates *H,* which can be bolted in suitable positions on the drums.

A fourth type of cylindrical cam has a rib instead of a groove, and it is

Courtesy Jones & Lamson Machine Co., Springfield, Vt.

Fig. 46. Barrel Cams on Automatic Lathe.   Drum *A* carries cams which control tool movements by shifting the "center bar" *F*.   Drum *B* carries cams which move the "front former" *G* (a translation cam) that is bolted to the "front former slide" *D*. As the follower *E* moves up the incline on *G*, the tools are moved into cutting depth. The cams on drum *C* operate the main drive clutch and automatically stop the machine at the end of its cycle; also they regulate the speed of the machine so that it moves slowly during cutting, fast during return and nonproductive motions.

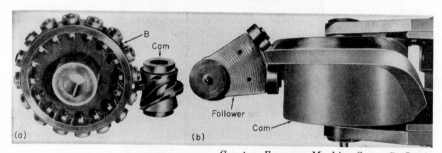

Courtesy Ferguson Machine Corp., St. Louis

Fig. 47. Roller Gear Drives.   In the form shown in (a), this mechanism is acting in the same manner as a Geneva drive; namely, the cam is driving the wheel *B* through $\frac{1}{16}$ turn (said to have 16 stops) with each engagement of the cam surface.   In form (b), the follower oscillates, and is used for winding electric-motor stators.

called a *roller gear drive* by its manufacturer, Fig. 47.   The manufacturer of these cams uses a modified trapezoidal motion (Fig. 34, zero initial acceleration) which makes them appropriate for the highest speeds. They are used where intermittent motion is desired, as in indexing (see § 50, Geneva mechanisms) and feeding machines.

53. Example—Theoretical Curve for Cylindrical Cam.   A reciprocating follower is to move toward the right for 12 in. with uniform motion in $1\frac{1}{4}$ turns of the cam; dwell for $\frac{1}{4}$ turn; return to the starting point with UARM in 1 turn; dwell for $\frac{1}{2}$ turn. Lay out the theoretical curve (a) on a projection of the cylinder, (b) on the developed surface.

Solution. As suggested by the statement of the problem, grooved cylindrical cams may have the cycle of operation continue for more than 1 revolution of the cam. In this event, the grooves necessarily cross and the shape of the follower

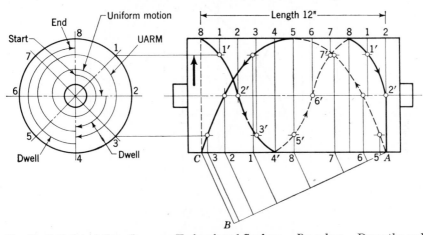

Fig. 48. Cylindrical Cam Curve on Undeveloped Surface. *Procedure.* Draw the end view and a side elevation. Divide the circular end view into a number of equal divisions, say 8, to represent equal time units, obtaining points 1, 2, . . . , 8. Divide the 12-in. stroke into 10 equal parts to give uniform motion for 10 time units ($1\frac{1}{4}$ rev.). Draw lines 1-1′, 2-2′, etc., from 1, 2, etc., in the end view and parallel to the cam's axis to obtain the intersections 1′, 2′, etc., which are points on the theoretical curve as laid out on the *surface* of the cylinder. The dwell for $\frac{1}{4}$ turn is represented by 2′A. Since the return motion occurs in 1 rev., there are 8 time units. (For UARM, there must be an even number of time units.) Divide the line of movement AC into 8 divisions to give UARM (1:3:5:7:7:5:3:1). Vertical lines from these divisions on AC and horizontal lines from 5, 6, etc., on the end view locate other points on the theoretical curve to point C. The follower is now in its initial position, but the cam is not. The dwell C8 of $\frac{1}{2}$ turn, represented by the dotted line, completes the curve on the cam.

in the groove must be elongated (somewhat the shape of a canoe as one looks straight down on it) so that it will satisfactorily pass the intersection of the grooves. Its length should be such that the leading end is across the intersection before the widest part of the follower reaches the intersection.

(a) Let the top of the cylinder in the plan view of Fig. 48 move away from you at the outset. Thus, the first part of the curve is visible; then it passes to the backside at 4′, front again at 8, backside again at A, front at 5, backside at C to the starting point. Since the cam turns at a uniform speed, equal arcs on the circumference represent equal units of time. The details of the procedure are given in the caption to Fig. 48.

(b) To lay out the curve on the developed surface, compute the circumference of the cylinder, which is one dimension of the rectangular developed surface, Fig. 49. The other dimension is the length of the cylinder. See the caption of

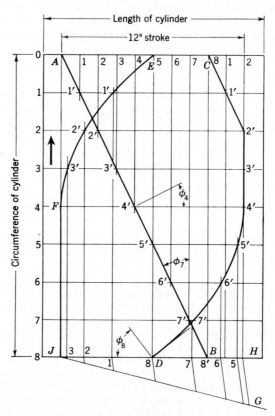

Fig. 49. Cylindrical Cam Curve on Developed Surface. *Procedure.* Draw the rectangle, circumference by length. Divide the circumference into a number of equal divisions (time units) as before, and divide the stroke into 10 equal parts for uniform motion. These are the divisions at the top. Dropping lines vertically downward to intercept the corresponding lines of equal cam angles locates the points $1', 2', \ldots, 8'$, then to $C2'$, which define that part of the theoretical curve. Notice that in the cylindrical form, the top and bottom elements are the same ones; hence, in the cylindrical form, $ABC2'$ is a continuous curve. The dwell $2'4'$ is vertical. For the return, divide the stroke into divisions for UARM as in Fig. 48, say along HJ, and project the points obtained vertically to find $5', 6' \ldots, D$ and $E,1',2',3'$; FJ is the dwell to the starting configuration.

Fig. 49 for details. If the undeveloped surface of Fig. 49 is wrapped about the cylinder, the curve is continuous. A template in this form serves to transfer the curve to the cylinder itself.

In order for the side thrusts not to be excessive, the maximum pressure angle must be held within limits, often stated as 30° maximum. See § 34. As before, the pressure angle is the angle between the direction of motion of the follower and a normal to the path, which is the same as the angle that the curve makes with a line parallel to the developed circumference; see ϕ_4. If the maximum desired pressure angle is exceeded, the remedy is to increase the size of the cylinder (motion remaining the same). It would appear in Fig. 49 that ϕ_8 is somewhat large.

54. Cylindrical Cam with Oscillating Follower. If the follower oscillates, its path on the developed surface is shown by circular arcs of radius equal to the radius of the oscillator—instead of straight lines as before. After dividing the circumference into equal time units, use the radius of the oscillator OA, Fig. 50, and draw the circular arcs shown from points

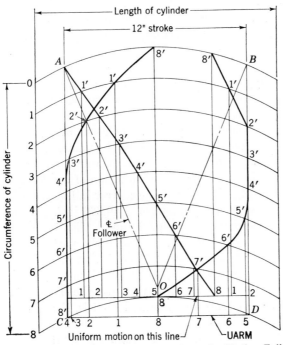

Fig. 50. Developed Surface, Cylindrical Cam; Oscillating Follower.

0, 1, . . . , 8 on the circumference. Then divide the path into the desired motion. The layout in Fig. 50 is for the motion given in § 53 and it is such that theoretically the component motion of the follower in an axial direction will be as given. If it is desired that the motion of the trace point *in the path* be as stated, the circular arc should be rectified and *then* divided into the required motion.

55. Miscellaneous Remarks on Cams. In case the pressure angle for a roller follower is too large and the cam cannot be made larger, perhaps it would be possible to use a lever (as a primary follower) with enough length to give a secondary follower near its end the required stroke (Fig. 31).

If the motion of the follower is to be continuous (without dwells), a linkage of cranks, connecting rods, and perhaps sliders will usually be a better solution than a cam and follower. Thus, cams are primarily functional where intermittent motion is involved. However, there are innumerable ways of providing intermittent motion (**47, 48**), including specially designed gearing. The roller gear drives, Fig. 47, could as properly be classified as gearing as cam mechanisms.

The limiting speed of operation of many machines is set by the cam design. If so, it will probably be found that the cam curves are such that infinite-jerk periods occur. Where this situation has existed in the past, a redesign of the cam curve to eliminate instantaneous changes of acceleration has permitted significant increases in the speed of the machine.

When one obstacle that limits speed has been eliminated, as by providing a better cam, one soon finds another obstacle at a somewhat higher speed. That is, something always sets a limit, or else the machine could be run infinitely fast.

When the new motions ($j \neq \infty$) are used, one must do more than just define the surface. The cams must be manufactured closely to the correct definition. On one test (**16**), it was found that an error of ± 0.0003 in. in the profile caused a variation of acceleration of $\pm 15,600$ ips². This error together with ± 5 sec. error in the cam angle resulted in a total fluctuation of acceleration of $\pm 46,800$ ips². With no error, the maximum was supposed to be 40,000 ips². This is to say that a small manufacturing error, especially in the early part of the curve, may nullify the designer's intentions. There is also sometimes an additional duty for the engineer, requiring a higher order of engineering analysis— modifying the cam curve in such a way that the elasticity and deflection of the follower system are compensated [that is, the final follower moves when and as desired (**3, 9**)].

Large master cams are helpful in obtaining accurate cams of the desired size. In the most precise cams, the radius of the surface should be calculated for every 0.5° to 1° of cam rotation (**52**). Accurate master cams (called leaders) are then made by taking cuts on a jig borer or a milling machine, holding the accuracy to, say, 0.0001 in. The milling operation leaves a series of shallow scallops on the cam surface, which are removed by careful filing. The master cam, or a copy of the master, is used on cam milling machines to guide the tool that cuts the production cams.   The roller and cam surfaces need to be hardened to withstand severe service.

The follower spring must be correctly designed to exert enough force to keep the follower and cam in contact.   This design involves dynamic-force analyses (**3, 27, 28, 29**).

The ratio of the diameter of the roller to the diameter of its pin axis should be at least 3 in order for rolling friction to produce enough turning moment to overcome the sliding friction between pin and roller (**52**).

There is no point in coupling two cycloidal curves together for rise and return, because the acceleration returns to zero at the end of the motion, an unnecessary event if the motion is to be continuous (without dwell). It would have the disadvantage of a higher pressure angle than necessary.

A harmonic motion is excellent for moving the follower to and from the end of a stroke *without* dwell.   But harmonic motion would not be preferred at high speed before or after a dwell ($j = \infty$).

An 8th-power polynomial (**9, 20**) has a nonsymmetric acceleration, with finite jerk at all times, which is advantageous in some circumstances.

56. Closure.   In any case where the follower has completed a round trip, the area between the vt curve and the $v = 0$ axis is zero—just as much negative area as positive ($\int v \, dt = \Delta s = 0$).   Also the area between the at curve and the $a = 0$ axis is zero.

In any case where the follower makes one stroke, $v = 0$ at each end, the area between the at curve and the $a = 0$ axis is zero ($\int a \, dt = \Delta v = 0$).

It is always the path of the follower which is divided into the required motion, with the same number of divisions as there are equal time units (angles).   The cam usually, but not necessarily, turns at constant speed, so that equal angles represent equal durations of time.

The reader has probably observed that none of the examples in this chapter were laid out to a large enough scale to give good results.

Unfortunately, smooth cam operation is not simply a matter of a proper kinematic design.   At high speeds, a flexible follower system (and no system is rigid) results in the motion of a particular point on the system not being the motion as designed into the cam.   Thus it is sometimes necessary to consider the dynamic reaction of the system itself, the way it deflects and its deflection at various positions in the cycle, in order to have the proper point in the follower describe the desired motion.   See references **3, 9, 16, 17, 25**.

The possible and actual variations in the details and appearances of cams are enormous (true also of linkages to be studied next—see references **47, 48**), so that all that can be done in a textbook is to develop as much of the fundamental theory as feasible and leave the remainder to the enterprise, intelligence, and ingenuity of the reader.   Since the subject is so extensive that some engineers spend years of their life with a major effort devoted to a study of cams, this chapter has necessarily left many important things unsaid. If the reader desires a more extensive knowledge of the subject, the current literature, much of which is cited among the References (at rear), is probably the best source of new information. Rothbart's (**3**) is the most complete book on the subject.

PROBLEMS

Note. Where graphical solutions are called for, use $8\frac{1}{2} \times$ 11-in. paper, unless otherwise indicated or unless it is desired to double or triple the size for greater accuracy. Lay out full size unless otherwise specified. See *Important Note* 2, p. 28. Considering those plates which have "same-as" alternate movements, the instructor may choose to divide all alternatives among teams of students and compare solutions, especially the appearances of the motion diagrams.

Radial Follower

71. A cam is to lift a radial follower through 3 in. with UARM during a 180° turn; then there is a dwell for 30°, and a return to the starting point with UARM in the remaining angle. The base circle is 3 in. in diameter; the cam turns CC; use a 1-in. roller. Lay out the cam curve and determine the maximum pressure angle. Is this angle satisfactory?

72. The same as **71** except that the follower moves with SHM.

73. The same as **71** except that the return motion is SHM.

74. The same as **71** except that the motion is cycloidal.

75. Lay out the working curve for a radial follower to move *out* 1 in. during 150° turn of the cam with UARM, to dwell for 30°, to move *in* 1 in. during 150° turn with UARM, to dwell for 30°. Use a 2-in. base circle, a $\frac{1}{2}$-in. roller, and draw twice size. Cam rotates CC at 660 rpm. What is the maximum pressure angle on each stroke (graphically and algebraically)?

76. The same as **75** except that the motion is SHM.

77. The same as **75** except that the motion is cycloidal.

78. The same as **75** except that the outward movement is with $v = C$ and the return is with SHM.

79. Use plate No. 22, Wingren. The follower moves up 2 in. with UARM while the cam turns 120°; dwell for 30°; moves down with constant acceleration during 120° and then continues down to the starting point with constant deceleration during 90°. The cam rotates 300 rpm CL. (a) The condition to be met on the downward stroke where the magni-

tude of the acceleration changes is that the velocity at the end of the speeding-up period is equal to the velocity at the beginning of the deceleration period. Find the working curve. (b) Determine the maximum pressure angle on each stroke, graphically and algebraically.

Ans. (a) $s_2 = 0.857$ in.

80. The same as **79** except that the timing is: *out* during 140°, dwell 10°, return the same as before.

81. Use plate No. 22, Wingren. The follower moves 2.25 in. with SHM while the cam turns 90°; dwells for 60°; moves down with SHM during 120°; dwells until the cycle repeats. The cam rotates at 480 rpm. (a) Construct the working curve for the cam. (b) Determine the maximum pressure angle both graphically and algebraically.

82. The same as **81** except that the motion is cycloidal.

Flat-face Follower

83. (a) Find the theoretical and working curves for a disk cam with flat-face reciprocating follower. The face is perpendicular to the line of travel of the follower. The base circle is 3.25 in. in diameter, the stroke is 1.25 in., and the cam turns 2400 rpm. The outward movement is with UARM during 120° rotation; return is with UARM during 150° rotation; dwell 90°. (b) Determine the face lengths needed on each side of the trace point by both graphical and algebraic means.

84. The same as **83** except that the motion is SHM.

85. The same as **83** except that the motion is cycloidal.

86. The same as **83** except that the return motion is SHM.

87. Use plate No. 23, Wingren. The flat-face follower is to move outward $1\frac{1}{4}$ in. with harmonic motion during 165° of cam rotation, to return to the starting point during 165° with harmonic motion, to dwell 30°. The cam turns 450 rpm CC. (a) Find the working curve. Let the trace point be the tangent point of the face and the base circle. Measure the minimum values of a and b and specify dimensions to be used. (b) Compute the dimensions a and b.

88. The same as **87** except that the motion is UARM.

89. The same as **87** except that the motion is cycloidal.

90. The same as **87** except that the return motion is UARM.

91. A reciprocating follower has a flat face inclined at 75° with its line of travel as shown in Fig. P11. Let $D = 1.5$ in.

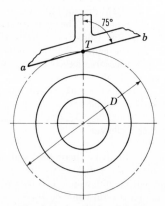

Fig. P11. Problems 91–94.

The lift is 0.7 in. in 40° of rotation; the cam rotates at 600 rpm. Use the trace point T for the theoretical curve. The outward motion is SHM; there follows a 120° dwell; return with SHM during 120°; dwell 80°. Rotate the cam in the most advantageous direction. (a) Find the working curve. Draw twice size. (b) Measure the minimum lengths of face Ta and Tb needed. If you do not find a satisfactory working curve, suggest ac-

tions which could be taken to improve the situation (using flat face). On word from the instructor, revise the cam design to find a true working curve.

92. The same as **91** except that the motion is UARM.

93. The same as **91** except that the motion is cycloidal.

94. The same as **91** except that the rise occurs during a 90° cam angle followed by a 70° dwell.

Offset Follower

95. (a) Find the working curve for a disk cam whose reciprocating follower is offset $\frac{3}{4}$ in. to the right of the center of the camshaft. The diameter of the base circle is 3 in. and the diameter of the roller is $\frac{1}{2}$ in. Rotate the cam for best results; let $\omega = 100$ rad./sec. The follower is to move outward $1\frac{3}{4}$ in. during 150°, to dwell 30°, to return during 180°. Use cycloidal motion. What is the maximum pressure angle? See **232.**

96. The same as **95** except that the motions are UARM.

97. The same as **95** except that the motions are SHM.

98. The same as **95** except that the outward motion is cycloidal and the return is SHM.

99. The same as **95** except that the offset is toward the left.

100. Use plate No. 25, Wingren. A reciprocating follower, offset to the right as shown, is to undergo a DRDR motion and it is important that the jerk be finite at all times. Therefore, it is decided to use cycloidal motion. The stroke is 2 in. The outward movement occurs during 120° of rotation; dwell 60°; return during 120°; and dwell 60°. The cam turns at 900 rpm in the better sense for the offset given. (a) Find the working curve. Use a horizontal line through the center of the roller as the base line for the displacement diagram. (b) What are the maximum pressure angles on forward and return strokes? See **232.**

101. The same as **100** except that the motion is UARM.

102. The same as **100** except that the motion is SHM.

103. The same as **100** except that the outward motion is cycloidal and the return is SHM.

104. The same as **100** except that the outward motion occurs during 180° with no dwell at the "top" (return in 150°; dwell 60°).

105. Use plate No. 24, Wingren. A reciprocating follower, with offset to the left as shown, moves up 2 in. with cycloidal motion during a 165° turn of the cam; returns with modified uniform motion during 165° (accelerates for 30°, $v = C$ for 105°, decelerates for 30°); dwells 30°. The speed of the cam is 780 rpm. (a) Construct the working curve. Use a horizontal line through the center of the roller as the base line for the displacement diagram. (b) Determine the maximum pressure angle on each stroke. See **232.**

106. The same as **105** except that the motion is cycloidal on both strokes.

107. The same as **105** except that the motion is SHM on both strokes.

108. The same as **105** except that the motion is UARM on both strokes.

Algebraic Computations Only

109. A follower moves 4.5 in. with constant speed while the cam turns 120° at the rate of 36 rpm. Determine the speed of the follower, and its acceleration and jerk.

110. A follower moves 3 in. from rest with constant acceleration during 80° of cam rotation. The cam turns 20 rpm. Compute (a) the acceleration, (b) the maximum speed, (c) the speed when $\theta = 60°$, (d) the displacement when $\theta = 60°$.

111. From its maximum speed, a follower comes to rest with $a = C$ after a displacement of 6 in. and during a 100° rotation of the cam. The cam turns 40 rpm. Let the point of maximum speed be the origin for s and θ, and compute (a) the acceleration, (b) the speed when $\theta = 60°$, (c) the maximum speed, (d) the displacement when $\theta = 60°$.

112. A follower is to move 4 in. during 150° of cam rotation. The speed of the cam is 60 rpm. If the motion is UARM, determine (a) the maximum acceleration, (b) the maximum speed.

113. The same as **112** except that the motion is SHM. Also (e), compute the acceleration at $\theta = 100°$.

Oscillation Roller Follower

114. The oscillating follower in Fig. P12 is to move through an angle $\delta = 25°$. Roller diameter is $\frac{1}{2}$ in., and rotation is CL at 540 rpm. For the purposes at hand, it will be accurate enough to lay out the motion on the chord subtending the 25°. The outward motion is with constant speed during 90° of rotation, preceded by a constant acceleration during $22\frac{1}{2}°$, followed by constant deceleration for 45°; dwell 45°; return with constant acceleration during 45°, constant speed for 90°, constant deceleration for $22\frac{1}{2}°$. (a) Construct the working curve. (b) Estimate the maximum pressure angle (angle of force vector with the direction in which roller center is moving).

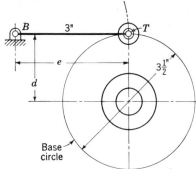

Fig. P12. Problems 114–117.

115. The same as **114** except that both movements are with cycloidal motion.

116. The same as **114** except that the return is with UARM.

117. The same as **114** except that the center of oscillation B is located by $e = 3.25$ in. and $d = 1.5$ in. (no change in length of arm BT).

118. Use plate No. 26, Wingren. The

trace point A on the oscillator moves through an arc 1.25 in. long with SHM during 165° of cam rotation, dwells 30°, returns to the starting point in 165° with SHM. The cam turns 450 rpm CC. Find the working curve.

119. The same as **118** except that the return is with modified uniform motion, accelerations and decelerations occurring during 30° turns.

120. The same as **118** except that the return is UARM.

121. The same as **118** except that the upward movement occurs during 180° and the dwell is shortened to 15°.

122. The same as **118** except that the upward movement is cycloidal.

123. Find the working curve for the oscillating follower shown in Fig. P13 if the slider A moves down 1 in. with UARM during 180° rotation of the cam, dwells for 60°, and returns to the starting point with UARM during the remaining angle. Let the cam rotate in the better direction at 420 rpm. Lay out full size.

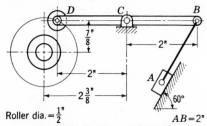

Roller dia. $= \frac{1}{2}''$

$AB = 2''$

Fig. P13. Problems 123–126.

124. The same as **123** except that during the outward movement, the follower moves with modified uniform motion, the velocity being constant for 100°.

125. The same as **123** except that the motion is to be harmonic.

126. The same as **123** except that the motion is to be cycloidal.

127. Use plate No. 29, Wingren. The slider C, designed to feed blank license plates into a press, moves to the right 4 in. (note given scale) with UARM while cam turns 180° at a speed of 180 rpm CC. Then there is a 10° dwell, the return to the starting point with UARM, and a 20° dwell before the cycle repeats. De-

termine the working curve and the maximum pressure angle (as measured from a perpendicular to AB) on each stroke. (*Note.* If desired, the instructor could assign **127**, **129**, and **130** to three teams of students, after which results could be compared and a final choice made.)

128. The same as **127** except that the dwells are eliminated and the rightward motion is with modified uniform motion during 210°, the velocity remaining constant during 150°, and the return with UARM during 150°.

129. The same as **127** except that the motion is SHM.

130. The same as **127** except that the motion is cycloidal.

131. Use plate No. 29, Wingren. The slide C moves 3 in. (note scale given) to the right and returns. The rightward motion is UARM during 180° CL rotation of the cam. The return is SHM during 150°; then the follower dwells until the next movement. (a) Find the working curve and the maximum pressure angle on each stroke. (b) If the speed of the cam is 120 rpm, determine the maximum values of v and a.

Oscillating Flat Follower

132. With the system shown in Fig. P14, a cam is to be laid out to lift the link C a distance of 2.5 in. with modified uniform motion (a-v-a) during a 180° turn, to lower C with harmonic motion during 120°; then C dwells for 60°. Use a flat face with T as the trace point. The dimensions in inches are: $D = 3$, $a =$

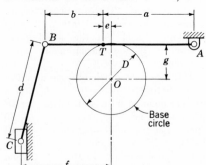

Fig. P14. Problems 132, 133.

4.5, $b = 2.5$, $d = 4.5$, $c = \frac{1}{2}$, $f = 3.5$, $g = 1.4$.   Draw full size.

133. The same as **132** except that the return motion is cycloidal.

Geneva Mechanism

134. (a) Design a Geneva mechanism. Fig. 38, to connect shafts which are 4 in, on centers. The driven member with radial slots makes $\frac{1}{8}$ turn for each revolution of the driver, which rotates at a constant speed of 600 rpm. Provide a positive stop for each dwell of the driven. Complete your sketch, using a $\frac{1}{2}$-in. driving pin, a $\frac{3}{4}$-in. driving shaft, a 1-in. driven shaft. (b) Construct an *st* diagram for the displacement of a point on the outer circumference of the driven member.

Tangential Cam

135. A valve cam on an internal combustion engine has a profile similar to that in Fig. 16, and it has a $\frac{7}{8}$-in. roller. The radii of the two circular arcs are 0.80 and 0.53 in., respectively, and their centers are 0.64 in. apart. Construct a displacement diagram for this cam.

Positive-motion Plate Cams

136. Lay out a constant breadth cam similar to that in Fig. P15 that will move a follower outward $1\frac{1}{2}$ in. with UARM during 135° of CC cam rotation,

then allow a dwell of 45°; $D = 4$ in. What is the dimension a?

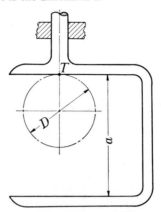

Fig. P15. Problems 136–140.

137. The same as **136** except that the motion is SHM.

138. The same as **136** except that the motion is cycloidal.

139. The same as **136** except that the middle part of the stroke is with constant velocity, the speeding up and slowing down occurring in 30° each with constant acceleration.

140. The same as **136** except that a constant-diameter cam is to be used with the center of one roller at T; use $\frac{5}{8}$-in. rollers.

141. Positive drive for both strokes of a follower is to be obtained with con-

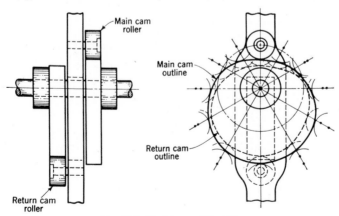

Fig. P16. Problems 141–143.

jugate cams similar to those in Fig. P16. It is necessary for the follower to move 2 in. with SHM during 180°, dwell for 60°, and return to the starting point during 120°. Use a $4\frac{1}{2}$-in. base circle, 1-in. rollers, and a center distance between rollers of 6 in. Lay out the required profiles for CC rotation.

142. The same as **141** except that the motions are UARM.

143. The same as **141** except that the motions are cycloidal.

Cylindrical Cams

144. (a) Lay out the theoretical curve on the undeveloped surface of a 2-in. cylindrical cam. The top of the cam in the plan view turns toward the observer. The reciprocating follower is to move to the left 3 in. with SHM in 1 rev., to dwell for $\frac{1}{2}$ rev., to return to the starting point with UARM in $1\frac{3}{4}$ rev.; dwell. (b) Lay out the foregoing curve on the developed surface.

145. The same as **144** except that the first motion is cycloidal.

146. The same as **144** except that the middle 1 rev. during the return is to be with constant velocity. The acceleration and deceleration occur during $\frac{3}{8}$ rev. each, making the total angle for the motion the same as before.

147. A cylindrical cam is to move a pivoted roller follower through an angle of 50°. The follower moves from left to right, Fig. P17, in such a manner that

the motion in the direction of the axis is UARM during 135° of cam rotation, dwells 30°, returns to the starting point with SHM (axially) during 180°, dwells. Cylinder diameter is 21 in.; roller diameter is 5 in.; radius of oscillating-follower arm is 7 in. Lay out the theoretical curve on the developed surface.

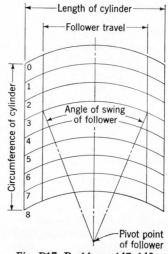

Fig. P17. Problems 147–149.

148. The same as **147** except that the motions are reversed; out with SHM, in with UARM.

149. The same as **147** except that the outward motion is cycloidal.

Larger Paper Needed

150. (a) Lay out the working curve

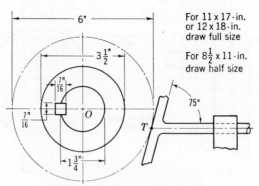

Fig. P18. Problems 150, 151.

for a plate cam with flat-face follower as shown in Fig. P18; O (8, 7), LDH. The follower is to move 3 in. to the right (working stroke) with SHM in $\frac{3}{8}$ sec., to dwell for $\frac{1}{8}$ sec., to move 3 in. to the left with UARM in $\frac{3}{8}$ sec., to dwell 1 sec. The cam turns 60 rpm in the more advantageous direction. Use the trace point T where the face is tangent to the base circle. (b) Determine and dimension the length of flat face, allowing $\frac{1}{8}$ to $\frac{1}{4}$ in. more than theoretically needed.

151. The same as **150** except that the working stroke is with cycloidal motion.

152. Figure P19 shows diagrammatically a cam operated shear for cutting strips of metal by the jaws at D; C (6, 6), LDH. (a) Design a cam that will hold the jaws in the open position shown for $\frac{1}{4}$ rev.; close the jaws during $\frac{1}{2}$ rev. with a modified uniform motion; open the jaws in $\frac{1}{4}$ rev. with UARM. Locate the trace point T by drawing AB tangent to the base circle. (b) If the cam makes 60 rpm, how long will the jaws remain wide open? Show computations. Determine and dimension the length of contact surface on each side of point T.

153–190. These numbers may be used for other problems.

For 11 x 17-in. or 12 x 18-in. draw 3 in. = 1 ft

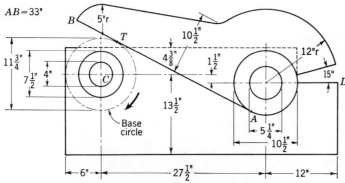

Fig. P19. Problem 152.

3

Centros and Velocities

57. Introduction. There are several graphical means of determining the velocities of points in a linkage system, given the velocity of one point, and all of them will not be discussed here. Mathematical analyses (**53**)

Courtesy The Heim Co., Fairfield, Conn.

Fig. 51. Links on Cartoning Machine. Notice the construction of the various links. The link ends circled are patented ends as shown in Fig. 55.

of linkages with four or more links are usually rather complex and have been traditionally avoided. However, because of the greater precision needed in some instances, mathematical analyses are made more often than formerly, and it is possible to obtain help from computers.

The first method to be described is the **centro method.** This chapter
will study the significance of centros and the methods of finding velocities
when a linkage's centros are known.   Also, for your general information,
we shall include a brief description of a few linkages, some of which have
been so useful that they have acquired names.

The concept of the centro is one of the most significant in kinematics.
To those dealing with problems on motion, it often comes in handy
because it unravels a confusing problem by providing a simple and direct
thinking process.   The best way to become acquainted with this concept is to solve a variety of problems finding centros and velocities.   But

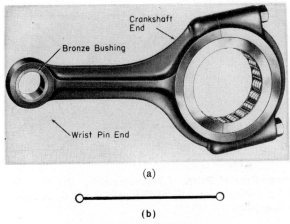

(a)

(b)

Courtesy F. E. Myers & Bro. Co., Ashland, Ohio

Fig. 52. Connecting Rod. The connecting rod in (a) is for a compressor. Note the
roller bearings for the crankshaft end—higher pairing between rollers and rod. The
more common construction is a sleeve bearing, but kinematically there is no difference, except as one might study the motion of the rollers themselves.

first, we should define a few terms and the manner in which linkages are
represented.

58. Pairs and Their Representations. Shorthand representation of
linkages like that shown in Fig. 51 is an evident advantage.   Hence, the
connecting rod of Fig. 52(a) is represented by a line with a small circle
at each end.   The small circle represents a point where there will be a
pin joint when the link is assembled.

A *pair* involves two links in such a manner that one link has a particular
motion relative to the other.   Figure 53(a) represents a *turning pair*
(also called a *hinge*), in which the joint is such that links B and C can
turn with respect to each other about the axis bc.   Either B or C may be
stationary, or perhaps neither is stationary.

When the line which represents the link is drawn through the circle, we shall mean that there is one solid piece. In Fig. 53(b), the link B may turn about a pin at the circle, but it is one rigid link. It could represent a portion of what we call a *bell crank*. In Fig. 53(c), the line CC drawn through the circle shows that there is no break in the link; it is one link C. As shown in (c), B turns about pin bc with respect to C; and C turns about pin bc with respect to B.

Sliding pairs, in which one link slides on another link that may be straight or curved, are represented as shown in Fig. 53(d), (e), and (f). The link C may or may not be stationary. The motion of B with respect to C in (d) and (e) may be plane motion (right and left, say) or motion

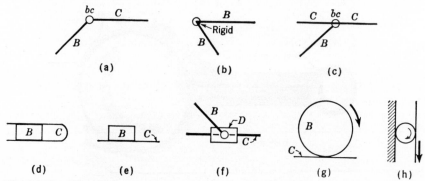

(a)   (b)   (c)

(d)   (e)   (f)   (g)   (h)

Fig. 53. Representations of Pairs. (a) Turning pair. (b) No pair. (c) Turning pair. (d) and (e) Sliding pairs. (f) Sliding pair D and C; turning pair B and D. (g) and (h) Rolling pairs.

in space if C has a motion perpendicular to the plane of the paper. Unless otherwise indicated, plane motion is implied in our discussions. Thus, when C is a straight link, the representation in (d) and (e) means that B has rectilinear motion with respect to C. See also Fig. 54.

A *rolling pair,* Fig. 53(g) or (h), is one in which two surfaces (usually a ball or cylinder) roll on one or other. One or both of the surfaces may be moving. In pairs such as (g) and (h), some sliding may occur unless, for example, the links have meshing teeth, as in gearing. (As we shall learn later, gear *teeth* slide on one another.) Thus actually, the bearings in the connecting rod, Fig. 52, roll and slide on their contacting surfaces.

A *spherical pair* is illustrated by a ball-and-socket joint, in which the motion of one link with respect to the other is not constrained to plane motion. The link end shown in Fig. 55 has a spherical bearing which permits nonplanar motion.

A *helical pair* is illustrated by a nut on a screw. Figure 56 shows a ball-bearing design that provides helical motion of a point on one member with respect to the other member. Other types of pairing can be defined (**31**), but the foregoing will serve our purposes.

Pairs might be classified as *open* or *closed*. A **closed pair** is one in which the construction keeps the pair intact without external force; examples: Fig. 53(a), (d), (h), and positive-motion cams. An **open pair**

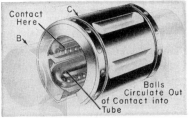

Courtesy Thomson Industries, Inc., Manhasset, N.Y.

Fig. 54. Ball-bearing Sliding Pair. An example of a ball system permitting reciprocating motion. The balls roll on the reciprocating surface, circulating through the tubes when they are out of contact. The visible rows of balls (and other rows out of sight) carry the load.

Courtesy The Heim Co., Fairfield, Conn.

Fig. 55. Rod End. An example of commercial ends available. The outside surface of the bushing is spherical and it fits in a spherical seat. A particular advantage of this construction is that a considerable and varying misalignment (intentional, perhaps) may be cared for without subjecting the link to bending.

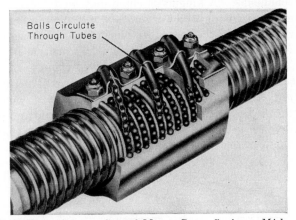

Courtesy General Motors Corp., Saginaw, Mich.

Fig. 56. Ball-bearing Screw. There is approximate rolling contact between the balls and the grooves in screw and nut. This action is made possible by providing the tubes through which the balls can circulate.

is one which requires gravity or springs to maintain the intended configuration; examples: Fig. 53(e) and (g) as drawn, plate cams with spring-loaded followers.

In this country, **lower pairing** is said to exist when the contact between the bodies is on a surface; examples: pin-connected joints, sliding pairs such as Fig. 53(d), (e), and (f). When the contact between the bodies is on a line or point (ideal rigid bodies), as between cylinders or balls and any other surface (except internal contact with a surface of the same radius), it is said to be **higher pairing;** examples: Figs. 53(g) and (h), 54, 55, and cam-and-follower pairs. With higher pairing, the stresses between the contact surfaces may be quite high. Hence, these surfaces are manufactured with a high hardness (great strength). The distinction between higher and lower pairs is of little kinematic significance.

It might be inferred from the discussion so far that a **link** is any rigid member (or effectively rigid with respect to the force on it) which exerts force on another member, moving or tending to move it. No body is actually rigid; all deform under load. However, for kinematic purposes, even a leather belt might be assumed to be rigid. (It is effectively so under constant load in tension.) A link which **rotates** about a center attached to the frame of a machine is called a **crank** or **rotor.** These same names are also used for a link that *oscillates* about an axis attached to the frame; but in addition it may be called an *oscillating link, oscillator,* or *lever.* The name **connecting rod** is used in general for a link joining two other links, as a crank to crank, crank to oscillator, crank to slider, etc.

59. Constraint. We shall be dealing only with *constrained* kinematic chains or mechanisms. A kinematic chain may be defined as any group of interconnected links for which significant relative motion is possible. In Fig. 57(a), three bars are joined by pin joints. With zero clearances at the connections and for ideal rigid bodies, no relative motion of the members E, F, and G is possible; this is a **locked chain** (avoid the name *kinematic chain* in this arrangement) and is used in structures such as bridge trusses. At the opposite extreme, in the group of connected links in Fig. 57(d), link B can move without causing a unique and predictable motion of any of the other links C, D, or E. When such a situation exists, we have an *unconstrained kinematic* chain or *unconstrained mechanism.*

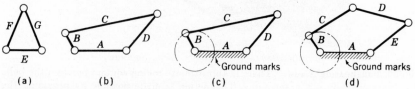

Fig. 57. Kinematic Chains. (a) Locked. (b) Constrained kinematic chain. (c) Mechanism. (d) Unconstrained mechanism. All these links are *bar links.*

In contrast, the motions of the links *relative to each other* in Fig. 56(b) are always related in a definable manner. A particular movement of B relative to A will induce movements of C and D relative to A which are always the same. Thus, Fig. 57(b) is a **constrained kinematic chain.** If one of the links in (b) is a part of the frame of the machine, say A,

it would be used as the **reference link,** and it would be designated
by the so-called **ground marks,** Fig. 57(c).   The reference link (frame)
may be stationary ($v = 0$ with respect to the earth) or it may be in
motion, as in the case of vehicles.   Even when the reference link is
moving, it is thought of as having zero velocity during a velocity analysis.
When a link of a kinematic chain is held stationary (or is part of a frame),
the chain becomes a **mechanism.**   Figure 57(c) is a **constrained mechanism.**   Whenever the word *mechanism* is used without a descriptive adjective, constraint is to be understood.   In the simple mechanisms with
which we shall deal, one can tell by inspection whether or not there is
constraint.   In more complex link arrangements, this is not always
practicable; and a mathematical test for constraint must be applied (**31**).

60. Centro.   A **centro** is two coincident points (hence, it can be referred
to as a point) and its significant property can be stated advantageously
in several ways: (1) *The coincident points in two different bodies have the
same linear velocities.*   (2) *The coincident points in two different bodies
have no relative motion (one does not move with respect to the other)*—perhaps
true for only an instant.   (3) *The coincident points in two different
bodies mark a point in one of the bodies about which the other body rotates,
permanently or momentarily.*   Any centro satisfies all of these statements,
but oftentimes one is more to the point than another.   It is probably
most useful to think of it as a point common to two bodies and having the
same linear velocity in each [statement (1)].   A more descriptive but too
cumbersome name for centro is **instantaneous center** of rotation, so
called because the point which is this center may remain in a particular
position in space only for an instant.   Other names in use for centro
include *instant center, virtual center,* and *rotopole.*

You have no doubt already observed the system we shall use in naming
centros.   Name the links with capital letters, A, B, C, . . . ; then the
centro common to links A and B will be called ab (lower-case letters);
the centro links for A and C will be ac; etc.

61. Location of Primary Centros.   When we use the expression *known
centro,* we shall mean one whose location is known; similarly an *unknown
centro* is one whose location is not known.   By *primary centros,* we mean
those which are easily spotted as being obviously in accord with the
definition.   All the primary centros must be known before all the others
can be found.   In Fig. 58(b), you might think of B as being stationary;
then it is easy to visualize that since C is moving in a straight line, its
center of rotation must be at infinity (a straight line is a limiting case of a
circle with infinite radius).

The logic for the case of *rolling* bodies (any shape), Fig. 58(c), is as
follows: When it is said without qualification that a body C is rolling, it is
meant that it is not sliding on any other body B.   If it is not sliding on B,

then the point on C in contact with B must at that instant be stationary with respect to the contact point (else it would slide over the contact point). If the contact point on C does not move with respect to the contact point on B, then these coincident points bc must be moving with the same velocity. Hence bc is a point common to the two bodies B and C having the same velocity in each and is thus the common centro. In accordance with this finding, it is often advantageous to think of the

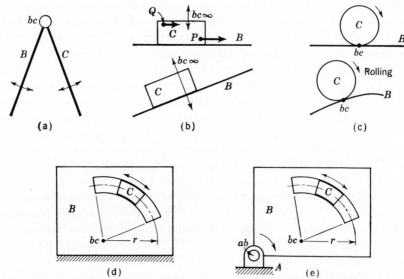

Fig. 58. Primary Centros. In (a), bc is the centro for B and C because it is the point in B about which C rotates and the point in C about which B rotates. For sliders C moving in paths on link B, the center of rotation of C with respect to B is at the center of curvature of the path. In (b), the path is a straight line and its center of curvature is at infinity; since a radius to a path is perpendicular to the path, bc is then at infinity in a direction (upward or downward) perpendicular to the straight link B. In (d) and (e) the path has a radius r and bc is as shown. In (c), if a body C *rolls* on another body B, which may or may not provide a straight path, the point of common velocity bc is the point of contact between B and C.

motion of a rolling wheel as a *series of minute rotations* about each element of the wheel as it comes into contact with the link on which it rolls.

If a wheel slips, the contact point on C has motion relative to the contact point on B, and therefore these two points do not have the same velocity and consequently the contact point cannot be the centro bc. If there is no rolling, we are back to cases in Fig. 58(b), (d), (e). If there is sliding *and* rolling, the centro is easier to locate after one has learned about relative velocities.

62. Kennedy's Theorem. Kennedy's theorem states that any three bodies moving relatively to each other have but three centros, *all of which lie on a straight line.* Let A, B, and C, Fig. 59, be three bodies with relative motion. There is a centro ab for A relative to B; another bc for B relative to C; and a third ac for links A and C. For three links, these are all the possible combinations of relative motions of two links. Choose the body A as the reference body. You may think of it as stationary if you wish, but this is not essential. With hinges at ab and ac, you spot these points immediately as centros for A and B and for A and C, respectively.

Now you must recognize, as you probably do, that *a point in a rotating body always moves in a direction perpendicular to its radius*, the direction of the velocity vector. The only motion that B can have relative to A is one of rotation about ab; and the only possible motion of C relative to A is rotation about ac. Thus the direction of motion of any point in either B or C relative to A must be in a direction perpendicular to its particular

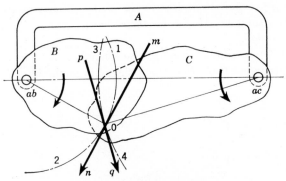

Fig. 59. Kennedy's Theorem.

radius. To complete the background for the final argument, recall that the centro bc must have a particular velocity (magnitude and sense) whether it is considered to be in B or C.

First, assume that bc lies at point O, Fig. 59. When O is considered to be in B, it has a radius ab-O and moves in the arc 1-2 in the direction mn. When O is considered to be in C, its radius is ac-O and it moves in the arc 3-4 in the direction pq. Thus, point O cannot be the centro bc because its velocity as a point in B is different from that when O is in C (even if the magnitudes of mn and pq are the same, their senses are different). We shall come to this same conclusion for any other position of O except a position on the line of centers ab-ac, inasmuch as the vectors mn and pq must not only have the same magnitude but also point in the same direction. It follows that the third centro bc must be on line ab-ac, and we have proved Kennedy's theorem that for *any three bodies moving relatively to each other, their three centros must lie on a straight line.* To find the

precise position of *bc* on line *ab-ac* in Fig. 59, we must know more of the
motions of *B* and *C* (see § 64, for example).

63. Number of Centros in a Kinematic Chain. The number of centros
in a constrained kinematic chain with *N* links is equal to the number of
possible combinations taken two at a time (see your algebra text); or

(a)   Number of centros $= \dfrac{N(N-1)}{2}.$

64. Centro for Two Rotating Bodies Which Slide in Contact. There is
one other centro, which might be classified as a primary centro (§ 61),
whose location is partly determined by definition. Kennedy's theorem
completes the requirements for a precise location. The basic principle
to be noted in this article is that if two bodies are sliding on one another,
their entire relative motion at that instant is in the direction of their
common tangent; and the radius to the center of that motion is perpendicular to the tangent; therefore, *their centro must lie somewhere along
the common normal.*

In Fig. 60, let the bodies *B* and *C* be hinged on *A* at *ab* and *ac* and in
contact at the point *P,Q* (*P* is the tangent point in *B*, designated P_b in

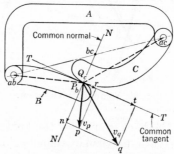

Fig. 60. Centro for Hinged Bodies in Sliding Contact.

Fig. 60, and *Q* is the tangent point in *C*). You may think of *A* as stationary if it helps, but this system of bodies could just as well be spinning
through space. We are referring all motions to *A*. Let *Pp* at a right
angle to the radius of P_b be the velocity of *P* (v_p) with respect to *A*; *Qq* at
a right angle to the radius of Q_c be the velocity of *Q* (v_q) with respect to *A*.
Now you should study the situation until you see that if the bodies are to
remain in contact, these velocities of *P* and *Q* must be of such magnitude
that their normal components, *Pn* and *Qn*, are equal. If *Pn* were greater
than *Qn*, the bodies would separate and contact would cease; if *Qn* > *Pn*,
the bodies mash into each other or overlap. Thus, for the condition
of continuous contact of rigid bodies, the total *relative* motion must

be in a tangential direction TT, and *the center of rotation bc of B rela-tive to C (and C relative to B) must be on the perpendicular NN to the surfaces of B and C at their point of contact*—which is what we wished to demonstrate.

To locate the specific point which is bc in Fig. 60, we use Kennedy's theorem, which says that ab, ac, and bc must be on a straight line. Hence, the intersection of NN and ab-ac is the centro bc.

While studying Fig. 60, we shall find it worthwhile observing that the sliding velocity between Q and P is the algebraic difference of the tangential velocities, $Qt - Pr = rt$. This value (rt) is the entire relative motion between Q and P since there is none in the normal NN direction

65. Example—Locating Centros. Locate all centros for the quadric chain $ABCD$ of Fig. 61.

Solution. This is the simplest problem, but study the explanations and technique carefully because they apply to all problems. First, always set up some means of keeping track of the total number of centros and of the ones which have been located. This is commonly done either by a *table of centros*, Fig. 61(a), or by the circle of Fig. 61(b).

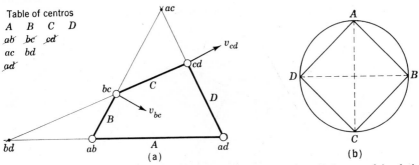

Fig. 61. Centros for Open Quadric Chain. This linkage is called a quadric chain because there are four links; it is "open" because the links do not cross. Notice that the velocity vector v_{bc} is that with respect to A; v_{cd} is also the velocity with respect to A. Hence, we have known the directions of the velocities of two points in C relative to A. Since the direction of motion is perpendicular to the radius, one could draw perpendiculars to v_{bc} and v_{cd}; their intersection would be the instant center for C relative to A (ac).

First, name all the links in the system and write down their names. Starting with the first name, say A in this case, combine it with each of the succeeding link names, using lower case—as ab, ac, ad—and writing these names in a column under A. (Avoid the use of script in centro names. Use engineering lettering.) Combine the name of the next link B with all links succeeding it, and so on, until there are no succeeding names for combining purposes, as D in this problem. This procedure gets onto paper the names of all the centros for the chain, which is its advantage. For all 4-link chains, the number of centros is, from equation (*a*) for $N = 4$,

$$\frac{N(N-1)}{2} = \frac{4(4-1)}{2} = 6 \text{ centros.}$$

We see that we have six names of centros in the table, which checks. (If the circle is used, arrange the link names symmetrically around the circle.)

The next step is to locate all the primary centros. In this case, they are all at turning pairs, *ab*, *bc*, *cd*, and *ad*. In the table of centros, mark out the names of those found. In Fig. 61, the foregoing names have been marked out. (If the circle is used, draw a line connecting the names of the links whose centros have been found. The line joining the two names indicates that that centro has been found. The names of the primary centros are those whose names are designated by the letters at the ends of the solid lines.)

The third and final step is to apply Kennedy's theorem repeatedly until the remainder of the centros are located. The plan is to find two straight intersecting lines which locate an unknown centro (at the intersection). As good an idea as any is to try to find the first unknown centro in the table, after all primary centros are spotted, in this case *ac*. Observe that *ac*, Fig. 61, is common to links *A* and *C*; hence if links *A* and *C* are considered in their relative motion with respect to any other link in the system, we shall have three links moving relatively to each other, whose three centros, one of which is *ac*, lie on a straight line (Kennedy's theorem). Therefore, *A* and *C* taken in turn with two other links

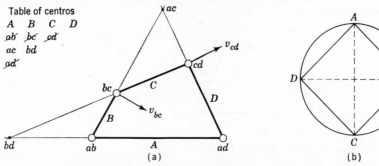

Fig. 61. Repeated.

provide two lines, each containing *ac*. Their intersection is *ac*. (If there are two parallel lines, their intersection is at infinity. If two lines are *coincident*, another intersecting or parallel line must be found.) Thus, consider the relative motion of *A* and *C* with *B* and with *D*. (These two lines are all that are available in a 4-link chain, but if there are 5 or more links, there will be more lines.) To spot the lines, write down the names of the groups of three links and then the names of their centros; thus:

(1)

A   *B*   *C*
ab   *bc*
ac

(2)

A   *C*   *D*
ac   *cd*
ad

It is helpful to circle the name of the centro being sought. (This part is scratch work in the solution of problems.) The centros of each of the foregoing groups are on a straight line. Therefore, we draw a line (1) through *ab* and *bc* of indefinite length, draw another line (2) through *ad* and *cd*, find their intersection (if there is one), and name it *ac*, Fig. 61. Mark out the name *ac* in the table to show at a glance that its location is known. (If the circle is used, draw a line from *A* to *C*—shown dotted, but there is no reason for you to dot it—which indicates that *ac* is known. From each letter on the circle, a line must go to each and every other letter when all centros are found.) This procedure is repeated in detail until all desired centros are known.

Since centro *bd* is common to links *B* and *D*, we use links *B* and *D* with *any* other two links in the system in order to find two intersecting lines. When there are only four links, only two lines are possible for *bd*: in this case the lines of centers for *ABD* and *CBD*. The reader should now go through the technique and find *bd*. After *bd* is found, mark it out on the table of centros (draw a line from *B* to *D* in the circle).

66. Remarks on Finding Centros.

When there are more than four links, it is not always possible to find the "first" unknown centro first. Among all the possible lines, only one may be found. In any case, apply the procedure given above; and if after trying all possible combinations, you do not have two intersecting lines, go to the next unknown centro and repeat the process. Keep trying the "next" unknown until one is located. Then perhaps you can start over and have no trouble. (It is not necessarily so that there will be no more trouble.)

If after going through the entire problem in this manner, you have been unsuccessful in locating an unknown centro, check the primary centros again (§§ 61, 64). Perhaps you have overlooked one.

The sequence of letters in the name of a centro is not significant, *ac* or *ca*. With practice, one learns to find centros rapidly in a complex linkage, by making mental search of the table of centros, or of the circle, for two intersecting lines. You may not get this much practice; so a systematic approach as outlined may do the most good in a given time.

Review the significance of a centro; *ac* is a point in link *A*, Fig. 61 or 62, about which *C* is rotating (momentarily, in this case); *ac* is also a point in *C* about which link *A* is rotating. It is also a point common to links *A* and *C* and has the same velocity in each. If link *A* is attached to the ground and has zero velocity, point *ac* at this instant also has zero velocity (but the centro *ac* will be somewhere else in the next instant if the links are moving!); thus, *ac* is then a point in *C* with zero velocity. Evidently, a centro is not necessarily physically in a link; one mentally extends the link to include the centro, if such an extension is appropriate. Review the definition of centro, § 60.

If you arrive at an impasse in the more complex problems, restudy the above example and this article for clues about what to do.

Engineers do not use heavy dots at centros. Those in Fig. 61 are for pedagogical purposes. Use pinpricks for accurate location. Use a sharply pointed pencil for precise definition. You might as well get into this habit now, since it is essential to accurate solutions of velocity and acceleration problems.

67. Example—Centros for Slider-crank Chain. Find the centros for the kinematic chain $ABCD$, Fig. 62.

Solution. Since the links have the same names as before (A, B, C, D), the names of the centros are the same as in § 65. See table of centros and circle, Fig. 62. The primary centros are at turning joints except for A and D. Since D

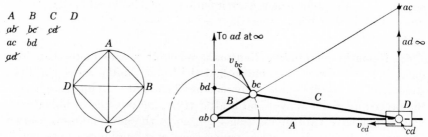

Fig. 62. Centros for Slider-crank Chain.

slides in a *straight line* on A (or A slides in D), the centro ad must be at infinity. *This is a primary centro and must be recognized.* The lines for the unknown centros are:

	For ac				For bd	
(1)		(2)		(1)		(2)
A C B		A C D		B D A		B D C
ac bc	and	ac cd		bd ad	and	bd cd
ab		ad		ab		bc

The student should go through the mental processes of the solution. Note that since ad is at infinity in a *vertical* direction, a line from ab to ad [line (1) for bd] is a vertical line. Said another way, the parallel lines ab-bd and cd-ac intersect at infinity where ad is.

68. Example—Centros in a Six-link Mechanism. Find the centros for the mechanism shown in Fig. 63. See Fig. 64 for an explanation of its use.

Solution. You should have observed in Figs. 61 and 62 that the relative location of the centros in a kinematic chain is independent of the link, if any, that is stationary. However, for determining velocities, one link needs to be the reference or ground link. Since this is an actual mechanism, the frame A (reference link) is conventionally marked. From equation (**a**) and for 6 links, we have $N(N-1)/2 = (6 \times 5)/2 = 15$ centros. The table of centros in Fig. 63 checks. The primary centros are marked out in the table of centros; and are

indicated by the solid lines in the circle. The significant observations are that *af*
is at infinity in a horizontal direction (*F* has rectilinear motion on *A*) and that *ae*
is at the center of curvature of the path of *F* on *A*.

The next step for a novice might best be a routine search for the remainder of
the centros, starting with the first unknown *ac*. Note that in a 6-link mechanism,
there are 4 possible lines intersecting at a particular centro; for example, the lines
of centers for the relative motions of links *A* and *C* combined with each of the
other 4 links, *B*, *D*, *E*, and *F*. One could (in scratch work) quickly write the
names of these lines, to wit, *ab-bc*, *ad-cd*, *ae-ce*, *af-cf*, and then compare these

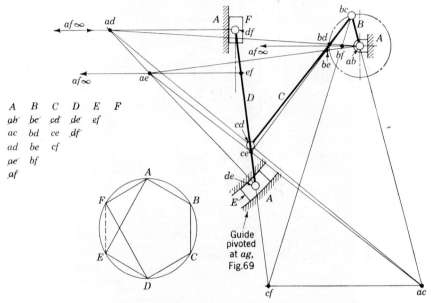

Fig. 63. Centros for Six-link Mechanism. The proportions of this linkage are
approximately those of Fig. 64; the path of *E* is somewhat steeper. If the path of *E*
has a constant curvature and remains fixed, the centro *ae* remains in the position
shown for all positions of the pumping mechanism; otherwise, it would be at the
center of curvature of the path at the point of contact between *A* and *E*.

names with those marked out in the table (or with lines drawn in the circle).
In this case, you will find only one line (*bc-ab*); that is, *ac* cannot be found until
other centros have been found. It happens that the next unknown *ad* can be
found (from *ae-de* and *df-af*), after which there should be no further trouble.

If the circle is used, one might notice that enough centros are known with
respect to *ef* to find it immediately; then *ad*; etc. If you should become skilled at
finding centros, *find them as far as possible from primary centros*, which are in
general more accurately located. Also, avoid depending on the intersection of
two lines that make a small acute angle with one another; try for lines whose
angles of intersection are closer to 90°. Since linkages of six or more links are
often made up of combinations of four-link chains and since you will no doubt

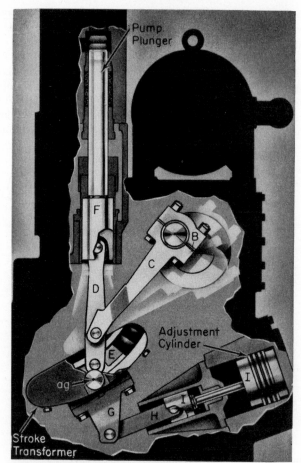

Fig. 64. Controllable-stroke Pump. Over and over again, inventors and engineers have worked on inventing or developing means of varying the stroke of pumps and compressors, because of the great demand for those which have a varying output and because, from the standpoint of power required, this system of controlling output is efficient.   Uses of such pumps would include: pumping feed water into a boiler, delivering oil to an oil burner, pumping fluids in pipes, etc.   The method of operation is seen by comparing Figs. 63 and 64.   As shown in Figs. 63 and 64, the pump plunger F reciprocates in a vertical direction.   The guide for E, moving about the center ag above, can be tipped to various slopes by the adjustment cylinder (not shown in Fig. 63) and its accompanying links.   For a particular output of the pump, the adjustment mechanism remains stationary.   (Figure 63 shows the centros for one position of the pump linkage only.)   If a motion analysis were made, it would probably be for several positions of the adjustment mechanism, these links being allowed each time to be stationary.   Notice, in Fig. 64, that if the path for E is tipped to a steeper position, the plunger has a greater stroke and pumps more fluid.   The steepness of the path for E can be reduced until the plunger does not move (zero output); the radius of curvature of the path is equal to de-df.

have a more intimate knowledge of the centros in four-link chains, a good technique is to isolate mentally the four-link chains and locate those centros first. The reader is advised to complete the thinking necessary in solving the remainder of this example. It is a matter of applying Kennedy's theorem over and over.

69. Relative Motion. In a constrained chain, the motion of any one link relative to any other is always the same; that is, motion is purely relative. We think of the motion of the moon with respect to the earth; of the earth with respect to the sun (plus the earth's motion about its own axis); but the sun is moving, and so far as is known there is no such thing in space as a stationary point. Hence, in the actual determination of velocities and accelerations, we must use some convenient reference body. Unless otherwise stated, that reference body is understood to be the earth (or the frame of a machine); so that when we speak of *the* velocity or of *absolute* (or *resultant* or *total*) velocity, we mean with respect to the earth (or frame—the context will indicate which).

Except for a body or point moving with constant velocity, the velocity that we find will be an instantaneous one; it will change in the next instant. Points on a driving crank usually move with approximately constant speed (velocity changes because direction changes), because driving motors and engines are more economically operated at some certain speed. But points on the other links (Fig. 64, for example), will have continuously varying speeds. When velocities are analyzed, we usually wish to know the whole pattern of variation for some point or points in the mechanism, a velocity diagram, say, plotted against either displacement or time. However, for learning purposes, you will gain more engineering experience by finding velocities for only one position of a particular mechanism, but for many mechanisms.

70. Velocities by Centro Method. The finding of velocities by the centro method is the most direct (if the centros are located) and is based on the simplest fundamental, namely: *the linear velocities of two points in a rotating body are proportional to their distances from their center of rotation and the directions are perpendicular to their respective radii.* That is all. To find velocities graphically, we need to use constructions which give intercepts proportional to the radii—using similar triangles, say. We note that each link has a centro common with the reference link; this centro has zero velocity with respect to the reference link and is the instantaneous center of rotation of a particular link in an absolute sense; and the velocities of all points in a particular link are proportional to and perpendicular to their respective radii. Inasmuch as the principle is so simple, the best procedure now is to illustrate its application with examples.

71. Example—Velocities by Line-of-centers Method. In the cross-link mechanism of Fig. 65, the crank *B* is turning at a uniform speed and the velocity of point 1 is given (represented by $\frac{9}{16}$ in.).   Find the velocity of point 2 in D by the line-of-centers (LOC) construction.   Also determine the angular velocity of D. (Point 2 might be at a turning pair with another link E attached there.   A link like D with three turning pairs is called a ***ternary link;*** with two turning pairs a ***binary link.***   A hinge with three links on the same axis is a ternary hinge.)

Solution.   If it is not already known, compute the velocity of point 1.   Then the first thing to do is to determine the *line of centers,* defined by the name of

1. The link in which the point 1 with known velocity lies (known point),

2. The link in which the point 2 with the unknown velocity lies (unknown point),

3. The stationary or reference link (ground link).

Thus, with 1 in B, 2 in D, and A the reference link, the line of centers is defined by the centros for links B, D, and A; or ab, ad, and bd.   Having found the centros in Fig. 65, we see that the line of centers is the horizontal line through the fixed

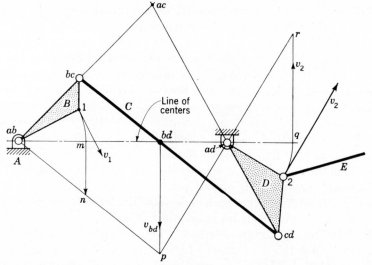

Fig. 65. Velocities—Line of Centers. Rotate "known point" 1 into LOC at m; find velocity of common centro bd; rotate "unknown point" 2 into LOC at q and find its velocity.

pins.   Construction is based on the line of centers (LOC) and any point outside of it is rotated into it.   Thus rotate 1 into the LOC and get point m, Fig. 65. Erect the perpendicular and lay out $mn = v_1$ to some velocity scale K_v.

Now bd, which we shall call the ***common centro,*** is in both B and D and has the same velocity in each.   This statement is the key to the LOC procedure.   Since bd is in B, its velocity is proportional to its radius as a point in B, which is ab-bd. Inasmuch as we know the velocity of at least one point in B, that is, v_1, we can find the velocity of any other point bd in B by a similar triangle construction. Thus drop the perpendicular to the line of centers bd-p; draw a line from ab (point of zero velocity in B) through the end n of the vector for v_1; the intercept by this line on bd-p is the velocity of bd in magnitude and direction.   In the similar triangles ab-m-n and ab-bd-p, Fig. 65, we know that ab-m is equal to the radius of 1 and mn represents v_1 to scale; also ab-bd is the radius of bd as a point in B. Therefore, if $mn = v_1$ to scale, then bd-p represents v_{bd} to the same scale, because this vector is proportional to its radius as mn is to its radius (mn/ab-$m = bd$-p/ab-bd).

Now we know the velocity of a point in link D, so that the velocity of any other point is easily found. Draw a line through p, the end of v_{bd}, and ad of indefinite extent. Rotate point 2 into the LOC to get q so that ad-q is equal to the radius of 2. Erect a perpendicular qr at q. The magnitude of the intercept on qr by p-ad is to scale the speed v_2 of 2, because in the similar triangles bd-p-ad and r-q-ad (parallel sides) we have $v_2/v_{bd} = ad$-q/ad-bd. The direction of motion is, as oft repeated, perpendicular to the radius; hence, the proper location of the vector v_2 is at point 2 as shown. Its magnitude is measured and found to be about 1.14 in.

Summarized briefly: No matter how many links the mechanism has, find first the velocity of the common centro from a known velocity, so that the velocity of a point in the "other" link is known; then find the unknown velocity. Use the important principle that the velocities of points in a rotating body are proportional to their radii.

The angular velocity of D is found by dividing the linear velocity of any point in it by the corresponding radius. If all dimensions were known, it could be computed from (§ 16)

$$\omega_D = \frac{v}{r} = \frac{v_2}{ad\text{-}2} \quad \text{CC.}$$

If the velocity v_2 is in feet per minute, then ad-2 should be in feet to get radians per minute.

72. Example—Velocities by the Link-to-link Construction.

Let this problem be the same as that of § 71 except for the method of construction. Find the angular velocity of link C (instead of D).

Solution. This construction, as the name implies, is one wherein the velocity of a point in an adjacent link is first found from the known velocity, then the velocity of a point in the next link is found, and so on, until one has found the velocity of a point in the link containing the point whose velocity is desired. The solution is in Fig. 66, where the vector v_1 has been laid out to a velocity scale K_v, pointing toward its base line (it does not make any difference on which side of the line the construction is placed). The procedure is as follows:

Rotate point 1 to find h, so that h and bc (a point in the adjacent link) are in the same straight line from the center of rotation ab. Erect a vector gh at h representing v_1 to scale. Draw line ab-g through the end of this vector, and find the intercept point s on the perpendicular erected at bc. This length bc-s is the velocity of bc in magnitude and direction because in the similar triangles ab-h-g and ab-bc-s

$$\frac{v_1}{v_{bc}} = \frac{r_1}{r_{bc}} = \frac{r_h}{r_{bc}} = \frac{ab\text{-}h}{ab\text{-}bc}.$$

Now we are "in" the adjacent link, on the way to link D in this case.

The construction characteristic of the so-called link-to-link (L-to-L) *method is shown next.* Rotate the velocity vector v_{bc} *into the radius* (either CL or CC), and obtain point t. Since we wish to know the velocity of a point in the next adjacent link, we want v_{cd}, because cd is the only point in C that is also in D. Therefore, draw tu parallel to bc-cd; the intercept cd-u on the radius ac-cd of cd is v_{cd} in magnitude. The reason for this is that the intercepts (bc-t and cd-u) on intersecting lines (ac-bc and ac-cd) made by parallel lines (tu and bc-cd) are proportional to the lengths (ac-bc and ac-cd) of the intersecting lines (measured from the intersection ac). This you recall from your geometry.

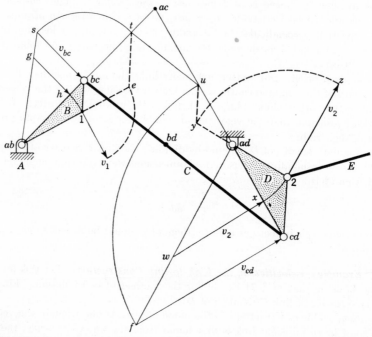

Fig. 66. Velocities—Link to Link. Link-to-link construction is to rotate velocity vector into radius, as s-bc to t-bc, and draw a line so as to intercept another vector on another radius, as u-cd.

Now rotate point u CC to get a vector cd-f, which is v_{cd} perpendicular to the radius of cd. Rotate 2 into the radius of cd to get x. Erect a perpendicular at x, draw the line ad-f, and obtain the intercept point w; the intercept $wx = v_2$ in magnitude to the same scale K_v as used for v_1. Its direction is perpendicular to the radius ad-2 of 2 (as shown, $2z$). As before, the proof of this construction is that, in similar triangles, wx/cd-$f = ad$-2/ad-cd; and the velocities of points in rotating bodies are proportional to their radii.

An alternate construction is shown dotted at 1 and 2. This dotted construction is the one typical of the link-to-link construction used at t to u, in which a *vector is rotated into a radius;* and you should study both positions 1 and 2 until you understand why this construction is correct. Rotate v_1 into radius ab-1 to e;

draw et parallel to 1-bc; then bc-t is v_{bc} in magnitude. At u, draw uy parallel to 2-cd; get y on radius 2-ad extended; swing vector 2y to its correct direction at z; 2z is v_2 to scale K_v in magnitude and direction. There is the advantage that the dotted construction is the shortest solution. There is also in this problem the question of the accuracy of drawing a line parallel to such a short line as 1-bc (or as 2-cd).

The angular velocity of C can be obtained from any known velocity and the corresponding radius. Thus if the actual dimensions and velocities were known, ω_C could be computed from either of two known velocities; in accordance with § 16, $v = r\omega$;

$$\omega_C = \frac{v_{bc}}{ac\text{-}bc} = \frac{v_{cd}}{ac\text{-}cd} \quad \text{CC.}$$

What you must be sure of in the link-to-link construction is that *intercepts on the radii* of the points are being obtained. It follows that the centro of a particular link must be known in order to be certain that you are dealing with intercepts on radii. In this problem, ac is the center of rotation of C with respect to the reference link A (that is, it has zero velocity relative to A), and therefore *all* points in link C have velocities proportional to their distances from ac with directions perpendicular to their respective radii (from ac). One may use this principle with several variations in the details of the graphical procedure.

The link-to-link method is often advantageous when one wishes to have many velocities of a certain point, as in obtaining a velocity diagram (§ 97). However, one cannot blindly draw lines parallel to kinematic lines representing links; but in each problem, a method of procedure must be decided upon with the location of significant centros (such as ac, Fig. 66) in mind. Moreover, there is an additional technique required in moving through a slider on another moving link. This situation is covered in each of the next two chapters.

Be certain that the newly found vector points in its correct sense. In most cases for velocities, the sense can be easily determined by imagining the linkage in motion. In any case, you should *try* to decide on this basis, because this would be the "common-sense check." However, if the given vector is rotated say counterclockwise (dotted construction v_1-e, Fig. 66), the final vector should be rotated clockwise (oppositely, y to z, Fig. 66) to compensate for the original rotation.

73. Examples—Velocities by Centro Method.

(a) In the five-link mechanism of Fig. 67, the cam B turns at 140 rad./min. and the radius to the point of contact 1 is $r = 1.1$ in. What is the velocity of point 2 in the rolling wheel? Use the LOC method. Determine the angular velocity of wheel E if ae-2 is 1.4 in. (b) The mechanism of Fig. 68 is much the same as the part $CDEA$ in Fig. 67. Given v_{cd} as shown, find v_2 by the link-to-link construction. *Note.* The mechanism of Fig. 67 is chosen as illustrative of several situations. The cam was omitted from Fig. 68 because we have not yet discussed how to "pass through" a sliding connection. (One could go from 1 to bc, which has the same velocity as cd, because both points are in link C, which has rectilinear motion. There are usually a number of roundabout procedures.)

Solution. (a) The linear speed of point 1 is

$$v_1 = r\omega = \left(\frac{1.1}{12}\right) 140 = 128.2 \text{ fpm.}$$

Let $K_v = 200$ fpm/in. The construction in Fig. 67 is highlighted by heavy dots and v_2 is found to be about 55 fpm. The reader will find it instructive to go through the motions of locating all centros. (Centro bd is missing because it would crowd the illustration.) Note that bc is on the normal to the cam profile at the point of contact (§ 64) and the line ab-ac. If the wheel E is rolling on A, its centro ae is at the point of contact between A and E.

To decide upon the line of centers, we think: the known point 1 is in B, the unknown point 2 is in E, and the ground link is A; therefore, the line of centers

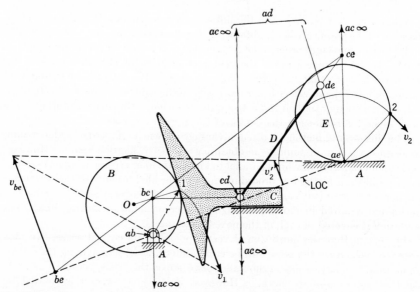

Fig. 67. Velocities—Line of Centers. For the rolling wheel E, ae is the centro with zero velocity; hence points in E rotate about ae. Points in B rotate about ab; be is the common centro.

for these three links is defined by ae-ab-be. The centros containing the name of the ground link A are momentarily or permanently stationary; the *common centro* be is made up of the names of the links containing the known and unknown points (1 and 2). One disadvantage of the LOC method is that it often happens that one of the needed centros is inaccessible or quite far away. For example, a small movement of the mechanism of Fig. 67 will put centro be out of reach. One can always get around this difficulty and still use the centro method by going through some intermediate link and its common centro, but there are other methods of finding velocities which many engineers prefer. (See the next chapter.)

The angular velocity of wheel E can be computed from $v = r\omega$, § 16, as follows:

$$\omega_E = \frac{v_2}{ae\text{-}2} = \frac{55}{1.4/12} = 472 \text{ rad./min.} = 7.86 \text{ rad./sec. CL}$$

(b) The link-to-link construction from C to 2, Fig. 68, is nicely defined by the centros ad and ae. Notice that ad is not needed for finding v_2; we only need to know that it is at the intersection of the vertical line cd-ac and the line ae-de, so that vectors can be rotated into and out of the radii as desired. However, to get

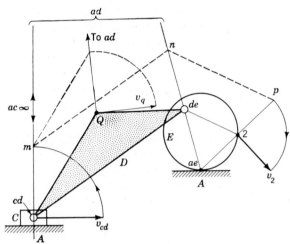

Fig. 68. Velocities—Link to Link. Find centros ad and ae; rotate v_{cd} into the radius ad-cd; mn is parallel to cd-de and np is parallel to de-2. To get the velocity of Q ($= v$-) by this method, ad must be located so that the radius ad-Q can be drawn.

the velocity of a point such as $Q(= v_q)$, the actual point ad is needed for the purpose of drawing the radius of Q to ad (to get an intercept on the radius).

74. Instantaneous Angular Velocities. We often have use for the angular velocity of a body, for example in computing normal accelerations, and equation (10), § 16, is applied ($v = r\omega$). The absolute angular velocity of any link is the absolute linear velocity of any point in the link divided by the point's instantaneous radius to the link's centro of zero velocity. In Fig. 67, the angular velocity of the wheel E is

$$(\boldsymbol{b}) \qquad \omega_E = \frac{v_2}{ae\text{-}2} = \frac{v_{de}}{ae\text{-}de} = \frac{v_{be}}{ae\text{-}be}, \text{ etc., CL.}$$

In this case, the reference link A is stationary (or part of the frame) and the angular velocity is called the *absolute* angular velocity. Be consistent with units.

The angular velocity ratio of two links in a system is obtained by division of angular velocities, if they are known, or by using the common centro as follows: In Fig. 67, the velocity of the common centro v_{be} is $r\omega = (ab\text{-}be)\omega_B$ as a point in link B, and $r\omega = (ae\text{-}be)\omega_E$ as a point in E. But of course these two values of v_{be} must be the same, or

$$v_{be} = r\omega = (ab\text{-}be)\omega_B = (ae\text{-}be)\omega_E,$$

from which

(c)
$$m_\omega = \frac{\omega_B}{\omega_E} = \frac{ae\text{-}be}{ab\text{-}be},$$

where m_ω stands for **velocity ratio,** which is defined as the angular velocity of the *driver* (*input link*) divided by the angular velocity of the *driven* link (*output link*). To illustrate further, in Fig. 60, $\omega_C/\omega_B = ab\text{-}bc/ac\text{-}bc$. Generalizing from equation (**c**), we may say that *the ratio of the angular velocities of any two bodies moving in a constrained system is inversely as the ratio of the distances of their common centro from their centers of rotation.* When the common centro lies between the centers of rotation of the bodies, the bodies rotate in the opposite sense; when the common centro lies outside the centers of rotation, the bodies rotate in the same sense. Those who are studying or have studied analytic mechanics should recall that for a given power transmitted, the torque T on the shafts is inversely as the angular velocities (power $= T\omega$).

75. Inversion. An inversion of a mechanism has occurred when a different link is held stationary (or attached to a frame). Consider a slider-crank mechanism $ABCD$. If, in Fig. 69(a), link A is held constant, the inversion is the common reciprocating engine mechanism (D

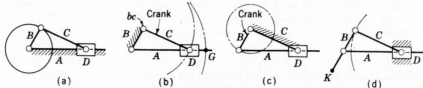

Fig. 69. Inversions of Slider-crank Mechanism. (a) Reciprocating engine. (b) Whitworth quick-return mechanism (Figs. 89 and 96). (c) Used for toy steam engines; oscillating cylinder D opens and closes ports. (d) Old-fashioned hand pump (D stationary, links B and C oscillate).

slides on A), characteristic of crankshaft-piston systems of automotive engines, gas compressors, etc. If link B is held stationary, Fig. 69(b), both links A and C can make complete rotations while A slides in D; C usually moves with constant speed, A with variable speed. Something must be done in the physical design to arrange for link A, say, to pass the pin bc (see Fig. 72). In the proportions shown, this is a Whitworth quick-return mechanism. If the link B is made significantly longer than C, this inversion becomes the oscillating-arm quick-return mechanism (Fig. 73). For links C or D, stationary, see the caption to Fig. 69.

76. Dead-center Positions. If the line of action of the force on a link as imposed by other links in the mechanism passes through the center of rotation of said link, the force then has no turning moment with respect to the center of rotation, and the mechanism is said to be on dead center. In the reciprocating-engine mechanism, there are two dead-center positions. Figure 70(a) shows the **head-end dead-center position** (in the

automotive industry and elsewhere, the ***top dead center,*** TDC); Fig.
70(b) shows the ***crank-end dead-center position*** (also called the ***bottom
dead center*** BDC).   Evidently, in order to start the engine with a force
F on the piston, the mechanism must be somewhat off of dead center.

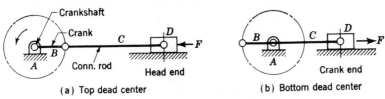

(a) Top dead center   (b) Bottom dead center

Fig. 70. Dead-center Positions.

77. Forms of Four-bar Mechanisms.   There are myriads of possible
combinations of links and linkages; hence only a meager assortment can
be mentioned here.   In addition to the few that have already been
discussed briefly, the ones taken up in the next few paragraphs are noteworthy at this time.

Some of the possibilities of the four-bar mechanism are indicated in
Figs. 71 and 72.   In Fig. 71(a), the link *D* oscillates while the crank *B*

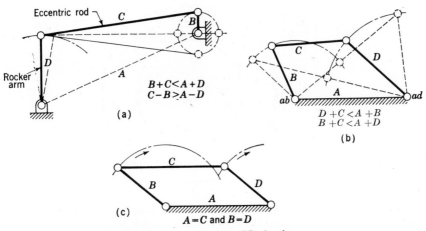

Fig. 71. Some Four-bar Mechanisms.

rotates.   When the crank is quite short, it becomes an eccentric [Fig.
36(c)]; the connecting rod for the eccentric is often called an ***eccentric rod;***
and an oscillator is sometimes called a ***rocker arm.***

In Fig. 71(b), complete rotation for either *B* or *D* is impossible.   The
limiting positions, shown dotted, are determined on one side by *B* and *C*
being collinear, on the other side by *C* and *D* being collinear.   See the
proportions on the illustration.

The ***parallel-motion mechanism*** has a number of uses.   With the
links *B* = *D* and *A* = *C*, the link *C* always moves so that it is parallel

to A.   A mechanism with these proportions can be so constructed that links B and D make complete revolutions.   One type of drafting machine consists of two parallel-motion linkages in series.

The **drag link** of Fig. 72 is proportioned so that both links B and D make complete rotations.   Notice that if a pin is placed at ad, there is interference between a bar link B and the pin ad (or bar link D and

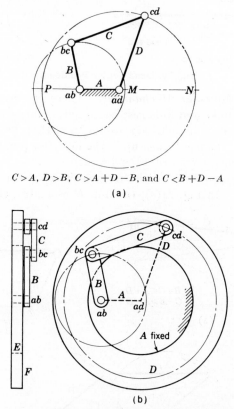

$C>A$, $D>B$, $C>A+D-B$, and $C<B+D-A$

(a)

(b)

Fig. 72. Drag Link.   The shortest link is always the stationary link; the sum of the shortest and longest links is less than the sum of the other two.

pin ab).   The construction of Fig. 72(b) is suggestive of what can be done to get around this difficulty—a stationary disk large enough to contain centers ab and ad with, say, ad at the center of the disk.   The large disk constitutes the bearing surface for link D, which now becomes a ring. The relative motions of points bc and cd are identical in Fig. 72(a) and (b).   The input is likely to be to crank B at constant speed; the output is no doubt to a link not shown but connected in some way to link D, say at cd.   The proportions can be adjusted to some extent, all the while staying within the limits specified in Fig. 72, so that one can obtain closely a desired movement of the output link, say a slider. This output slider may be at certain positions for certain crank angles, or it may have one slow stroke and one fast stroke, as in quick-return mechanisms, for which it is used. It is an interesting and useful mechanism, as a study of its output motions in various configurations reveals.

78. Quick-return Mechanisms. *Quick-return mechanisms* are useful in a great variety of situations. The reason for using this mechanism is that some link has a *working stroke*, the stroke during which the objective of the mechanism, such as cutting metal, is being accomplished; and then it has a return stroke, during which nothing is accomplished except the return of the mechanism to the beginning of its working stroke. Evidently, the output of the machine can be increased if its return stroke is speeded up. Quick-return mechanisms appear in several link forms, including noncircular gears (Fig. 84). We have already mentioned the Whitworth mechanism, Fig. 69(b), and the drag link, Fig. 72. The *oscillating-arm quick return*, Fig. 73, is another simple and advantageous form. Notice that this one is the same inversion of the slider crank as the Whitworth, Fig. 69(b); the difference is in the relative lengths of the links (which should be compared now). There are some minor variations

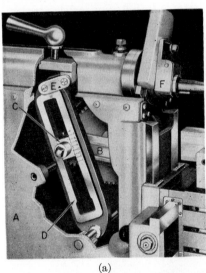

(a)

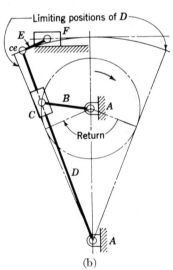

(b)

Courtesy South Bend Lathe Works, South Bend, Ind.

Fig. 73. Shaper with Oscillating-arm Quick Return. The links in the kinematic representation in (b) are named to correspond with (a). If the crank *B* turns with constant speed, the return stroke is made in less time than the working stroke (while the tool is cutting metal), because the return angle is less than the crank angle for the working stroke. The cutting tool moves with the slider *F*. The quick-return linkage consists of *A*, *B*, *C*, and *D*.

from the linkage shown in Fig. 73, some of which will be found in the problems.

79. Toggle Links. *Toggle links* are links arranged to give a large *mechanical advantage*—a large force produced by a small one. They come in various forms, Fig. 74. The toggle links C and D, Fig. 74(a), may be

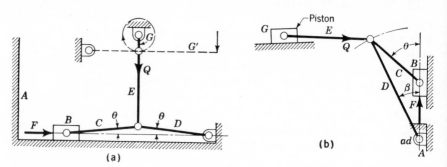

Fig. 74. Toggle Links.

moved by a rotating crank G, or by a hand-operated lever G' shown dotted. For the configuration of Fig. 74(a) only, the force $F = Q/(2 \tan \theta)$, in which it is seen that for small values of θ ($\tan \theta$ approaches zero as θ approaches zero), the force F may be quite large for a relatively small force Q in link E. The same general statement applies to Fig. 74(b)

Courtesy Allis-Chalmers, Milwaukee

Fig. 75. Jaw Crusher. For crushing rock or ore. There are other types of crushers.

and Fig. 75.   To get the proper relationship of force F and Q, one may use
the principles of analytic mechanics (2).   The arrangement in Fig. 74(b)
is used for a pneumatic riveter.   Of course, the toggle links C and D
in this arrangement can be driven by crank or hand lever, as the application requires.   The jaw crusher of Fig. 75, showing something of actual
construction, resembles the link arrangement of Fig. 74(a); but in Fig. 75,
the toggle links C and D, called toggle *plates* in this machine, are not in
a symmetric arrangement and they are in the form of a plate in contact
along the depth of the jaw (perpendicular to the paper); through an
approximately semicylindrical bearing surface, the crank has become an
eccentric, and the link B (jaw) has become an oscillator.   The distin-
guishing feature of toggle links is that, in action, the toggle links arrive at
or near a position of zero angle (θ about zero, plus or minus) with the line
of action of the force which they are to exert.

80. Straight-line Mechanisms.   A *straight-line mechanism* (**47, 54**)
is one in which some point in it moves in a straight line, or substantially
a straight line for some portion of the possible movement of the links.

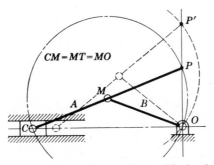

Fig. 76. Scott-Russell Straight-line Mechanism.

Such a motion of a point in a mechanism is useful where it is inconvenient
or impossible to provide plane guides and a slider link.   The Scott-
Russell, Fig. 76, and the Peaucellier, Fig. 77, provide theoretically
exact straight-line motion of the point P.   The required proportions of
the links are given in Figs. 76 and 77, which are positioned to trace a
vertical straight line.   No proof will be given here for Fig. 77, but note
in Fig. 76 that the link CP is always the diameter of a circle, passing
through O, and line CO is always horizontal.   Since the triangle COP
is inscribed within a semicircle, the angle COP is a right angle, and OP is
always vertical.

Many approximate straight-line motions are in use.   Those in Fig.
78 are seen to be modifications of the Scott-Russell.   Notice in Fig.
78(a) that instead of the slider C, point M moves in a circular arc.   If the

arc is small and the radius OM large, the arc closely approaches a straight line. Modifications similar to those of Fig. 78 are used for indicator mechanisms. The purpose of an indicator is to record the pressure changes of the fluid in the cylinder of a reciprocating machine. See Fig. 79.

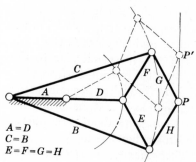

$A = D$
$C = B$
$E = F = G = H$

Fig. 77. Peaucellier Straight-line Mechanism.

Before machine tools capable of manufacturing a good plane surface were developed, a straight-line mechanism was often a necessity. Watt invented the one which goes by his name, Fig. 80, in order to guide the ends of the piston rods of his steam engine in a straight path. See Fig. 82. This feature contributed greatly to the success of his revolutionary invention.

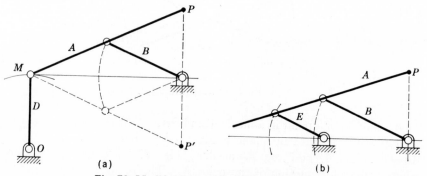

(a)   (b)

Fig. 78. Modifications of Scott-Russell Motion.

81. Parallel-motion Mechanism. A *parallel-motion mechanism* is a combination of links so constructed that when one point in it moves in a certain path, another point will move in a similar path, either larger or smaller (or the same size). In industrial use, this mechanism is called a *pantograph* and is widely used for copying purposes. A common feature of all pantographs, Fig. 81, is a parallelogram of links, $MNRS$, Fig. 81(b) and (c). The relative locations of the tracing point T and the pencil point P should be as indicated in Fig. 81, except as their relative posi-

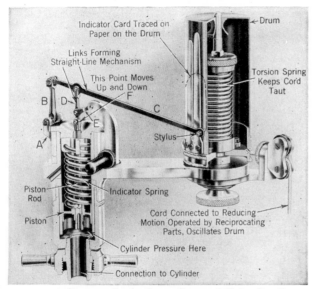

Courtesy Crosby Steam Gage and Valve Co., Boston

Fig. 79. Indicator. Steam or other fluid enters through the connection to the cylinder and acts on a small piston, whose motion is opposed by the indicator spring. The greater the pressure from the cylinder, the more the piston (and stylus) moves. The stylus, which traces the indicator card, is actuated through an approximate straight-line mechanism *BDEC*, a somewhat different arrangement from those shown in Fig. 78, but similar in principle to Fig. 78(a). The drum, through a cord, a reducing mechanism (Fig. 81), and a connection to the crosshead, is made to oscillate in phase with the movement of the engine piston. No indicator paper is shown on the drum.

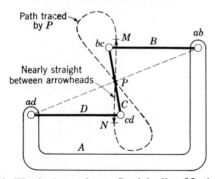

Fig. 80. Watt's Approximate Straight-line Mechanism.

tions are reversed. In Fig. 81(a), the outline produced by the point at *P* (which might, for example, be the flame of a cutting torch) is larger than the outline traced by *T*. If the positions of *T* and *P* are reversed, the produced (output) outline will be smaller than the traced (input) outline. For proof and the ratio of size, consider Fig. 81(a). Point *T* is located

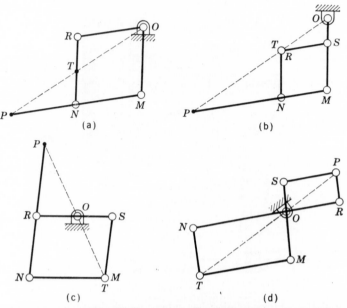

Fig. 81. Pantograph Mechanisms.

Courtesy Linde Air Products Co., Tonawanda, N.Y.

Fig. 81(e). Application of Pantograph. The hidden mechanism guides the torch in the same pattern traced.

on a straight line from P (which is chosen as desired on link MN) to the pivot O. From similar triangles PNT and PMO, we have

$$\frac{NT}{MO} = \frac{PN}{PM} \quad \text{or} \quad NT = \frac{MO}{PM} \, PN = \text{constant.}$$

This shows that for a particular position of P, point T is always at the same position of link NR, because NT is a function of constants MO, PM, and PN, and will follow a path similar to that of P. The ratio of size of the image produced by P to that by T is $OP/OT = MP/MN$. Thus, the ratio of sizes is determined in a given linkage by the location of P, and the linkage may be designed so that the size ratio (and locations of P and T) can be adjusted. In general,

$$\frac{\text{Size of image by } P}{\text{Size of image by } T} = \frac{\text{distance of } P \text{ from fixed point}}{\text{distance of } T \text{ from fixed point}}.$$

The distances OP and OT do not remain constant during a tracing operation, but their ratio does—as proved. Pantograph devices are used to enlarge or reduce engravings and maps, to guide cutting tools in accordance with a given pattern, etc. Watt's beam engine, Fig. 82, illustrates the use of his straight-line mechanism and a parallel-motion mechanism. In Fig. 82, link F has the same length as E, and $G = de\text{-}df$. The piston

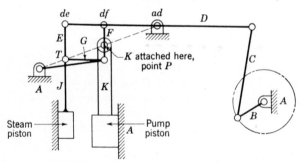

Fig. 82. Watt's Beam Engine. The straight-line mechanism consists of link F, that part of D between df and ad, and link H (not named), with point fk moving approximately in a straight line. The parallel motion (pantograph) consists of links E, F, G, and D (between de and df), with T ($= eg$) moving similarly to P ($= fk$).

rod K of the pump cylinder, shown broad to distinguish it from F, is connected to F at P (compare with Fig. 80). The piston rod J of the steam cylinder is connected to T, which moves in a similar but longer path than P by virtue of the parallelogram construction. Notice the absence of pin connections at the piston (lower ends of links J and K—which explains why the other ends of the links need to move in straight paths).

82. Elliptic Trammel. Excepting circular and straight paths, most points on linkages describe curves whose equations would be complicated and difficult to set up and use. However, it happens occasionally that one wishes to design a mechanism of such configuration that a certain point follows a predefined path, one which may or may not be simply expressed in equation form.   An example of such a mechanism is the *elliptic trammel,* Fig. 83, a device used to guide a pencil, pen, or cutting tool in the path of an ellipse (also used in other ways).   This mechanism consists of two sliders *D* and *C* and a connecting link *B*.   As *D* or *C* is

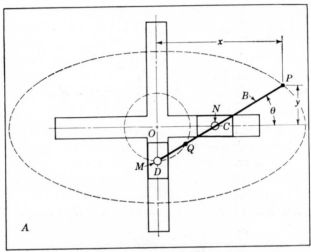

Fig. 83. Elliptic Trammel.

made to go from one extreme position in its straight path to the other extreme and back again, *any* point *P* on *B* traces an ellipse.   To prove this, note that in terms of the coordinates *x* and *y* of any point *P*,

$$\cos \theta = \frac{x}{MP} \qquad \text{and} \qquad \sin \theta = \frac{y}{NP}.$$

Square these terms and add them to get

$$\frac{x^2}{(MP)^2} + \frac{y^2}{(NP)^2} = \cos^2 \theta + \sin^2 \theta = 1,$$

which is the equation of an ellipse, origin at *O*, with major axis 2(*MP*) and minor axis 2(*NP*) (*NP* is the semiminor axis).   The mechanism is ordinarily made adjustable to get different lengths *MN* and *MP* and therefore different ellipses.   The point *Q* midway between *M* and *N* traces a circle.

83. Centrode.   A *centrode* is the path of a centro.   Recall that a centro is a point common to two bodies (really two coincident points). Hence, when we speak of the path of a centro, either we mean its "absolute" path (its path on the stationary link—or frame) or we must state which body the path is on.   If the centro's name is *bd* for example, there is a path of *bd* on *B* and another path of *bd* on *D* (since it is in both links). An interesting linkage for the purpose of studying centrodes is the cross-link chain of Fig. 84(a), in which $A = C$ and $B = D$.

In general, there is no problem concerning centrodes of centros at hinged joints; for example, the path of *bc* on link *A* is the circle *Q*. We might note that the centro *ac* is at the intersection of the lines *ab-bc* and *cd-ad* extended, then assume a number of positions of the linkage with link *A* stationary, find a series of positions of *ac*, draw a smooth curve through the points obtained, and find the path of *ac* on *A*. This path will be found to be a hyperbola (both branches) with foci *ad* and *ab*.

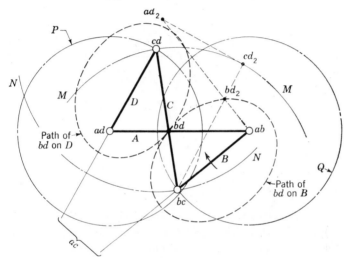

Fig. 84(a). Centrodes in Cross-link Chain.

To obtain the path of *ac* on *C*, hold the link *C* stationary, assume a series of positions of the linkage, and find the corresponding positions of *ac*. A smooth curve through these points also turns out to be the two branches of a hyperbola whose foci are at *cd* and *bc*, Fig. 84(a).

If *A* is the ground link and the linkage moves, *bd* simply moves from left to right to left along the straight line *ad-ab*. However, the paths of *bd* in *B* and in *D* are ellipses, as seen in Fig. 84(a). As before, the path of *bd* in *B* can be found by *holding link B stationary*, assuming a series of positions of the linkage, and locating the corresponding positions of *bd*. A second position is indicated dotted at bd_2. An easier and quicker way is to use tracing paper or cloth, on which is drawn link *B* with *bc* and *ab* accurately located. Also draw on the overlay the complete circle *MM*, which is the path of *cd* on *B*; that is, for all positions of *B*, *cd* will be somewhere on circle *MM*. Also note that with *A* stationary, *cd* is always on the circle *P* on the paper. These are the clues to the overlay procedure.

This overlay is hinged at *ab*, turned to many different positions quickly, for each of which *bd* is located on the overlay. After one complete turn of the overlay, draw a smooth curve through the various positions of *bd*, which in this case is an ellipse. Starting again at the initial position, but with link *D* stationary and pivot point *ad* and with complete circle

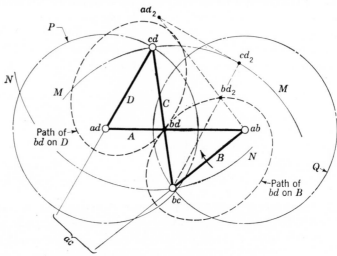

Fig. 84(a). Repeated.

NN on the overlay (the path of *bc* on *D*), circle *Q* on the paper (the path of *bc* on *A*), a series of positions of *bd* on *D* can be located as before. This path too is an ellipse, equal to the first one found. The foci of the ellipses are *bc* and *ab* for *B*, *ad* and *cd* for *D*. The major diameter is equal to the length of *A* = *C*.

Courtesy Illinois Gear & Machine Co.,
Chicago

Fig. 84(b). Elliptical Gears.

Since *bd* is a point common to *B* and *D* with the same velocity in each, there is no sliding at *bd*; or in other words, these ellipses are rolling on one another, because their tangent point is always *bd*. It follows that the kinematics of links *B* and *D* is the same whether *B* and *D* are two rolling ellipses (or elliptical gears) in continuous contact, or whether they are two cranks in a cross-link mechanism as shown. It is perhaps evident that the linkage needs some means of getting it past dead center; this means may or may not be the inertia of the driven members. Friction ellipses would drive for only half a revolution; toothed gears drive for the 360°, Fig. 84(b).

In § 74, it was shown that the angular velocity ratio of two rotating bodies is inversely as the distances of their common centro from the axes of rotation. In the position shown, Fig. 84(a), $\omega_B/\omega_D = ad\text{-}bd/ab\text{-}bd$. Imagining B rotating clockwise, you can see that bd is moving leftward, and therefore, for a constant speed of B, that the speed of D is increasing. It will be a maximum when bc reaches the line of centers $ab\text{-}ad$; and it will be a minimum when bc again reaches the line of centers 180° away. Thus, it is seen why both the cross-link mechanism and the rolling ellipses [the pitch surface of elliptical gears, Fig. 84(b)] can be and are used for quick-return mechanisms. Notice that with the links of Fig. 84(a) in the uncrossed configuration, the linkage becomes a parallelogram.

We can state in a general way that if two links, call them E and F, have constrained relative motion, the paths of the centrodes of these links (ef on E and ef on F) roll (pure rolling) on one another with the same kinematic relationship as exists between links E and F. This property of centrodes is used to obtain special rolling curves.

84. Closure. As a basic approach to velocities, no method is better than the centro procedures in which one uses repeatedly the simple fact that points in a rotating body have velocities proportional to their radii. If one can find the centros, the thought processes are easy and direct. Since centros are centers of rotation, a velocity vector for a particular point is always perpendicular to a line connecting the point and the centro. The reverse reasoning is sometimes useful: given a velocity vector at a point, a centro lies on a line through the point and perpendicular to the vector. The absolute velocity vector of any point is perpendicular to a radius from a *ground centro*—where, by *ground centro*, we mean the centro of the link involved and the ground link (or frame). Check all solutions against this rule. (Other methods of finding velocities have their advantages.)

No matter how complicated a body's motion may be, it may be said to be a series of infinitesimal rotations about an infinite number of centers. Test this statement against a rolling wheel and a connecting rod.

Review §§ 71 and 72 and note that in the line-of-centers construction, one first determines the LOC and then bases all construction on it with vectors erected perpendicular to it. Points are rotated into the LOC, as needed; but the common centro, whose velocity must always be found, is always on the LOC. In the link-to-link construction, the velocity vectors are rotated into the radii, and the construction is such that intercepts on radii are proportional to velocities.

Most of our dealings are with linkages in which each link moves in plane motion in parallel planes. However, three-dimensional motion of linkages is sometimes needed (**56, 57**). Time and space do not permit an investigation of such mechanisms here, but the finest examples are automotive (and other vehicle) steering mechanisms.

In solving velocity problems for numerical answers, put the answers along the respective vectors in the solution.

PROBLEMS

Note. Unless otherwise stated, lay out all the mechanisms given below full scale on $8\frac{1}{2} \times 11$ paper. The locations given are approximate. Most go on the paper upright. Some should have the long dimension horizontal (abbreviated LDH)— but it should be evident if the over-all dimensions given seem too much for $8\frac{1}{2}$ in. In many situations, there is little choice inasmuch as centros and construction lines frequently are off the paper in practice. Dimensions are placed alongside the link when convenient and are in inches unless otherwise indicated. Always reproduce the mechanism about in the configuration shown. Usually a satisfactory location for the drawing is given as is Q in **193** to **208**, indicated as follows: "Q (2, 6)," which means that Q can be located at 2 in. from the left edge of the paper and 6 in. from the bottom edge. By permission of the instructor, the student may locate his own figures and have more than one mechanism to the page.

Show all necessary construction lines, drawn in lightly with a hard, sharp pencil. Be sure the intended length of a vector is clear.

Centros

191. Use plate No. 3, Wingren. Find the centros for all mechanisms shown.

192. Use plate No. 4, Wingren. Find the centros for all mechanisms shown.

193–208. Find all centros. See captions for problem number.

Velocities, Miscellaneous

209–222. In each of these problems, the velocity of point J in the driving link is $v_j = 180$ ips ($K_v = 200$ ips/in.); J_b means consider J in link B, etc. Find the velocity of any point lettered $M, P, Q,$ or R. Find the angular velocity of the link named C, if any. If appropriate, find the angular velocity ratio of the links containing J and P. (a) Use the LOC construction. (b) Use the link-to-link construction (except Figs. P27, P28, P31, P32).

209. Use Fig. P20.
210. Use Fig. P21.
211. Use Fig. P22.
212. Use Fig. P23.
213. Use Fig. P26.
214. Use Fig. P27 (no link-to-link).
215. Use Fig. P28 (no link-to-link).
216. Use Fig. P29.
217. Use Fig. P30.
218. Use Fig. P31 (no link-to-link).
219. Use Fig. P32 (no link-to-link).
220. Use Fig. P33.
221. Use Fig. P34.
222. Use Fig. P35.

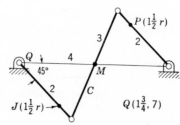

Fig. P20. Problems **193, 209.**

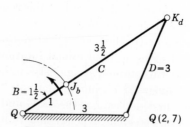

Fig. P21. Problems **194, 210.**

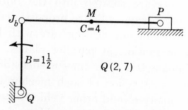

Fig. P22. Problems **195, 211.**

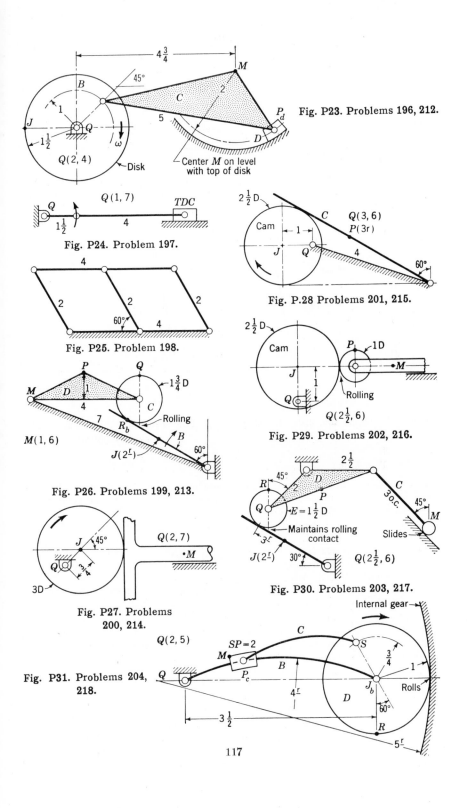

Fig. P23. Problems 196, 212.

$Q(1, 7)$

Fig. P24. Problem 197.

Fig. P25. Problem 198.

Fig. P.28 Problems 201, 215.

Fig. P26. Problems 199, 213.

Fig. P29. Problems 202, 216.

Fig. P27. Problems 200, 214.

Fig. P30. Problems 203, 217.

Fig. P31. Problems 204, 218.

117

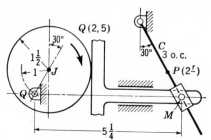

Fig. P32. Problems 205, 219.

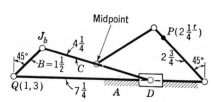

Fig. P34. Problems 207, 221.

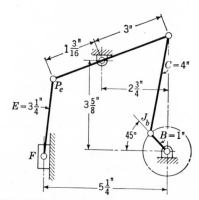

Fig. P33. Problems 206, 220. Rocking-beam Engine or Pump.

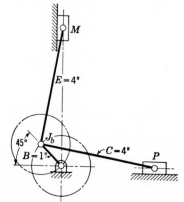

Fig. P35. Problems 208, 222. Right-angle Cylinders.

Velocities, Four Links

223. Use plate No. 3, Wingren. In Fig. 1, let the velocity of bc be represented by a $\frac{3}{4}$-in. vector ($K_v = 400$ fpm/in.) and find the velocity of the mid-point of link D by the LOC method.

224. The same as **223** except that the link-to-link construction is used.

225. Use plate No. 3, Wingren. In Fig. 4, the velocity of the center of the cam is a 1-in. vector ($K_v = 300$ ips/in.). Find v_A by the LOC method.

226. Use plate No. 3, Wingren. In Fig. 3, $v_{ab} = 75$ fpm CC ($K_v = 100$ fpm/in.). Find the velocity of the right end of link B and of its mid-point by the LOC method.

227. The same as **226** except that the link-to-link construction is used.

228. Use plate No. 3, Wingren. In Fig. 5, $v_{ad} = 400$ fpm CL ($K_v = 400$ fpm/in.). Find the velocity of the outer end of link C and of its mid-point by the LOC method.

229. Use plate No. 3, Wingren. In Fig. 7, the velocity of the center of wheel A is 380 fpm leftward ($K_v = 400$ fpm/in.). Find v_{ab} and v_D by the LOC method.

230. The same as **229** except that the link-to-link construction is used.

231. Use plate No. 3, Wingren. In Fig. 8, let the points of contact in A and B be named P and Q, respectively. Let the top corner of link B be called M. Given the velocity of the center of circle A as 80 fps CL ($K_v = 100$ fps/in.), find v_P, v_Q, and v_M by the LOC method.

232. In § 35, it was shown that the pressure angle in cams was a function of certain linear velocities, equation (14).

A similar statement can be made for cams with offset followers: offset = e, Fig. P36. Perhaps the easiest way to find the proper equation is by considering velocities from the viewpoint of centros. In Fig. P36, imagine the follower B as making point (or line) contact with the cam C (center of roller on theoretical curve), ignoring the roller. The centros

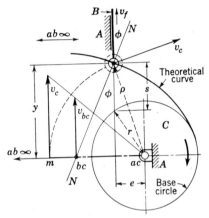

Fig. P36. Problem **232.**

are then as shown, $v_c = \rho\omega$ is the velocity of the contact point on the cam C, and $v_{bc} = v_f$ = the velocity of the follower. Set up a proportionality between these velocities and their radii (along ac-m) and show that (on upstroke)

$$\tan\phi = \frac{v_f - e\omega}{y\omega},$$

in which $y = (r^2 - e^2)^{1/2} + s$; the base circle radius is r and s is the displacement of the follower. The negative sign in the equation for $\tan\phi$ changes to a positive sign on the downstroke.

233. Find the centros for the cam and follower shown in Fig. P37. Sketch about as shown. Now noting that the speed v_f of the follower is the same as v_{bc}, show that the length e needed to make contact as shown is given by

$$e = v_f/\omega,$$

where ω is the angular velocity of the cam. The maximum value of e needed

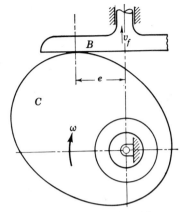

Fig. P37. Problem **233.**

occurs when v_f is a maximum. See equation (15), § 38.

234. Find all the centros for the Scotch yoke in approximately the configuration shown in Fig. 15. Assume that the velocity of Q is represented by some convenient vector and show the construction for finding the piston velocity by the LOC method.

235. Use plate No. 1, Wingren. (a) Find all centros. (b) Link B is moving toward the right with $v_B = 30$ fps ($K_v = 40$ fps/in.). Find v_x and v_z, the velocities of points X and Z, by any convenient construction.

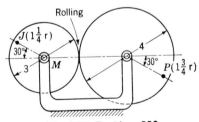

Fig. P38. Problem **236.**

236. Two gears as shown in Fig. P38 run together; locate M (2, 5). The velocity of point J is represented by a vector 2 in. long. Show a graphical construction for finding v_p.

237. In the drag-link mechanism of Fig. P39, $\omega_B = 10$ rad./sec. CC. Locate M (2, 3) LDH; lay out half size; $K_v = 4$ fps/in. (a) Locate all centros. (b) Find v_h, v_q, and v_p by the LOC method. (c) Find ω_C. (d) What is the velocity ratio, links B and D?

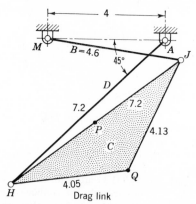

H   Drag link

Fig. P39. Problems 237, 238, 273, 274.

238. The same as **237** except that the link-to-link procedure is used in (b).

239. Two rolling wheels are connected by link C, Fig. P40; locate M $(1\frac{1}{2}, 2\frac{1}{2})$. The momentary angular velocity of B is 5 rad./sec. CC; $K_v = 2$ ips/in. (a) Locate all centros. (b) Find v_p, v_r, and v_q by the LOC method. (c) Find ω_C, ω_D, and the angular velocity ratio of the wheels.

240. The same as **239** except that the link-to-link construction is used in (b). Lay out $\frac{3}{4}$ size.

241. (a) Find all centros for Fig. P41; M (3, 4). (b) Given $v_B = 200$ ips rightward ($K_v = 200$ ips/in.) and $\theta = 30°$, find v_q, v_p, and ω_C.

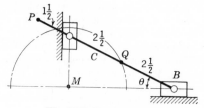

Fig. P41. Problem 241.

242. (a) Find all centros for Fig. P42; Q (3, 6). (b) Given $v_q = 450$ fpm ($K_v = 400$ fpm/in.) and $\theta = 45°$, find v_p and ω_C.

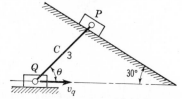

Fig. P42. Problem 242.

243. The linkage shown in Fig. P43 is an approximate straight-line mechanism. Link D oscillates as C slides in it; M (6, 6). (a) Find all centros. (b)

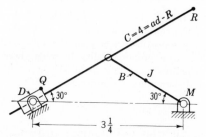

Fig. P43. Problem 243.

The mid-point J of link B has a velocity represented by a vector 1 in. long. What are the velocities of points R and Q?

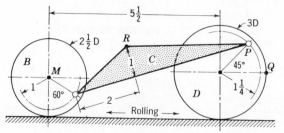

Rolling

Fig. P40. Problems 239, 240.

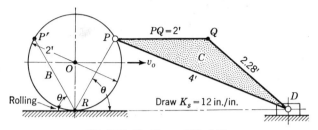

Fig. P44. Problems 244, 245.

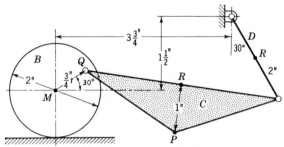

Fig. P45. Problems 246, 247.

Use the LOC method. (c) What are the angular velocities of links C and D if $K_v = 200$ fpm/in.?

244. The center O $(2\frac{1}{2}, 2\frac{1}{2})$, Fig. P44, of the rolling wheel (which oscillates to and fro) is moving with $v_o = 900$ fpm $(K_v = 400$ fpm/in.$)$; $\theta = 60°$. (a) Find v_p, v_q, and v_D by the LOC method. (b) Find v_p, v_q, and v_D by the link-to-link construction. (c) What is ω_C?

Ans. $v_q \approx 1385$ fpm,
$\omega_C \approx 210$ rad./min.

245. The same as **244** except that the position of the wheel is such that P is in the position P'.

246. The wheel M $(2, 5)$ in Fig. P45 rolls toward the right at 10 ips $(K_v = 5$ ips/in.$)$. (a) Find all centros. (b) Determine v_q, v_p, and v_r, using the LOC. (c) Determine ω_C and ω_D.

247. The same as **246** except that the link-to-link construction is used in (b).

248. An epicyclic gear train is shown in Fig. P46; M $(5\frac{1}{2}, 5)$. The arm rotates 10 rps CL, carrying with it gear C, which rolls on the stationary gear A. (a) Locate all centros. (b) Determine v_p, v_q, and v_r. (c) What is ω_C?

249. The same as **248** except that the stationary gear A is an internal gear, Fig. P47; M $(4, 5)$.

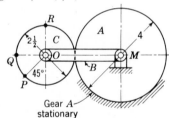

Gear A
stationary

Fig. P46. Problem 248.

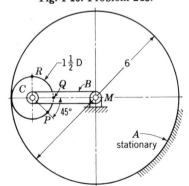

Fig. P47. Problem 249.

250. In the epicyclic gear train of Fig. P48, gear C drives, rotating at 2000 rpm CC; M (4, 5). (a) Locate all centros. (b) Determine v_r, v_p, and v_q. (c) What are ω_B, ω_D, and the velocity ratio C to B?

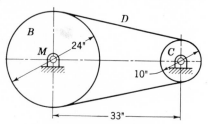

Fig. P50. Problem 252.

252. (a) In the belt drive of Fig. P50, locate all centros. There is no slipping; M (2, 7). (b) If C rotates 1000 rpm CL, find the velocity of the centro bc and from this find the velocity of a point on the periphery of B. Is the result what you would expect?

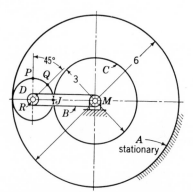

Fig. P48. Problem 250.

251. The cam in Fig. P49 turns at 900 rpm CL; M ($2\frac{1}{2}$, 5). (a) Find all centros for the position defined. (b) Using the LOC method, find v_J and v_R. (c) What is ω_C?

Velocities, Five Links

253. In Fig. P51, there is no slipping between the two friction wheels B and C; M ($2\frac{1}{2}$, $4\frac{1}{2}$), LDH. The driver B turns 200 rpm ($K_v = 1000$ fpm/in.). For $\theta = 30°$, find v_q and v_p (a) by the LOC method, (b) by the link-to-link construction. (c) Determine ω_D in radians per second.

254. The same as **253** except that $\theta = -30°$, $v_J = 100$ ips, and $K_v = 100$ ips/in.

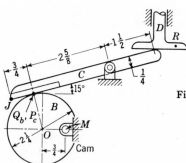

Fig. P49. Problem 251.

Fig. P51. Problems 253, 254.

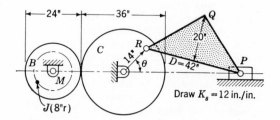

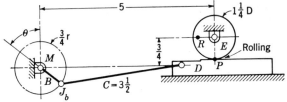

Fig. P52. Problems 255, 256.

255. The arrangement in Fig. P52 is used to cause gear E to oscillate through a large angle; M ($1\frac{1}{2}$, 6). Let $\theta = 30°$. (a) Find all centros. (b) Find v_p in link D and v_r by the LOC method. (c) Determine ω_C and ω_E.

256. The same as **255** except that the link-to-link construction is used in (b).

257. The 1-in. crank B in Fig. P53 turns at 20 rps CL; M (4, 6). (a) Locate all centros. (b) Let T be at a radius of about 2 in. and find its velocity by the LOC method, considering point R in B; $K_v = 60$ ips/in. (c) The same as (b) except that R is used as a point in C. (d) What is ω_E? CL or CC?

$\theta = 30°$ and locate all centros. (b) Let R be at a radius of 3 in. and find its velocity by the LOC method, considering point bc in B; $K_v = 400$ fpm/in. (c) The same as (b) except that bc is used as a point in C. (d) What is ω_E?

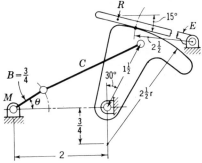

Fig. P54. Problems 258, 259.

259. The same as **258** except that $\theta = 120°$.

260. In Fig. P55, the bodies B and C remain in contact. The radius of the part RS of cam B is 1 in., as shown. Draw that part of the mechanism not completely defined approximately as shown; M (2, 5) LDH. (a) Find all centros. (b) The velocity of J is 300 fpm ($K_v = 200$ fpm/in.) rightward. Use

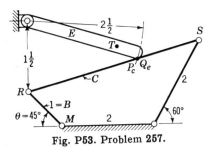

Fig. P53. Problem 257.

258. The $\frac{3}{4}$-in. crank in Fig. P54 turns at 1440 rpm CC; M ($2\frac{1}{2}$, 5). (a) Let

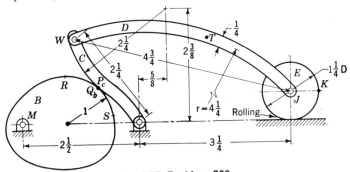

Fig. P55. Problem 260.

the LOC method and find v_t, v_p in C, and v_q in B, each as a separate problem. Consider point J in either D or E as convenient. For certain choices, some of the construction is off the sheet. (c) What is the angular velocity ratio of links B and E? What is the angular velocity of D?

Velocities, Six Links

261. Use plate No. 6, Wingren. Locate another point Q in link E on a perpendicular at the mid-point and $\frac{3}{4}$ in. away from (above) E. (a) Find all centros. (b) Given $v_J = 80$ fpm ($K_v = 60$ fpm/in.), B rotating CC, find the velocities of points cd, K, Q, and link F by the LOC method. (c) Determine ω_E and ω_C in radians per minute.

262. The same as **261** except that the link-to-link construction is used.

263. Use plate No. 16, Wingren. (a) Locate all centros. (b) Locate the end positions of link F. (c) Given the velocity of bc as 200 fpm ($K_v = 200$ fpm/in.), find v_F and v_{ed}. Use the LOC method. (d) What are ω_D and ω_E?

264. The crank B in Fig. P56 turns at 300 rpm; M (2, $4\frac{1}{2}$). (a) For the position shown, find all centros. (b) If the actual length of B is 2 ft. (and other links in proportion), use the LOC method and find v_J, v_F, and v_Q; $K_v = 3000$ fpm/in. (c) What are ω_D and ω_E? (d) Locate

the ends of the stroke of F and dimension the stroke ($\frac{1}{2}$ in. = 1 ft.).

265. The same as **264** except that the two short sides of D are each 1 in. and D makes a complete revolution.

266. In Fig. P57, locate the significant points as shown; M (2, 5). (a) Find all centros. (b) The velocity of crank B at the instant shown is 200 rpm CC. Find the velocities of points J, R, and link F. (c) What is the angular velocity ratio of links C and D? What are ω_C and ω_E?

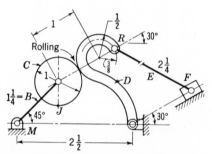

Fig. P57. Problems 266, 267.

267. The same as **266** except that the link-to-link construction is used for the velocities.

268. The crank in the quick-return linkage of Fig. P58 turns at 170 rpm CL and is actually 4 in. long (other links in proportion); M ($2\frac{1}{2}$, 5); $K_v = 300$ fpm/in. (a) Locate all centros. (b) Find the velocity of the slider F by the LOC construction and ω_D.

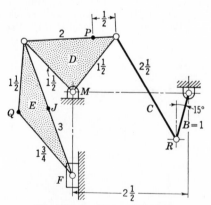

Fig. P56. Problems 264, 265.

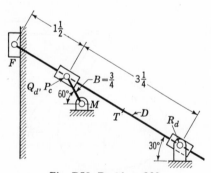

Fig. P58. Problem 268.

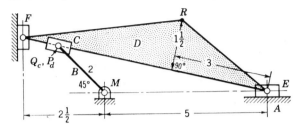

269. The slider F carries the working tool in the quick-return mechanism of Fig. P59; M ($4\frac{1}{2}$, $5\frac{1}{2}$), LDH. (a) Locate all centros. (b) Let the crank B rotate at 300 rpm CC; $K_v = 600$ fpm/in. Its actual length is 6 in. Find the velocities of links F and E and point R. (c) What is the angular velocity of D?

270. The same as **269** except that the absolute velocity of R is 350 fpm, and in (b) the problem is to find v_E, v_F, and ω_B.

Centrodes

271. For the cross-linked quadric chain of Fig. P60, find the centrodes of ac on links A and C; M (5, 4) on 11 $\times$ 17-in. paper, LDH; draw full size.

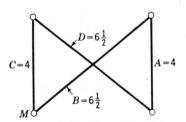

Fig. P60. Problems 271, 272.

272. The same as **271** except that the centrodes of bd on B and D are desired.

273. For the drag-link chain of Fig. P39, determine rolling curves that will roll on each other with the same relative motion that B and D have in the given chain. It is all right in this problem to represent link C by the line JH only.

274. The same as **273** except that rolling curves which will reproduce the relative motion of links A and C are desired.

275. Use plate No. 1, Wingren. Name the ground link D. Find rolling curves

such that one curve rolling on the other reproduces the same motion of points ac and bc as in the given mechanism (centrodes of cd on links C and D, respectively). What are the shapes and sizes of the resulting curves?

Larger Paper Needed

276. The linkage of Fig. P61 is used as a means of obtaining a piston stroke (of F) that is greater than the crank throw; M ($5\frac{1}{2}$, $2\frac{1}{2}$), LDH. Lay out the dimensions shown half size. The crank rotates 1600 rpm CC; $\theta = 45°$. (a) Find the stroke of the piston. (b) Locate all centros. (c) Find v_q and v_F by the LOC method; $K_v = 2000$ fpm/in. (d) Solve (c) by the link-to-link construction. (e) What are the instantaneous angular velocities of the connecting rods?

277. The same as **276** except that $\theta = 135°$.

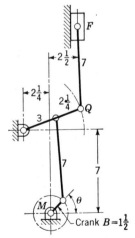

Fig. P61. Problems 276, 277.

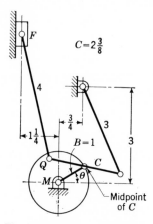

Fig. P62. Problems 278, 279.

279. The same as **278** except that $\theta = 210°$.

280. In Fig. P63, the two friction wheels B and G roll on each other, with B turning 75 rpm CL; M (7, 4) on 11×17 paper, LDH; draw full size. Let $\theta = 45°$ and $\beta = 30°$. (a) Locate all centros. (b) Find v_p and v_Q by the LOC method; $K_v = 40$ fpm/in. (c) What are the angular velocity ratios ω_F/ω_B and ω_E/ω_B? (d) What are the instantaneous angular velocities of links C, E, and F?

281. The same as **280** except that $\beta = 60°$ and $\theta = -45°$.

282. The same as **280** except that the link-to-link construction is used in (b).

278. The mechanism of Fig. P62 results in a stroke of the piston F greater than twice the crank throw; M (8, 4), LDH. Let $\theta = 30°$. (a) Find the length of stroke of F and compare with the stroke of a simple slider-crank mechanism with the same length of crank B. (b) Find the centros. (c) If B turns 30 rpm CL, find v_q and v_F by the LOC method; $K_v = 10$ fpm/in. (d) The same as (c) except use the link-to-link construction. (e) Determine the instantaneous angular velocities of links C, D, and E.

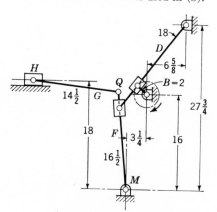

Fig. P64. Problem 283.

283. A mechanism for a hack saw is shown in Fig. P64, in which the crank B is driven at 83 rpm CL by a silentchain drive from an electric motor. The saw blade moves with slider H. Locate M (8, 3), LDH. Use a space scale of 3 in. = 1 ft.; $K_v = 100$ fpm/in. (a) Locate all centros. (b) Using the LOC method, find the velocities of Q and H. (c) What is the angular velocity ratio of links D and F (from centros)? (d) What are the angular velocities of links D and F?

284–290. These numbers may be used for other problems.

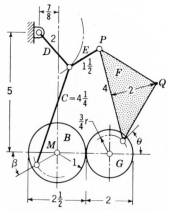

Fig. P63. Problems 280–282.

4

Relative Velocities

85. Introduction. If one has time to learn only two methods of finding velocities in mechanisms, the centro and the relative velocity methods are probably the best choices. The principle of centros provides economical thought processes in many situations; relative velocities are economical and more revealing in other situations, and in general provide a quicker solution to velocity problems. An added virtue of relative velocities is that the method of thinking through a solution is the same as thinking through a solution of relative-acceleration problems. Hence, we shall not only be learning something useful in its own right, but also useful in preparation for knowledge to be gained in the future.

As before, by *absolute velocity* or *total velocity* we shall mean the velocity relative to the earth, or if a machine is moving relative to the earth, we shall mean relative to the frame unless the context indicates that velocities relative to the earth are meant. The reader is intuitively familiar with relative velocities from his own movements. If one is in a vehicle moving along a straight road at 70 mph, he realizes that if another vehicle is moving in the opposite direction at 70 mph, the relative velocity of the vehicles is 140 mph. If they happen to be moving on the same straight line, the consequences are likely to be unpleasant eventually. Other examples are the relative speed of a river to a person moving across a bridge, the absolute velocity of a person walking across the deck of a moving ship, the velocity of a missile to intercept a moving target. We shall be interested in applications to mechanisms.

86. Relative Velocity of Coincident Point in Two Bodies, Plane Motion. Let a link A, Fig. 85, be moving with plane motion in space, and let a slider B move along A. Concentrate your attention on position 1, where we have two coincident points P in the path which is link A and Q

in the slider B; P and Q are particular points fixed with reference to their respective links. We shall take the point on the ground link at P, Q as the reference point, so that the displacements of P and Q may be thought of as the absolute displacements. (Some other origin in the ground could be used if desired, but the result would be the same.) After some time interval Δt, the links arrive at position 2; P in path A has moved from P_1 to P_2 (the displacement Δs_p as shown); Q in B has moved to position Q_2. In this position, Q has an absolute displacement of Δs_q from its original position and it has moved along the link a distance $\Delta s_{q/p}$, Fig. 85. In longhand, this symbol $\Delta s_{q/p}$ means the change of displacement of point Q with respect to point P. If the displacement of

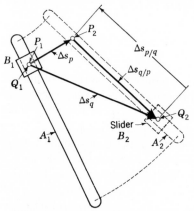

P with respect to Q is desired, it should be written as $\Delta s_{p/q}$ and this vector points in the opposite sense, Fig. 85. Since displacements as well as velocities must be with respect to some other body (frame of reference), we would be more precise to write the absolute displacements of P and Q as $\Delta s_{p/g}$ and $\Delta s_{q/g}$, say, in which the subscript g implies that the reference link is the ground. There will be no trouble, however, if the reader keeps in mind that if the symbol does not include a reference body, the ground link (or frame) is understood. This

Fig. 85. Relative Displacement.

simplifies the symbolization somewhat. There will be times when the inclusion of a point in the ground link in the symbol will be helpful to the thinking processes.

In Fig. 85, the relation of the vectors is seen to be

(**a**) $\qquad \Delta s_q = \Delta s_p + \!\!\!\rightarrow \Delta s_{q/p} \qquad$ or $\qquad \Delta s_q \rightarrow \Delta s_p = \Delta s_{q/p},$

where the symbols $+\!\!\!\rightarrow$ and $\rightarrow$ mean a *vector sum* and a *vector difference*, respectively. Since these displacements occur during time interval Δt, we may divide the terms of (**a**) by Δt and obtain a relation of average velocities (also by differentiation of displacements with respect to time); thus

(**b**) $$\frac{\Delta s_q}{\Delta t} = \frac{\Delta s_p}{\Delta t} + \!\!\!\rightarrow \frac{\Delta s_{q/p}}{\Delta t}.$$

Let Δt approach zero; then the time interval becomes infinitesimal, and the corresponding displacements are so small that *even if curved paths are being followed*, the displacements are coincident with the paths, and the various terms of equation (**b**) become instantaneous velocities. The *relative velocity equation* can then be written,

(17) $$v_q = v_p \rightarrowtail v_{q/p} \qquad \text{or} \qquad v_{q/g} = v_{p/g} \rightarrowtail v_{q/p},$$

in which G, g can be any reference point. Notice that the vector v_q is the resultant of the vectors v_p and $v_{q/p}$. This relation (17) has been derived for coincident points on two bodies having relative motion and it says that the absolute velocity of Q (or with respect to G) is equal to the absolute velocity of P (or with respect to G) vectorially plus the velocity of Q relative to P. In a particular situation, equation (17) is an instantaneous relationship. It will help the thought processes to observe the various ways in which the relative velocity equation can be written. Notice the sequence of subscripts.

(*c*) $v_q = v_p \rightarrowtail v_{q/p},$ $v_q \rightarrow v_p = v_{q/p},$

(*d*) $v_p = v_q \rightarrowtail v_{p/q},$ $v_p \rightarrow v_q = v_{p/q}.$

Read the symbol $v_{q/p}$, for example, as the velocity of Q with respect to P; $v_{p/q}$ represents the velocity of P with respect to Q. Now we are in a position to "move through" a sliding-contact connection so far as velocities are concerned.

87. Example—Velocities at a Sliding Connection. The $4\frac{3}{4}$-in. crank B of a Whitworth quick-return mechanism [Fig. 69(b)] turns at 600 rpm. For the configuration shown in Fig. 86, what is the velocity of a point P_d in link D?

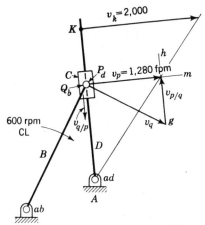

Fig. 86. Relative Velocities at Slider.

Solution. Consider two coincident points Q_b in B and P_d in D. Since the radius of Q_b and angular velocity n_b are known, the velocity of Q_b is computed as

$$v_q = 2\pi r n = 2\pi \left(\frac{4.75}{12}\right)(600) = 1492 \text{ fpm}$$

in a direction perpendicular to the radius of Q_b.   Therefore, use some convenient scale $K_v = 1600$ fpm/in. and lay off v_q as shown, at right angles to B.   From the previous article, we know that equation (17) holds for instantaneous velocities as follows,

(17) $v_p = v_q +\!\!\!+ v_{p/q},$

and the next thing to do is to complete a velocity polygon that agrees with this vector equation.   Since the point P_d is in link D which is rotating about ad, we know that the direction of v_p is perpendicular to its radius.   Therefore, draw a line Pm of indefinite extent normal to ad-P.   Now considering the relative motion of P_d and Q_b, we note that C slides on D and that D is straight (if D were

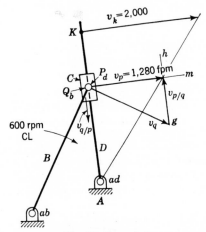

Fig. 86. Repeated.

curved, the direction of the relative velocity would be along a tangent); it follows that the relative velocity vector must be in the direction of the path.   Thus, from the end of the vector v_q, draw gh parallel to D.   The intersection of Pm and gh defines the relative velocity polygon.   Since the senses of the vectors must be correct, check with equation (17), which says that vectors v_q and $v_{p/q}$ must add to give v_p.   Checking Fig. 86, you see that this is true.   Scale v_p and find $v_p = 1280$ fpm.

The velocity $v_{q/p}$ is equal and opposite to $v_{p/q}$.   Of course, both of these vectors "act" at point P, Q.   When a vector is in some position other than where it actually acts and is pointing in its correct sense, it is sometimes called a **free vector**, as $v_{p/q}$ in Fig. 86.

With the velocity of one point P_d in link D known, the velocity of any other point in D is quickly found by the principle of the previous chapter ($v_1/v_2 = r_1/r_2$). For example, a radial line through the end of v_p intercepts v_k on a perpendicular at K; $v_k = 2000$ fpm.   Perhaps K is a hinge for additional links; if so the construction shown would be a literal link-to-link procedure through a slider, and would be convenient in finding a velocity diagram (§ 99).   But see succeeding articles of this chapter.

88. Relative Velocity of Two Points in a Rigid Body, Plane Motion.

Equation (17) can be easily generalized by using it as the definition of the relative velocity; thus, let the velocity of Q relative to P be defined by $v_{q/p} = v_q \rightarrow v_p$. Considering this equation with respect to two points fixed in a rigid body M (that is, the distance between the points is a fixed amount r), we see in Fig. 87 that the only possible motion that Q may have relative to P is one of rotation, or one in a direction perpendicular to the line joining P and Q. No matter how complicated the plane motion of body M may be, it is only necessary to know the angular velocity ω of M in order to compute a relative velocity for equation (17);

$$(e) \qquad v_{q/p} = r\omega \qquad \text{and} \qquad \omega = \frac{v_{q/p}}{QP},$$

an equation that can be used to compute $v_{q/p}$ when r and ω are known or ω when $v_{q/p}$ and r are known; r is always the distance PQ between the points P and Q. If ω is counterclockwise, the relative velocity vectors point as shown by the dotted vectors in Fig. 87. While equation (17) can be said to hold for any two points that have relative motion, it is not always easy to determine either the magnitude or direction of the relative velocity vector. However, most engineering problems fall into either of the two categories described: coincident points in constrained motion (with respect to some frame of reference or axes) or two points fixed in a moving rigid body. A vector equation can be used to determine two unknowns: two magnitudes, two directions, or one of each.

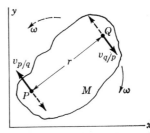

Fig. 87. Points Fixed in Rigid Body. The solid vectors show the relative velocities for clockwise (CL) rotation, the dotted vectors for counterclockwise (CC) rotation.

89. Velocity Polygons.

One may draw a velocity polygon at any convenient position. In Fig. 86, it was convenient to our purpose to start the polygon at P, Q. However, for more extensive problems (more links), there will often be some advantage in having the polygon entirely separate from the linkage. Thus, we choose some convenient point, called the **velocity pole** and designated O_v, and then proceed with the construction of a velocity polygon, a method often called the **velocity-polygon method**. As usual, the reader will obtain a better understanding of the application of (17) by studying some examples.

90. Example—Cam Mechanism.

An eccentric cam B, Fig. 88, with an eccentricity of 3 in., rotates at 900 rpm CL and drives a flat-face oscillating follower.

For the position shown in Fig. 88, find the velocity of point M and of point N in C. What is the sliding velocity at P, Q? Compute the angular velocity of C.

Solution. Since the radius ab-T is 3 in., the velocity of T (perpendicular to its radius) is computed as

$$v_t = 2\pi \left(\frac{3}{12}\right)(900) = 1417 \text{ fpm}, \quad \nearrow$$

where the small arrow indicates its general direction. Choose a convenient pole O_v and lay out $v_t = O_v t$ to scale, say $K_v = 1000$ fpm/in., and name its end t. In the velocity polygon, use lower-case letters to correspond to the points in the mechanism whose names are capital letters, as t for T. This is important for easy

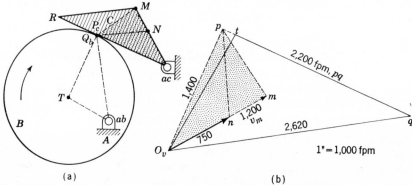

(a)   (b)

Fig. 88. Velocity Polygon, Cam Mechanism. *Construction.* Draw $O_v t = v_t$ to scale; $O_v q$ perpendicular to ab-Q; tq perpendicular to TQ; $O_v p$ perpendicular to ac-P; qp perpendicular to TQP; $O_v m$ perpendicular to ac-M; pm perpendicular to PM; pn perpendicular to PN. Arrowheads are not necessary because all absolute vectors point away from the pole O_v. Except for thinking purposes, arrowheads might be meaningless on relative velocity vectors; $v_{p/m}$ points oppositely from $v_{m/p}$. The location of vector $v_{p/m}$ is at P; the location of vector $v_{m/p}$ is at M.

identification of the various vectors in the polygon. The construction of the velocity polygon is a link-to-link procedure, but of course a different construction from that previously given that name and it is not so called now. Since v_t in B is known and since velocities in link C are desired, we first find the velocity of the coincident point Q in B, then of the coincident point P in C, then of any other point in C as desired, all by using repeatedly the relative velocity equation. Therefore, the first thing to do is to write the relative velocity equation for points T and Q,

(f) $v_q = v_t \leftrightarrow v_{q/t}.$

The vector v_t is already plotted. Remember the following rule for velocity polygons:

> *All absolute velocity vectors are laid out from the pole O_v.*

In equation (f), we know v_t completely and the directions of the other two vectors.

(If desired, we could compute $v_{q/t} = r\omega = (TQ)\omega$, § 88, knowing the magnitude of TQ, but there is no need.)   Since the direction of v_q is perpendicular to the radius ab-Q, draw a line $O_v q$ of indefinite extent from O_v in this direction.   Since the direction of Q relative to T is perpendicular to the line TQ joining the points, draw through the point t a line tq of indefinite extent perpendicular to TQ.   The intersection of lines $O_v q$ and tq at q completes the polygon for equation (f); $O_v q = v_q$ to scale, $tq = v_{q/t}$ to scale.

Next we complete the construction for the relative velocity equation

$$v_p = v_q \mathbin{+\mkern-8mu+} v_{p/q},$$

in which v_q and the directions of v_p and $v_{p/q}$ are known.   Since C rotates about a fixed point ac, the direction of v_p is perpendicular to ac-P; v_p is an absolute velocity, so draw $O_v p$ of indefinite extent from O_v normal to ac-P.   If the bodies B and C remain in contact, their relative motion is in the tangential direction (the normal velocities must be the same, § 64).   Therefore through q, draw qp perpendicular to the normal TQ to the point of contact (that is, parallel to the tangent at the point of contact).   The intersection of $O_v p$ and qp locates p and defines the magnitudes of $v_p = O_v p$ and $v_{p/q} = qp$, Fig. 88.   Now we are in link C and can use the principle of § 88.   The equation for P and M is

$$v_m = v_p \mathbin{+\mkern-8mu+} v_{m/p},$$

in which v_p and the directions of v_m and $v_{m/p}$ are known.   Draw a line $O_v m$ of indefinite extent perpendicular to the radius of M, which is ac-M.   From the end of vector $O_v p$, draw pm perpendicular to PM, which is the direction of the relative velocity of P and M.   The intersection of $O_v m$ and pm locates m and defines the magnitudes of $v_m = O_v m = 1200$ fpm (one of the required answers) and $v_{m/p} = pm$.

In finding v_n, suppose we write the equation

$$v_n = v_m \mathbin{+\mkern-8mu+} v_{n/m}.$$

Since points N and M are in the same radial line, their directions are the same; $O_v n$ coincides with $O_v m$ and no intersection is obtained.   Hence v_n may be found from some other known point, say P (but see below).   Therefore, we write

$$v_n = v_p \mathbin{+\mkern-8mu+} v_{n/p},$$

and draw a line from p perpendicular to PN, intersecting $O_v m$ already drawn and locating n; $v_n = O_v n = 750$ fpm.

The sliding velocity at P, Q, which is $v_{p/q}$ or $v_{q/p}$, $= pq = 2200$ fpm, the velocity of Q relative to P.   The angular velocity of C can be determined from any of the known velocities and their corresponding radii.   Use v_p and ac-$P = 4.95$ in.   Then for $v_p = 1400$ fpm, Fig. 88,

$$\omega_C = \frac{v_p}{r_p} = \frac{1400}{(60)(4.95/12)} = 56.6 \text{ rad./sec. CL.}$$

Notice that the shaded triangles PM-ac and pmO_v are similar because their corresponding sides are perpendicular (pmO_v was drawn that way).   For this reason, pmO_v is called the **velocity image** of that part of link C that is PM-ac. Notice that the orientation of the image of C is that it is rotated 90° *in the same sense as* ω_C. [You could complete the image of C and perhaps learn something by locating r in Fig. 88(b).] The ends of the velocity vectors of all points within PM-ac will fall within the triangle pmO_v. Also, the point n could have been located by proportion, the easiest way; $O_v n/O_v m = ac$-N/ac-M. It probably occurs to you that a technique of applying the relative velocity principle could be worked out such that lines are drawn parallel to links and radii (instead of perpendicular). It can and has been done (**36**), but we shall not go into it.

The velocity polygon has some characteristics which are sometimes advantageous. You have observed that all absolute vectors start from O_v, a point of zero velocity and the velocity image of the frame. Any line between points of the velocity polygon is to scale the relative velocity of those two points in accordance with our relative velocity equation. For example, if needed, a line mq may be drawn and its length would be $v_{m/q}$ and $v_{q/m}$ to scale.

Some of the lines in the vector polygon, Fig. 88(b), are shown dotted in order to help you follow the descriptions, but in your own solutions, you will find it quicker to use solid lines—made with a *sharp* lead, and with each point (q, k, etc.) named as soon as you find it.

91. Example—*Whitworth Quick Return*. A 5-in. crank on a Whitworth quick-return mechanism turns at 300 rpm. For the configuration of Fig. 89,

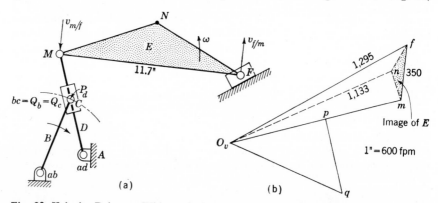

Fig. 89. Velocity Polygon, Whitworth Quick Return. *Construction.* $O_v q \perp$ to ab-Q; $O_v p \perp$ to ad-P; pq parallel to path at P; $O_v f$ parallel to path; $O_v m/O_v p = ad$-M/ad-P; $mf \perp$ to MF; $mn \perp$ to MN; $fn \perp$ to FN; draw $O_v n$.

determine the velocities of N and link F. What is the angular velocity of E (always include sense)?

Solution. The velocity of the end of crank B, which is the centro bc now called Q, is

$$v_q = 2\pi r n = 2\pi \left(\tfrac{5}{12}\right)(300) = 785 \text{ fpm.}$$

Choose a pole O_v, lay out this vector perpendicular to ab-Q, and name its end q. We may write a relative velocity equation for Q and N, or Q and F, or even Q and M, but you will find in each case that you do not know the direction of the relative velocity vector; hence, these equations do no good. We must go through P because we know the direction of motion of P relative to Q and therefore there are only two unknowns in the relative velocity equation,

$$v_p = v_q \mathbin{+\mkern-8mu+} v_{p/q}.$$

We know v_q and the directions of v_p and $v_{p/q}$. For the absolute v_p, draw $O_v p$ of indefinite extent perpendicular to its radius ad-P; draw qp in the direction of the relative motion of Q and P, namely, parallel to link D; the intersection p of qp and $O_v p$ defines the velocity polygon for the foregoing equation.

If the equation for M and P is written, $v_m = v_p \mathbin{+\mkern-8mu+} v_{m/p}$, again you will find that intersecting lines are not obtained. Moreover there is no point other than P of known velocity in D (as there was in § 90 when we used P instead of M in finding v_n). The way out of this difficulty is to notice that $v_m/v_p = ad$-M/ad-P and to find m on $O_v p$ extended so that $O_v m/O_v p = ad$-M/ad-P. One may measure the linkage dimensions and compute $O_v m$ or use a graphical construction (Fig. 20 at Obd and § 177) which may be less accurate.

After m has been determined, we use

(*g*) $$v_f = v_m \mathbin{+\mkern-8mu+} v_{f/m};$$

$O_v f$ is in the direction of motion of F and mf is perpendicular to MF as the only possible relative motion of these two points (§ 88). We find $v_f = 1295$ fpm. Now notice in the following that we cannot find v_n directly from v_m alone.

(*h*) $$v_n = v_m \mathbin{+\mkern-8mu+} v_{n/m},$$

wherein we know v_m and the direction of $v_{n/m}$. There remain three unknowns, the magnitude and direction of v_n and the magnitude of $v_{n/m}$, but only two unknowns can be found from a vector equation. Since we now know v_f, the thing to do this time is to write the equation for N and F,

(*i*) $$v_n = v_f \mathbin{+\mkern-8mu+} v_{n/f},$$

and solve these two equations (*h*) and (*i*) simultaneously. Part of this solution is in dotted lines in Fig. 89(b). From equation (*h*), draw mn of indefinite extent perpendicular to MN, which is the direction of the relative motion of M and N. From equation (*i*), draw fn perpendicular to FN for the same reason. Checking the intersection n of these lines reveals that point n satisfies both equations (*h*) and (*i*), and $O_v n = v_n = 1133$ fpm to scale in the sense $O_v n$. The actual location of the vector is naturally at point n. This situation, as regards N, arises often and in general one needs to know completely or to be able to find the velocities of two other points in the same link (or know the absolute angular velocity of the link—see below).

The angular velocity of link E can be found from v/r, § 88;

$$\omega_E = \frac{v_{f/m}}{FM} = \frac{350 \text{ fpm}}{11.7/12} = 359 \text{ rad./min. CC.}$$

To be sure of the sense, think through a relative velocity equation, equation (*g*) in this case. We see from (*g*) that $v_{f/m}$ points upward, which means that F is moving upward with respect to M and therefore the angular motion of E must be counterclockwise.

Before leaving this example, *notice that the shaded triangle mnf is the velocity image of link E; $O_v m$ is the velocity image of the line that is link D;* etc. Also observe that the links $ADEF$ constitute an ordinary slider-crank mechanism, whose velocity polygon is $O_v mf$.

92. Example—Oscillating Cylinder. The 1-in. crank of the oscillating-cylinder engine of Fig. 90 turns 600 rpm CL; $RQ = 4$ in.; $RM = 2.4$ in. Determine the speed of point M and the piston speed for the given configuration. Also

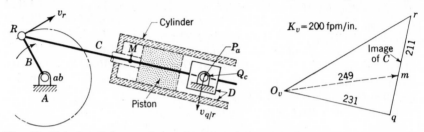

Fig. 90. Velocity Polygon, Oscillating Cylinder. *Construction.* $O_v r \perp$ to ab-R; $O_v q$ parallel to C; $rq \perp$ to RQ; point m from $rm/rq = RM/RQ$.

find ω_C. The piston and cylinder are suggested by the dotted lines, where it is noted that the physical dimensions of C may be such as not to include the axis of oscillation P (ad). The kinematic representation shown is equivalent, however.

Solution. The velocity of R, the end of the crank, is

$$v_r = 2\pi r n = 2\pi \left(\frac{1}{12}\right) (600) = 314 \text{ fpm}, \quad \nearrow$$

which is of course the velocity of one point in link C. Now from equation $v_m = v_r + v_{m/r}$, you will note that there are three unknowns (magnitude and direction of v_m, magnitude of $v_{m/r}$) and that therefore the value of v_m cannot be found directly. This is similar to previous situations, but this time the problem is to find another point on C whose velocity is known in direction. The only such point is Q_c, which is coincident with the axis of oscillation P. Since P (in A or D) is stationary, or at least in the reference link, the velocity of Q relative to P is an absolute velocity and is seen to be at any instant in the direction of sliding of Q past P, which is the direction of RQ. Thus, we can draw the polygon for

(*j*) $$v_q = v_r \mathbin{+\!\!\!\!+} v_{q/r},$$

in which $O_v r$ is perpendicular to ab-R, $O_v q$ is parallel to C, and rq is perpendicular to RQ in the direction of the relative velocity of Q to R; v_q is seen to be the piston velocity (which is the sliding velocity of the piston in the cylinder). We measure $v_q = 231$ fpm.

Since rq is the velocity image of RQ, the point m is located by proportion, $rm/rq = RM/RQ$. An idea that should be noticed in this example is that the actual physical dimensions of a link (piston link) may be *imagined* to be extended in any manner to include a coincident point which would involve only two unknowns in a relative velocity equation.

An examination of the relative velocity equation for Q and R and the polygon O_vrq shows that $v_{q/r}$ points downward at Q. Since $v_{q/r} = 211$ fpm and $RQ = 4$ in. $= \frac{4}{12}$ ft., we get the angular velocity of link C as

$$\omega_C = \frac{211}{\frac{4}{12}} = 633 \text{ rad./min. CL.}$$

93. Example—Rolling Wheel. Let the mechanism be the same as in Fig. 68, p. 101. Given $v_c = v_p$ as represented by a vector 1.5 in. long, to find v_q and v_2 [as in § 73(b), for purposes of comparing methods], and also to determine the velocity image of the rolling wheel E. Show how to compute ω_e from relative velocity. See Fig. 91.

Solution. Since we are starting with a vector v_p twice as long as in Fig. 68, the other vectors should be twice as long, except for errors inherent in the small scale used and in the reproduction processes. Observe that in solving this problem by the relative velocity method, we must recognize the centro ae; that is, we must know that all points in E are momentarily rotating about ae.

Choose a convenient pole O_v, Fig. 91, and lay out $O_vp = 1.5$ in. toward the right, which is the direction of motion of P. Since the center of rotation ae of

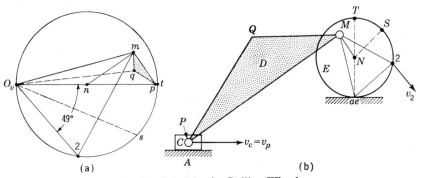

Fig. 91. Velocities for Rolling Wheel.

M is known, v_m can be found next from

$$v_m = v_p \nrightarrow v_{m/p}.$$

Accordingly, O_vm is drawn of indefinite extent perpendicular to ae-M; pm is drawn perpendicular to PM; the intersection is m in the velocity polygon; $v_m = O_vm$. Because neither the direction nor magnitude of v_q is known, a single relative velocity equation with P or M will have three unknowns. Therefore, both relative velocity equations must be used (as in finding v_n, Fig. 89).

$$v_q = v_p \nrightarrow v_{q/p} \qquad \text{and} \qquad v_q = v_m \nrightarrow v_{q/m}.$$

In accordance with these equations, draw pq from p of indefinite extent perpendicular to PQ; draw mq from m perpendicular to MQ; their intersection marks q and $v_q = O_vq$ in magnitude and direction.

To get the velocity image of the wheel E, find $v_n = v_m + v_{n/m}$ by drawing mn perpendicular to MN. Since N moves horizontally (perpendicularly to its radius ae-N), the intercept $O_vn = v_n$. One could find the velocities of several points S, T, etc., on the circumference of the wheel and draw a smooth curve through these points. On the other hand, we could note that all points on the circumference have the same relative velocity in magnitude with respect to the center of the circle N; that is, in $v_s = v_n + v_{s/n}$, $v_{s/n}$ is the same length of vector for any location of S on the circle. Therefore with a radius of $v_{s/n} = v_{2/n}$ (where $v_2 = v_m + v_{2/m}$) and with the center at n, draw a circle O_v2st as shown. The area of this circle is the velocity image of wheel E. You might note that the maximum velocity occurs at the top of the wheel, point T $(v_t = 2v_n)$; that the circle goes through O_v. The point O_v is the velocity image of point ae $(v_{ae} = 0)$. Also observe that the velocity images of D and E would be parallel to their respective links if the links were rotated 90° in the sense of their angular velocities.

The angular velocity of E may be computed from $v_{2/m} = m\text{-}2$ in the polygon, pointing downward; thus,

$$\omega_E = \frac{\text{velocity } m\text{-}2 \text{ in Fig. 91(a)}}{\text{distance } M\text{-}2 \text{ in Fig. 91(b)}} \text{ CL.}$$

94. Relative Angular Velocity. Angular velocities may be added and subtracted vectorially. A vector representing an angular velocity is perpendicular to the plane of motion and conventionally points in the direction in which a nut on a right-hand thread moves when turned in the same sense as the angular velocity. When the bodies move in the same plane or parallel planes, this vector sum reduces to an algebraic sum. It is easy to show this for the case of parallel-plane motions by using relative velocity equations.

In Fig. 92, let the angular velocity of body B with respect to A be ω_2 CL; then let A rotate with angular velocity ω_1 CL. For the coincident points P_b in B and Q_a in A (link A imagined extended to include Q), we may write, § 86,

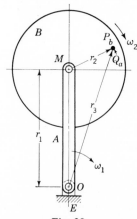

Fig. 92.

$$(k) \qquad v_p = v_q + v_{p/q}.$$

The radius of Q_a is r_3, which is seen to be the vector sum of r_1 and r_2; $r_3 = r_1 + r_2$. The velocity of Q_a is then

$$(l) \qquad v_q = r_3\omega_1 = (r_1 + r_2)\omega_1.$$

Since the angular velocity of B with respect to A is ω_2 and since point M is the point in A about which B is rotating, we have $v_{p/q} = r_2\omega_2$. Using this expression and equation (l) in equation (k), we find

(m) $$v_p = (r_1 \twoheadrightarrow r_2)\omega_1 \twoheadrightarrow r_2\omega_2$$
$$= r_1\omega_1 \twoheadrightarrow r_2(\omega_1 \twoheadrightarrow \omega_2).$$

Now we know that

(n) $$v_p = v_m \twoheadrightarrow v_{p/m}.$$

Inasmuch as $v_m = r_1\omega_1$ in equation (n), it is reasonable to conclude by a comparison of (m) and (n) that the last term in (m) is the same as the last term in (n), or

(o) $$v_{p/m} = r_2(\omega_1 \twoheadrightarrow \omega_2).$$

That is, the angular velocity of body B about the axis M (an axis does not rotate) is the vector sum of the angular velocity ω_1 of A and the angular velocity ω_2 of B relative to A. If the reference body E is the earth, then the sum of these angular velocities $\omega_1 \twoheadrightarrow \omega_2$ is the absolute angular velocity ω_B of the body B. Dropping the vector notation because the planes are parallel, this statement is written in equation form as

(p) $\qquad \omega_B = \omega_1 + \omega_2 = \omega_A + \omega_{B/A} \qquad$ or $\qquad \omega_{B/A} = \omega_B \rightarrow \omega_A,$
$\qquad\qquad$ [PARALLEL PLANES]

in which the signs are algebraic. For example, if $\omega_A = 100$ rad./min. CL and $\omega_{B/A} = 50$ rad./min. CL, then $\omega_B = 150$ rad./min. CL (and the velocity of P relative to the axis M is $v_{p/m} = 150r_2$). If $\omega_A = 100$ rad./min. CL and $\omega_{B/A} = 50$ rad./min. CC, the absolute $\omega_B = 100 - 50 = 50$ rad./min. CL because the CL rotations exceed the CC rotations. This principle is sometimes helpful in solving relative velocity (and relative acceleration) problems. The determination of the absolute velocity of P for a situation such as Fig. 92 would probably be easiest by equation (n); hence, it is important to understand the meaning of $v_{p/m}$.

95. Closure. The basic ideas of this chapter have already been stated so many times, that a repetition here is unnecessary. However, after the reader has worked a number of problems, it would be a good idea to study again §§ 86 and 88, noting the limitations, especially in § 88. Always start solutions by writing appropriate relative velocity equations and always name the points in the velocity polygons consistently because it is too difficult to understand the polygon otherwise. There is no quicker way of gaining engineering experience and know-how in this field than by thinking through the examples in detail (a pencil and paper help the thought processes). Let your mind go and think of relationships not put into words.

In practice, problems must be solved correctly and the right answer obtained, else there is something to pay. To be sure an answer is right, independent checks can be made by one or more other engineers and solutions compared until agreement is reached. One of the finest ways of checking solutions is to solve the problem by different methods—definitely the best way during the learning process. Velocity problems are particularly suited to this technique because of the several different ways in which velocities can be found. The reader can use the methods of this and the previous chapter. Practicing engineers often use the principles of two or more methods in the same problem. When numerical answers are called for, be sure to put the numbers adjacent to the appropriate vector.

PROBLEMS

Note to Students. Unless otherwise stated, lay out all mechanisms given below full scale on $8\frac{1}{2} \times 11$ paper in the upright position. In some cases, the long dimension of the paper should be horizontal, abbreviated LDH. Dimensions are placed alongside the link when convenient and are in inches unless otherwise indicated. Usually, satisfactory locations of points are indicated as follows: O_v (2, 6), which means that the velocity pole O_v is located 2 in. from the left *edge* of the paper and 6 in. from the bottom *edge*.

Unless instructed otherwise, work these problems by the methods explained in this chapter; that is, do not use your knowledge of centros unless the problem so suggests. As an act of courtesy to those who must check your work and *for your own later convenience*, always letter the points in the polygon in the conventional manner. As a regular practice, place the magnitude alongside certain vectors in the polygon: (a) those given or computed and (b) those representing the velocities asked for in the problem. Be sure the intended length of a vector is clear.

Miscellaneous and Four Links

291. (a) From a train moving at 60 mph is thrown a parcel whose speed relative to the train is 30 fps in the forward direction but making 30° with the line of travel. What is the absolute speed of the parcel? (b) The same as (a) except the relative direction of the parcel is 30° from the trailing part of the train. Consider all velocity vectors to be in the horizontal plane.

Ans. (a) 115.1 fps, $\theta = 7.49°$.

292. The center of the wheel A, Fig. P65, is moving toward the right at 10 fps, but excessive torque causes the wheel to slip so that its angular velocity is 6 rad./sec. Determine (a) the absolute velocity of the point of contact with the ground, (b) v_J, and (c) the location of the centro ag, where G is the ground.

Ans. (a) 1 fps, (b) 17.5 fps at 21.3°.

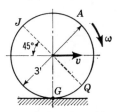

Fig. P65. Problems 292, 293.

293. The same as **292** except that $\omega = 12$ rad./sec. CL.

294. A 72-in. driving wheel on a locomotive connects with the connecting rod on a 30-in. crankpin circle, Fig. P66. If the train is moving at 45 mph, find the absolute velocities of the crankpin when it is in the positions P and Q. The wheel rolls.

Ans. 87.6 fps at 12.8°, 57.4 fps at 335.5°.

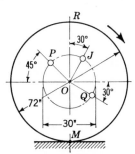

Fig. P66. Problem 294.

295. An iceboat is headed due north. A 40-mph wind blows in a direction N 30° W. The plane of the sail is at an angle of 30°, as measured clockwise, with the direction of motion of the boat. (a) If the velocity of the wind relative to the boat is parallel to the plane of the sail, what is the speed of the boat? Assume that the frictional resistance is negligible. (b) If the relative velocity of the wind to the boat is perpendicular to the sail, what is the boat's speed? Which is the better way to trim the sail when maximum speed is desired? (c)

Should the plane of the sail make a larger
or smaller angle with the direction of the
boat in order to increase the speed of the
boat? Justify your answers.

Ans. (a) 69.3 mph; (b) 23.1 mph;
(c) smaller.

296. The crank in the Scotch yoke
connection, Fig. P67, turns 60 rpm CL.
The piston's stroke is 18 in. What is
the piston velocity for $\theta = 60°$?

Ans. −4.08 fps.

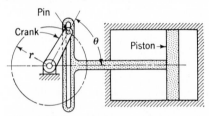

Fig. P67. Problem 296. Scotch Yoke.

297. In the drag-link mechanism of
Fig. P68, $\omega_B = 10$ rad./sec. CC; $K_v = 3$
fps/in.; M (2, 3) LDH; O_v (8, 4). Lay
out half size. (a) Find v_h, v_q, v_p, and
the velocity image of link C. (b)
What is ω_C? (See **237.**)

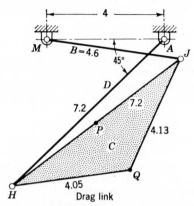

Fig. P68. Problem 297.

298. Two rolling wheels are connected
by link C as shown in Fig. P69. The
momentary angular velocity of wheel B
is 5 rad./sec. CC; $K_v = 2$ ips/in.;
M $(1\frac{1}{2}, 1)$; O_v $(7, 7)$. (a) Find v_r, v_q, and
the velocity image of C. (b) What is
ω_C? (See **239.**)

299. The wheel M (2, 5) in Fig. P70
rolls toward the right at 10 ips; $K_v = 3$
ips/in.; O_v (2, 3). Find v_p, v_r, and ω_C.
(See **246.**)

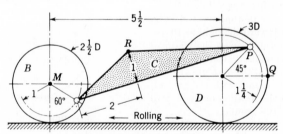

Fig. P69. Problem 298.

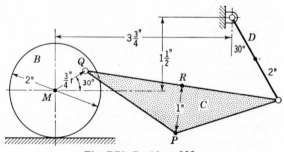

Fig. P70. Problem 299.

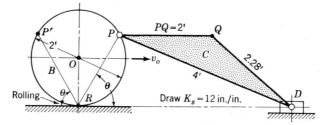

Fig. P71. Problem 300.

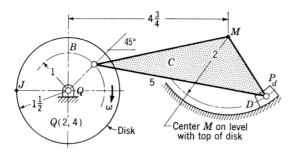

Fig. P72. Problem 301.

300. The center O $(2\frac{1}{2}, 2\frac{1}{2})$, Fig. P71, of the rolling wheel has an instantaneous velocity of $v_o = 900$ fpm ($K_v = 300$ fpm/in.); $\theta = 60°$; O_v $(1\frac{1}{2}, 9)$. (a) Find v_p, v_q, and v_D and the velocity image of C. (b) What is ω_C? (See **244**.)

301. In Fig. P72, let $v_j = 180$ ips; $K_v = 60$ ips/in. Find v_m, v_p, ω_C, and the velocity images of links B and C. (See **212**.)

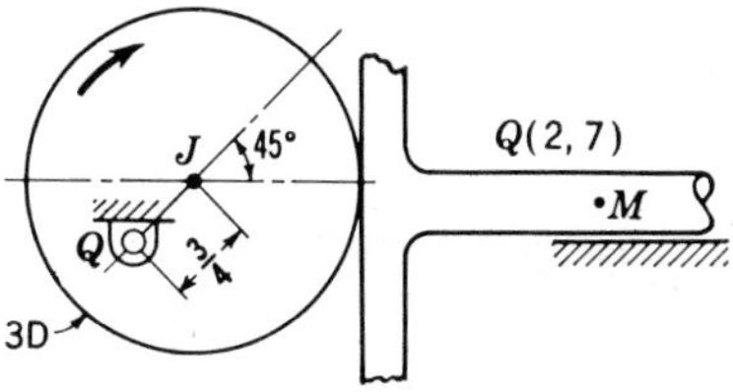

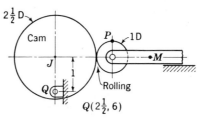

Fig. P73. Problem 302.

302. In Fig. P73, let $v_j = 180$ ips; $K_v = 60$ ips/in. Find v_m, the sliding velocity of the cam on the follower, and the velocity image of the cam. (See **214**.)

303. In Fig. P74, let $v_j = 180$ kps; $K_v = 60$ ips/in. Find v_p, the sliding velocity of the cam on the follower, the velocity image of the cam, and ω_C. (See **215**.)

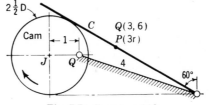

Fig. P74. Problem 303.

304. In Fig. P75, let $v_j = 180$ ips; $K_v = 60$ ips/in. Find v_m, v_p, the velocity images of the cam, roller, and reciprocating portion of the follower. (See **216**.)

Fig. P75. Problem 304.

305. An airplane pilot is to go from field A to a field B which is 100 mi. away in a direction S 40° E from A. A 30-mph wind is blowing from S 20° W. If the plane's airspeed is 200 mph, what straight course should the pilot set and how long is required for the trip?

Ans. S 32.54° E, 32.8 min.

306. In Fig. P76, the velocity of J is $v_j = 180$ ips. Let $K_v = 100$ ips/in.; O_v (4, 7). (a) Construct the velocity polygon, including the velocity images of links C and D. What is v_p? (b) Determine ω_C and ω_D. (See **213.**)

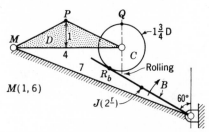

Fig. P76. Problem 306.

307. The cam in Fig. P77 turns 900 rpm CL; M ($2\frac{1}{2}$, 6); O_v (3, 2); $K_v = 200$ fpm/in. (a) Construct the velocity polygon including numerical values of v_j and v_r and the velocity image of the cam. (b) What is ω_C? (c) What is the sliding velocity at P? (See **251.**)

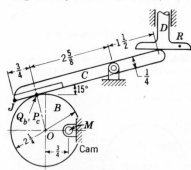

Fig. P77. Problem 307.

308. In Fig. P78, links D and C are pinned together and D is constrained to move in a vertical direction; B slides through C; v_b represents the velocity of a point in link B coincident with Q which is the center of the pin. If $\omega_B = 20$ rad./sec. CL, $MQ = 6$ in., and $\theta = 30°$, what is the velocity of Q?

Ans. $v_q = 11.52$ fps.

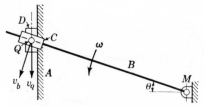

Fig. P78. Problem 308.

Five Links

309. The 1-in. crank B in Fig. P79 turns at 20 rps CL; M (4, 6); O_v (3, 3); $K_v = 40$ ips/in. (a) Construct the velocity polygon. Find v_q and the sliding velocity of link E on C. (b) Determine ω_C. (See **257.**)

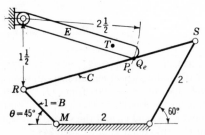

Fig. P79. Problem 309.

310. In Fig. P80, let $v_j = 180$ ips; $K_v = 60$ ips/in. Find v_r, v_p, and v_m, and draw the velocity images of the roller E and link D. Determine ω_E, ω_C, and the sliding velocity at M. (See **217.**)

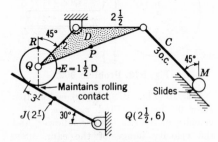

Fig. P80. Problem 310.

311. In Fig. P81, let $v_j = 180$ ips; $K_v = 60$ ips/in. Find v_p, ω_C, and the sliding velocity between cam and follower. (See **219.**)

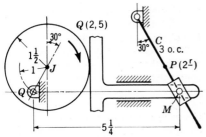

Fig. P81. Problem **311**.

312. In Fig. P82, let $v_j = 180$ ips; $K_v = 60$ ips/in. Find the velocity of P_c in link C, the velocity image of the small gear D (locate r on it), and ω_C. (See **218**.)

313. In Fig. P83, the bodies B and C remain in contact. The radius of the part RS of cam B is 1 in. Draw that part of the mechanism not completely defined approximately as shown; M (2, 5) LDH; O_v (5, 1); $K_v = 100$ fpm/in.; $v_j = 300$ fpm rightward. (a) Construct a velocity polygon including numerical values of v_t, v_w, v_p, and v_q. Show the velocity image of wheel E. (b) What are ω_D and ω_B? (c) What is the sliding velocity at P, Q? (See **260**.)

314. In Fig. P84, there is no slipping between the friction wheels B and C; M ($2\frac{1}{2}$, 5) LDH; O_v (6, 3); $K_v = 40$ fpm/in. (a) Construct the velocity polygon including v_r, v_q, and v_p. Show

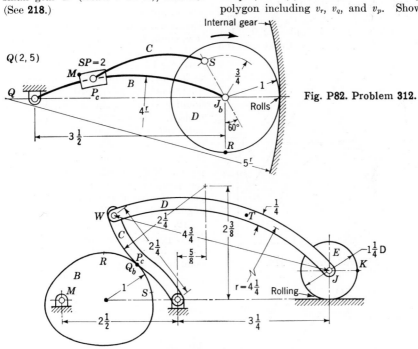

Fig. P82. Problem **312**.

Fig. P83. Problem **313**.

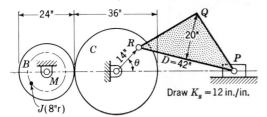

Fig. P84. Problem **314**.

the velocity images of B and C. (b) What is ω_D? (See **253**.)

Six or More Links

315. In Fig. P85, let $v_j = 180$ ips; $K_v = 60$ ips/in. Find v_p and ω_C. (See **221**.)

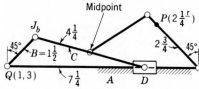

Fig. P85. Problem 315.

316. The crank B in Fig. P86 turns at 300 rpm CC; M (2, 6); O_v (2, 3). The actual length of the crank is 2 ft. Let $K_v = 1000$ fpm/in. (a) Construct the

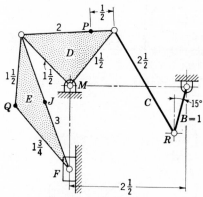

Fig. P86. Problem 316.

velocity polygon including v_p, v_q, and v_F. (b) Determine the angular velocities of C and E. (See **264**.)

317. Use plate No. 6, Wingren. Locate another point Q in link E on a perpendicular at the mid-point and $\frac{3}{4}$ in. away from (above) E. Name additional points as follows: centro $bc = R$; $ce = P$. Given $v_j = 80$ fpm with ω_B CC; $K_v = 30$ fpm/in.; O_v (1, 5). (a) Construct the velocity polygon, including the velocity image of link E (with point Q), and noting the magnitudes of v_D, v_p, v_q, and v_F. (b) Determine ω_E and ω_C in radians per minute. (See **261**.)

318. The mechanism of Fig. P87 represents approximately part of the linkage for operating the shutters for a camera. The crank B, driven by a torsion spring, is the input and the shutter blade moves with (is) link H. The complete linkage includes five additional shutters. In analyzing this mechanism, the relative angular velocities for several links are desired. The dimensions shown are proportional to the actual dimensions. Lay out full size for $8\frac{1}{2} \times 11$; M (2, $6\frac{1}{2}$); O_v (3, $4\frac{1}{2}$). (Double size for 11×17 and place velocity polygon alongside.) For the position shown, draw the velocity polygon, using $v_p = 1.5$ in., and find the angular velocity ratio ω_H/ω_B.

319. Use plate No. 15, Wingren. Name points in the quick-return mechanism as follows: centro $bc = Q$; coincident point in $D = P_d$; centro $ef = W$, coinci-

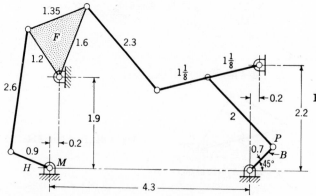

Fig. P87. Problem 318.

dent point in $D = M_d$. The arm B is driven at 96 rpm CL. (a) Find the velocity of slider F by the velocity polygon. What is ω_D in radians per minute? Use $K_v = 60$ fpm/in.; O_v (1, $10\frac{1}{2}$). What is the sliding velocity of link D in C; in E?

   Ans. $v_F \approx 372$ fpm.

   320. Use plate No. 7, Wingren. Link B rotates 100 rpm CC. The points involved are as follows: P (P_b) is the point bc in B; Q (Q_d) is the coincident point in link D; let centro de be R; erect a perpendicular to link E at point L and locate point M on it 1 in. (actual measurement) above L; let centro ef be T. Use $K_v = 100$ fpm/in.; O_v (5, 3). (a) Draw the velocity polygon and find the velocities of M and L and the velocity of slider C on D. (b) Determine the angular velocity of link D and of link E. (c) Also find v_F by the link-to-link construction.

Relative Angular Velocity

   321. In Fig. P88, the arm B of the epicyclic gear train rotates 10 rps CL, carrying with it gear C, which rolls on the stationary gear A. Draw the velocity polygon and the velocity image of C, noting the magnitudes of v_p, v_q, and v_r. What are ω_C and the relative angular velocity $\omega_{C/B}$? Locate M ($5\frac{1}{2}$, 7); O_v (4, 1); $K_v = 10$ fps/in. (See **248.**)

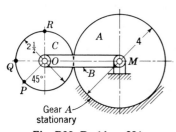

Fig. P88. Problem 321.

   322. The same as **321** except that the stationary gear A is an internal gear, Fig. P89; M (4, 7); $K_v = 6$ fps/in. (See **249.**)

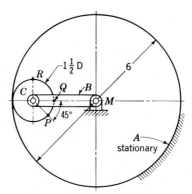

Fig. P89. Problem 322.

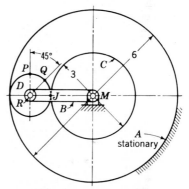

Fig. P90. Problem 323.

   323. In the epicyclic gear train of Fig. P90, gear C drives, rotating at 2000 rpm CC; M (4, 7); O_v (4, $3\frac{1}{2}$); $K_v = 600$ fpm/in. (a) Construct the velocity images of gears C and D and insert magnitudes of v_j, v_p, and v_q. (b) Determine ω_C, ω_B, $\omega_{D/B}$, and $\omega_{D/C}$. (See **250.**)

Larger Paper Needed

   324. The same as **276** except that the linear velocities are to be found by a velocity polygon.

   325. The same as **278** except that the linear velocities are to be found by a velocity polygon.

   326. The same as **280** except that the linear velocities are to be found by a velocity polygon.

   327. The same as **283** except that the

linear velocities are to be found by a velocity polygon.

328. The 3-in. pinion, Fig. P91, turns at 100 rpm, driving the 10-in. gear. The arm B is integral with the gear and therefore has the same angular velocity. The link C is adjustable along the length of B, which provides a means of changing the length of stroke. Locate M (8, 6); space scale $K_s = 2$ in./in. (a) Find the length of stroke of F. What is the ratio of the time of the slow-speed stroke to the time of the quick stroke? (b) Construct the velocity polygon (use the property of velocity images, $FP/FR = fp/fr$, etc.). Record the numerical value of v_F. (c) Determine ω_D and ω_E.

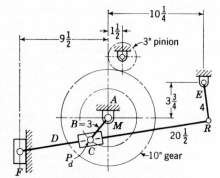

Fig. P91. Problem 328.

329–350. These numbers may be used for other problems.

5

Velocity Diagrams

96. Introduction. It often happens, as in our study of cams, that the engineer wishes to know, not just the velocity or acceleration of a point in a mechanism when the mechanism is in a particular position, but how the velocities and/or accelerations of the point vary throughout a whole cycle of events. A picture of the variation of velocity is called a ***velocity diagram,*** and a picture of the variation of acceleration is called an ***acceleration diagram;*** these diagrams may be drawn against either displacement or time as the abscissa. Usually, a relatively simple procedure can be devised for finding velocity diagrams, but for acceleration diagrams, one can seldom do better than to complete the acceleration polygons (Chapter 6) for several configurations of the linkage and plot the results.

97. Velocity Diagram of a Slider-crank Mechanism. One faced with the problem of obtaining velocity diagrams can sometimes find graphical procedures which reduce the time and effort. A case in point is shown in Fig. 93, which can be taken as the kinematic representation of a 4 × 4-in. automotive engine turning at 4200 rpm CC; (connecting-rod length)/ (crank length) = $C/B = 4$. If the stroke is 4 in., the radius of the crank B is 2 in.; hence, the speed of the crankpin bc is

$$v_{bc} = 2\pi rn = 2\pi \left(\frac{2}{12}\right)\left(\frac{4200}{60}\right) = 73.3 \text{ fps.}$$

The obvious construction and one that is satisfactory and takes little time is the link-to-link construction, § 72. By this method, the velocity vector $v_{bc} = v_p$ is rotated into its radius to get point E, Fig. 93; draw $E2''$ parallel to C until it intersects the vertical line from $2'$; then the height

149

$2'$-$2''$ represents the speed of cd ($v_{cd} = v_D$) because the grounded centro (point of zero velocity) in C is ac. If this procedure is used to find the entire velocity diagram, it would be a good plan to draw a circle of radius eE with center at e and extend the radial lines until they intersect this circle (as is done in Fig. 95), so that all of the rotated positions of the vector are obtained at once.

A somewhat shorter procedure when the crank moves at a constant speed is to draw the crank circle to scale to represent v_{bc} and use $C = 4v_{bc} = 4B$. The result represents the mechanism to some space scale

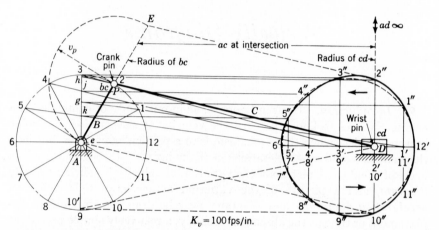

Fig. 93. Velocity Diagram (vs) of Slider-crank Mechanism. The appearance of the velocity diagram with v plotted against t can be determined by laying off equal horizontal spaces (that is, $12'$-$1' = 1'$-$2' = 2'$-$3'$, etc., instead of as shown), plotting the ordinates ($1'$-$1''$, $2'$-$2''$, etc.) on these equally spaced divisions, and drawing a smooth curve through the points obtained. Notice that the maximum velocity of D occurs when the crank and connecting rod are perpendicular to each other, between positions 2 and 3 (and 9 and 10). To define the diagram accurately, find more points, especially between 12 and 1, 2 and 3, 5 and 6.

with which we are unconcerned. Divide the circle into a convenient number of equal parts (it is not necessary that the parts be equal; it is only a convenience), obtaining points 1, 2, 3, ... Locate the corresponding positions of the wrist pin cd (piston D) at $1'$, $2'$, $3'$, . . . by setting the dividers for the distance bc-cd and using centers at 1, 2, 3, ... Then draw the lines $1'$-$1''$, $2'$-$2''$, etc., perpendicular to the path of motion at the various positions of the wrist pin and draw the various positions of the connecting rod C, 1-$1'$, 2-$2'$, 3-$3'$, . . . , extending these lines as necessary to intersect the vertical center line of the crankshaft at points g, h, . . . , k; draw $g1''$ horizontally to intersect $1'$-$1''$ and obtain point $1''$ on the diagram; repeat with $h2''$, etc. For proof, consider position 2. Swing vector v_{bc} downward into the radius at e, the center of the circle; draw $e10''$ parallel to C. The intercept on $2'$-$2'' = 10'$-$10''$ is proportional to radius cd-ac and therefore represents the speed of cd. But $e10''2'h$ is a parallelogram, and $eh = 10'$-$10''$; thus eh represents $v_{cd} = v_D$ to the same scale that $e2$ represents the crankpin speed, which proves that the horizontal line $h2''$ intercepts the vertical at $2'$ at a point on the velocity diagram. Since the diagram is symmetric with respect to the horizontal center line, points $7''$, $8''$, etc., may be obtained with the dividers by setting off $7'$-$7'' = 5'$-$5''$, $8'$-$8'' = 4'$-$4''$, etc.

If the line of travel $12'$-$6'$ of the piston should not pass through the center of the crankshaft e, the diagram is not symmetric about the line of piston travel and points on the diagram must be individually found for both strokes of the slider. In this case, the special construction of the previous paragraph does not apply. If one were satisfied to have points $1''$, $2''$, . . . , $5''$ below $12'$-$6'$, a series of lines $e10''$ all drawn from e parallel to the corresponding positions of the connecting rod intercept points on the diagram. This construction has some advantages and is also applicable when the center line of the cylinder does not pass through the center of the crankshaft. Evidently under this plan, points $7''$, $8''$, . . . , $11''$ would fall above $12'$-$6'$.

The motion of the piston D, Fig. 93, departs somewhat from harmonic motion because of what we call the *angularity of the connecting rod;* that is, when the crank has turned through $90°$ from head-end dead center, the piston has moved more than half of its stroke to $3'$, because of the movement of the connecting rod to an angular position with respect to the center line of the cylinder. The velocity diagram of the slider of a Scotch crosshead, Fig. 15 (this mechanism is equivalent to one with a connecting rod of infinite length), would be harmonic; the vs diagram for SHM is a circle (see Fig. 13) shown dotted in Fig. 93 for comparison. Also compare the vs curve of Fig. 93 with the vt curve of Fig. 14 for harmonic motion.

98. Polar Velocity Diagrams. When the speed of a point in a mechanism is depicted by intercepts on radial lines, the resulting diagram is called a **polar velocity diagram**. Two ideas are illustrated by Figs. 94 and 95. In Fig. 94, the circle is a polar velocity diagram for the constant-speed crankpin, as is the circle in Fig. 93. Now if we should wish to compare the speed of the piston (or any other point in the mechanism) with the speed of the crankpin, we could lay out on the radial lines representing crank positions distances that are equal to the speed of the piston when the crank is in said position. For example, lay off 0-$1''$, Fig. 94, equal to $1'$-$1''$, Fig. 93; 0-$2'' = 2'$-$2''$, etc. In the particular case of the problem of Fig. 93, given points g, h, . . . , k in Fig. 94 from Fig. 93, we could swing arcs $g1''$, $h2''$, etc., to get $1''$,

2″, etc., Fig. 94.   A smooth curve through these points is the velocity diagram of the piston on polar coordinates.   We notice that for a short

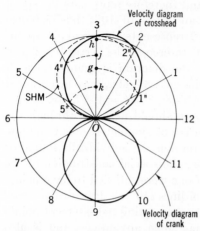

period between positions 2 and 3, the piston speed is a little greater than the crankpin speed. The velocity diagram of the slider in a Scotch crosshead plotted in this manner turns out to be a circle, shown dotted in Fig. 94.

Sometimes we wish to know the pattern of velocity of the end of a crank rotating or oscillating with variable speed, as in Fig. 95. If desired, the velocity of such a point can be found for a number of positions and laid off from a straight line, divided either into equal time units or into distances moved during equal

Fig. 94. Polar Velocity Diagrams.

time units.   Also the diagram may be plotted as a polar velocity diagram as in Fig. 95.   Again, the construction is by the link-to-link idea.   Instead

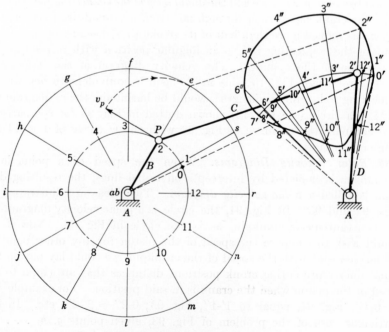

Fig. 95. Polar Velocity Diagram.

of rotating the vector v_p into each radius, draw a circle of radius ab-e, which becomes the polar velocity diagram of the crank as measured from the crank circle; then the rotated vector positions are e, f, g, etc.

Having drawn the mechanism in some position, say 2, draw an arc $0'$-$7'$, which is the path of pin cd. With dividers set at radius $B + C$ and center at ab, locate $0'$, one end of the path of D; with dividers set at $C - B$ and center at ab, locate the other end of the path of D (practically coincident with $7'$). Then with dividers set at the length of C and with centers at 1, 2, 3, etc., locate the various positions $1'$, $2'$, $3'$, etc., of the pin cd. Draw the radial lines through $1'$, $2'$, $3'$, . . . and find the intercepts $1''$, $2''$, $3''$, . . . on these radial lines by the method illustrated at position 2 ($e2''$ is parallel to the corresponding position of C). The zero line for this polar diagram is the arc $0'$-$7'$. It may be interesting and useful to note that since the radius to the pin cd is constant, this diagram is also an angular velocity diagram ($\omega = v/r$).

99. Velocity Diagram—Whitworth Quick-return Mechanism. To find the velocity diagram of a Whitworth mechanism, one must first get the mechanism, which is a design problem of importance to the engineer planning to use this configuration of links. However, for the purpose of explanation, we shall assume the linkage as shown in Fig. 96. The

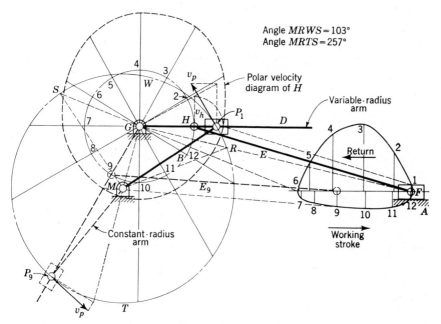

Fig. 96. Velocity Diagram for Whitworth Mechanism. The ends of the stroke of F are found by using radii of $E + (GH)$ and $E - (GH)$ with center at G. The slider F could be a cutting tool, a feeding device, etc.

process of finding any velocity diagram involves assuming the linkage
in a number of positions and finding the velocity of the desired point
at each position. If all of these positions are shown complete, the
drawing may become dense with lines. Hence, only two positions are
shown in Fig. 96, which should make the method clear. The construction shown must be completed at each of the positions 1 to 12, but
it is not necessary to actually draw all the lines. After completing the
work for one or two positions, you will see ways to economize in drawing
lines. The relative velocity principle is applied at P each time to get the
velocity of P_d in D; then the velocity of H, a point in both links D and E,
is found; then the velocity of F is found by the link-to-link construction
(the light dotted lines parallel to the positions of link E). Either the
circle of radius GH or the circle of radius B may be divided into equal (or
unequal) parts; we have divided circle G into 12 equal angles. (Some
prefer to divide the stroke of F into a certain number of equal parts.)
When the corresponding positions of F are found, draw vertical lines
through these positions, naming them with the number of the position.

It is seen that the return stroke is made while the constant-speed arm
B rotates through the angle $RMSW = 103°$; the working stroke occurs
during angle $RMST = 257°$. Since the constant-radius arm B rotates
at constant speed, these angles are proportional to time. The return
stroke occurs in less time than the working stroke and therefore the
velocities on the return are greater than on the working stroke, as seen
from the diagram, Fig. 96. The ratio of the times is

$$\frac{\text{Time of working stroke}}{\text{Time of return}} = \frac{257°}{103°} = 2.49,$$

which incidentally is rather high unless the speed is low (accelerations
may be high). The alternative to the procedure given above is to draw
a series of velocity polygons to find v_F; this would also require having the
links in a number of positions.

100. Closure. As you have no doubt observed, this chapter intro-
duces no new principles. It is of interest because it is indicative of the
type of detail studies that are made in practice. It also implies that if
you are faced with the problem of repeated solutions involving a given
mechanism, you might find it profitable to give thought in search of the
shortest way to the desired answers. The biggest problem in practice is
often the one of finding the proportions of the lengths of links to perform
a desired function. This sort of problem, which is basically a design
problem, is often attacked empirically by trial and error. However,
there is help to be had in the literature on this phase.

PROBLEMS

See *Note to Students* on p. 141.

351. Use plate No. 13, Wingren. The crank B in the Whitworth quick-return mechanism turns at a constant speed of 5 rad./sec. CC. Construct the velocity diagram of the slider F; $K_v = 5$ fps/in. Dimension the length of the stroke of F and complete the ratio of the times of the working to the return strokes.

352. The same as **351** except that plate No. 14, Wingren, is used and $\omega_B = 13$ rad./sec. CL.

353. The same as **351** except that plate No. 15, Wingren, is used.

354. The same as **351** except that plate No. 16, Wingren, is used.

355–361. In Figs. P92 to P98, inclusive, let the constant angular velocity of crank B be 100 rpm and also let 1 in. = 100 fpm be the velocity scale. Draw the figure on $8\frac{1}{2} \times 11$-in. paper so that slider F moves parallel to the 11-in. edge. (a) Plot the velocity diagram of slider F. Determine the length of stroke of F. In each problem, draw conclusions from the shape of the velocity diagram

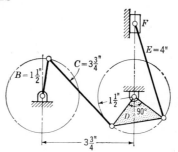

Fig. P94. Problem **357**. Crossed-link Mechanism.

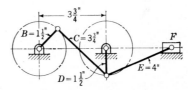

Fig. P95. Problem **358**. Crossed-link Mechanism.

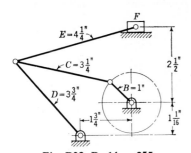

Fig. P92. Problem **355**.

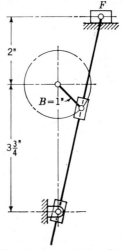

Fig. P96. Problem **359**. Oscillating-arm Quick Return.

regarding the suitability of the arrangement as a quick-return mechanism.

(b) Estimate the average velocity of the working (slow-speed) stroke and calculate the rpm of B to give an average cutting speed of 50 fpm.

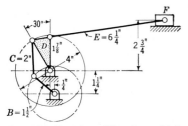

Fig. P93. Problem **356**. Drag Link.

(c) If the velocity diagram appears unsuited for quick-return mechanisms, keep the basic linkage but redesign by changing the direction of motion of F so that an appropriate shape of the velocity diagram is obtained.

$D_p = 3.5$ in., $D_g = 22$ in., what is the maximum speed on the working stroke; on the return stroke? The radius of Q, that is, the length of link B, is adjusted to give the desired stroke of the tool.

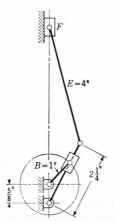

Fig. P97. Problem 360. Whitworth Mechanism.

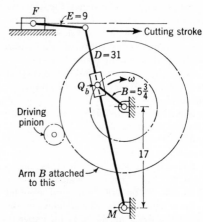

Fig. P99. Problems 362.

Larger Paper Desirable

363. The linkage of Fig. P100 will give a variable speed to the slider F; M (4, 6)

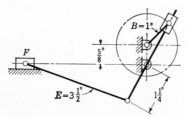

Fig. P98. Problem 361. Whitworth Mechanism.

362. The mechanism in Fig. P99 represents a shaper; the slider F carries the cutting tool and the working stroke is toward the right. Lay out to scale 1 in. = 5 in.; M (5, 1); let the crank length represent the constant speed of Q_b. Draw the polar velocity diagram of Q_b and find the velocity diagram of the tool. Let the curve for the cutting-stroke speeds be above the line of travel of F. What is the ratio of the time of the working stroke to the time of the return stroke? If the pinion turns 220 rpm,

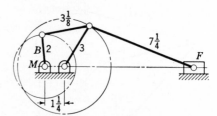

Fig. P100. Problem 363.

for 11 × 17 paper. Let the crank length represent the speed of the crankpin P. Its actual length is 2 in. and it turns 45 rpm CL. (a) Determine the length of stroke and draw the velocity diagram of link F. (It may be better to divide the angle of motion of the 3-in. link, or the stroke of F, into equal parts rather than use equal time units.) (b) What is the ratio of the times of the slower to the faster strokes? (c) Is this arrangement of these links a good one for quick-return purposes? Criticize and suggest improvements.

364. The mechanism of Fig. P101 represents a hack saw. The arm *B*, turning at 83 rpm CL, is driven via a silent chain by an electric motor. The saw blade moves with the slider *H*. Lay out to scale 3 in. = 1 ft.; *M* (8, 3). Let the crank length *B* represent the speed of the crankpin and construct a polar velocity diagram for the saw blade *H*. (Try dividing the angle of oscillation of link *D* into equal parts, say six, instead of using equal time units.) Determine the ratio of the time of the working stroke to that of the return stroke. Estimate the average cutting speed. (Same mechanism in **283**.)

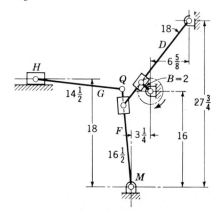

Fig. P101. Problem **364**.

365–370. These numbers may be used for other problems.

6

Relative Accelerations

101. Introduction. A knowledge of velocities is necessary to a study of accelerations; hence, the previous chapters. A knowledge of accelerations is necessary to a force analysis of mechanisms and machines, inasmuch as a force (and a dynamic reaction) is involved whenever a mass has an acceleration (Newton's law, $F = ma$); hence, the need for the next subject for study. A force analysis is usually needed for the rational design of machines; this last step is beyond the scope of this book. It also happens that accelerations sometimes need to be known for purely kinematic reasons.

In finding velocities needed for the solution of acceleration problems, we shall use the relative velocity principle and the velocity polygon, Chapter 4. Our aim will be to make clear the fundamentals of the relative acceleration equation in its simplest form, rather than to develop short-cut graphical solutions. The applications of this chapter are for points fixed in a rigid body, or are problems which can be reduced to this condition. The rigid body, (or a series of rigid bodies) however, may have any type of plane motion.

102. Review of Basic Acceleration Equations. We recall that linear acceleration is the time rate of change of linear velocity, where, by change of velocity, we mean a change of either magnitude or direction or both. Like velocities, accelerations are always relative. When unstated the implied reference body is the earth or the frame of the machine, whose acceleration is taken as zero. If a point is moving in a curvilinear path, the change of velocity in a direction tangent to the path is the change in speed and is the *tangential acceleration* a^t. If a point is moving in a curvilinear path, the change in velocity due to the change in direction is called the **normal acceleration** a^n and is found to be (§ 21)

$$(12) \qquad\qquad a^n = \frac{v^2}{r} = r\omega^2 = v\omega,$$

in which r is the radius of curvature of the path at the location of the point, v is the linear speed of the point, and ω is the angular velocity of the body in which the point lies (or the angular velocity of the radius to the point).

The angular acceleration is the time rate of change of angular velocity, or for *constant acceleration*,

$$(6) \qquad\qquad \alpha = \frac{\Delta\omega}{\Delta t}.$$

Thus, using $v = r\omega$ with r constant we have the tangential acceleration as

$$(a) \qquad a^t = \frac{\Delta v}{\Delta t} = \frac{\Delta r\omega}{\Delta t} = \frac{r\Delta\omega}{\Delta t} = r\alpha; \qquad a^t = r\alpha,$$

in which a^t is constant or an instantaneous value, v is the speed of the point, and α may be an instantaneous angular acceleration.

When a point in a body moves in a curved path, it is usually necessary to work with the tangential and normal components of its acceleration. Thus the **total** or **resultant** acceleration of such a point is

$$(18) \qquad\qquad a = a^t + a^n = [(a^t)^2 + (a^n)^2]^{\frac{1}{2}}$$

when the path is attached to the reference (stationary) link.

103. Relative Acceleration for Points Fixed in a Rigid Body. Since the acceleration of a point is the time rate of change of its velocity, we may with some logic write the relative velocity equation (17), p. 129, in the form

$$\Delta v_q = \Delta v_p + \Delta v_{q/p},$$

and then divide each term by Δt;

$$(b) \qquad\qquad \frac{\Delta v_q}{\Delta t} = \frac{\Delta v_p}{\Delta t} + \frac{\Delta v_{q/p}}{\Delta t}.$$

Accordingly, the terms of this equation are, respectively, the average acceleration of point Q, the average acceleration of point P, and the average acceleration of Q relative to P, each during the same time interval Δt. Using acceleration symbols, and going to the limit with $\Delta t = dt$, equation (b) becomes

$$(19) \qquad\qquad a_q = a_p + a_{q/p},$$

and in this form we may think of each acceleration as being an instantaneous value. That is, the accelerations may be varying, but at any particular instant, the accelerations of Q and P are related as in (19). This equation is deceptively simple looking, because for some situations of P and Q, the problem may be difficult if not impossible to solve. In this book, we shall limit its application to two points fixed in a rigid body. If a point is in a rotating body, we find it convenient to use the normal and tangential components of the resultant acceleration, perhaps $a^n_q \nrightarrow a^t_q$ for a_q and $a^n_p \nrightarrow a^t_p$ for a_p. Fortunately, the relative acceleration vector can also be resolved into normal and tangential components. If body M in Fig. 97 has an angular velocity ω in either CL or CC sense, the normal acceleration of any point Q in the body relative to any other point P is

$$a^n_{q/p} = r\omega^2 = \frac{(v_{q/p})^2}{r} = \frac{(v_{q/p})^2}{QP},$$

where $r = PQ$ is the distance between the points and ω is the *absolute* angular velocity of the body (frame of reference is attached to the earth). The vector is always directed from Q toward P. Likewise $a^n_{p/q}$ is always directed from P toward Q. The sense of the tangential component $a^t_{q/p} = r\alpha$ depends on the sense of α, and a moment's reflection on the vectors $a^t_{q/p}$ and $a^t_{p/q}$ in Fig. 97 reveals the manner of deciding. Quite

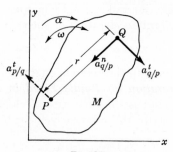

Fig. 97.

commonly, the practical situations are such that the sense of $a^t_{q/p}$ is defined by the vector polygon of equation (19); then the absolute angular acceleration of the body containing points P and Q is computed from

(*c*) $$\alpha = \frac{a^t_{q/p}}{QP},$$

with its sense determined by inspection. In accordance with the foregoing discussion, we can now write equation (19) in the form

(19) $$a^n_q \nrightarrow a^t_q = a^n_p \nrightarrow a^t_p \nrightarrow a^n_{q/p} \nrightarrow a^t_{q/p},$$
$$[P \text{ AND } Q \text{ FIXED IN BODY}]$$

in which one or more terms may be zero. Also, in many problems, it is inconvenient to find these components of either P or Q. All the tangential accelerations are of the form $r\alpha$. If a velocity polygon has been constructed so that any desired v is at hand, the normal accelerations are most conveniently found from the v^2/r form. Some examples will help explain the application of (19).

104. Example—Rotating Disk. The disk in Fig. 98 has an instantaneous angular velocity of $\omega = 3$ rad./sec. CL when the angular acceleration is $\alpha = 4$ rad./sec.^{2} CC. If its radius is $r = 24$ in., what is the absolute acceleration of a point P on its circumference?

Solution. The tangential acceleration is (24 in. = 2 ft.)

$$a^t = r\alpha = (2)(4) = 8 \text{ fps}^2,$$

directed so as to match the CC sense of α; see Fig. 98. The absolute normal acceleration is always directed toward the absolute center of rotation, PO in Fig. 98, and its value is

$$a^n = r\omega^2 = (2)(3^2) = 18 \text{ fps}^2.$$

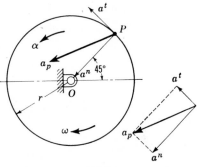

Fig. 98.

The resultant a_p is found graphically in Fig. 98, or it may be computed from $[(a^t)^2 + (a^n)^2]^{1/2}$; $a_p = 19.6$ fps^2.

105. Example—Accelerations in Rolling Wheel. The center of a 4-ft. tractor wheel, Fig. 99, has a velocity of $v_q = 10$ fps and an acceleration of $a_q = 15$ fps^2,

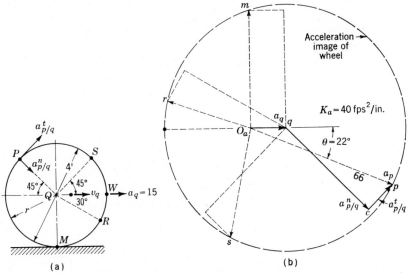

(a)

(b)

Fig. 99. Rolling Wheel. It will be helpful to notice certain features of the accelerations of points in a rolling wheel. The total acceleration of M, the point in contact with the ground, is the normal acceleration $a^n{}_{m/q} = r\omega^2$, because, in

$$a_m = a_q \mathbin{+\!\!\!+} a^t{}_{m/q} \mathbin{+\!\!\!+} a^n{}_{m/q},$$

we know that a_q is equal and opposite to $a^t{}_{m/q}$ (rolling).

both directed toward the right. What is the absolute acceleration of the point P if the wheel is rolling?

Solution. The first step is to write the relative acceleration equation for P relative to Q, since the motion of Q is known; thus

$$(d) \qquad a_p = a_q \rightarrowtail a^n_{p/q} \rightarrowtail a^t_{p/q}.$$

Since the total acceleration of Q is known, there is no point in putting down its

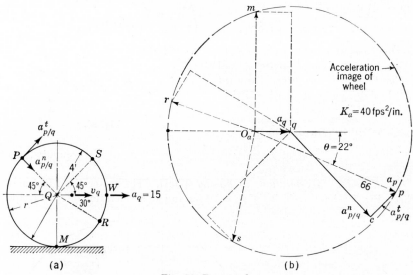

Fig. 99. Repeated.

components. Since the absolute acceleration of P is desired, we do not break a_p into its components.

The angular velocity of the rolling wheel is found from

$$(e) \qquad \omega = \frac{v_q}{MQ} = \frac{v_q}{r} = \frac{10}{2} = 5 \text{ rad./sec. CL.}$$

Then the normal acceleration of P to Q is [see equations (**b**) and (19)]

$$(f) \qquad a^n_{p/q} = (PQ)\omega^2 = r\omega^2 = (2)(5)^2 = 50 \text{ fps}^2.$$

Note. Whether the wheel rolls or not, $v_{p/q} = (PQ)\omega = r\omega = (2)(5) = 10$ fps for an angular velocity of 10 rad./sec. Only in case of rolling is $v_{p/q} = v_q$ in magnitude. Then $a^n_{p/q} = (v_{p/q})^2/r = 100/2 = 50$ fps^2, as before. Equation (**e**) and equation (**g**) below are true only for rolling; equations (**f**) and (**h**) are true whether or not there is some slipping.

The angular acceleration of the wheel is

$$(\mathbf{g}) \qquad \alpha = \frac{a_q}{MQ} = \frac{a_q}{r} = \frac{15}{2} = 7.5 \text{ rad./sec.}^2 \text{ CL.}$$

From equation (**c**), we get

$$(\mathbf{h}) \qquad a^t{}_{p/q} = (PQ)\alpha = r\alpha = (2)(7.5) = 15 \text{ fps}^2,$$

which is seen to be the same as a_q. Now write equation (**d**) with the known quantities substituted;

$$a_p = 15 \nleftrightarrow 50 \nleftrightarrow 15.$$

In your solutions, this line is placed *directly underneath the acceleration equation* (**d**) *so that each number is readily identified.* Also use short arrows as shown to indicate the general direction of each vector whose direction is known. Next choose a convenient scale and acceleration pole O_a, Fig. 99(b), and plot the known vectors; $O_a q = a_q = 15$, $qc = a^n{}_{p/q} = 50$, $cp = a^t{}_{p/q} = 15$. In accordance with equation (**d**), this vector sum is equal to a_p, shown dotted $O_a p$, Fig. 99; $a_p = 66$ fps². (After laying out the vector polygon, always check back against the acceleration equation to see that equation and polygon match.) In the acceleration polygon, name the ends of the *absolute* vectors with the lower-case letter of the name of the point, as p and q, Fig. 99. Since acceleration polygons may be dense with lines, it is important to be generous in naming the vectors. In most cases, it will be worthwhile to name each vector, as $a^n{}_{p/q}$, $a^t{}_{p/q}$, etc., Fig. 99(b).

If the accelerations of S, R, and M are found, as shown with light dotted lines, it is seen that a circle with center at q and radius qp goes through s, r, and m. This result is logical since the acceleration of any point x on the circumference is $a_x = a_q \nleftrightarrow a_{x/q}$, and a_q and $a_{x/q}$ are the same for all positions of x on the circumference. This circle bounds the *acceleration image* of the wheel.

106. Example—Acceleration in Four-bar Mechanism. The crank B, Fig. 100, is turning 70 rpm CC at the moment that its angular acceleration is 10 rad./sec.² CL (slowing down). The lengths of the various links are as shown in Fig. 100 ($A = 2.55$ ft., for example). Find the accelerations of M and Q, the angular velocity ω_C, and the angular acceleration α_C.

Solution. It generally saves time in linkages to construct the velocity polygon first. The velocity of P is

$$v_p = 2\pi Bn = 2\pi(2)\left(\frac{70}{60}\right) = 14.66 \text{ fps.}$$

With this velocity in link C known, we find v_m next because its direction is known to be perpendicular to D. Use the equation

$$v_m = v_p \nleftrightarrow v_{m/p},$$

and a scale of 1 in. = 12 fps. Lay out $O_v p = v_p = 14.66$ from a convenient velocity pole O_v; draw $O_v m$ perpendicular to D, and pm perpendicular to PM. This locates point m in the velocity polygon and defines v_m. Then draw mq

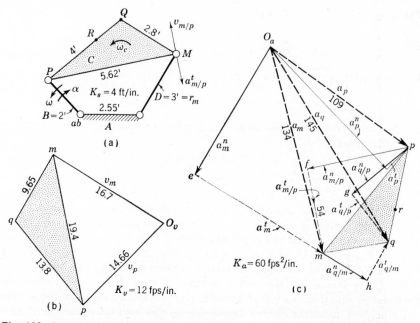

Fig. 100. Accelerations in Four-bar Mechanism. *Procedure.* For the acceleration polygon, choose a scale and pole O_a; lay out $a^n_m = 93$ and draw em perpendicular. Then lay out $a^n_p = 107.3$ and add $a^t_p = 20$ to this vector; this locates point p; $a_p = 109$ fps^2. Add vector $a^n_{m/p} = 67$ to a_p to get point f; draw fm perpendicular at f. The intersection of em and fm locates point m; $a_m = 134$ fps^2. Add $a^n_{q/p} = 47.6$ to a_p and raise perpendicular gq. Add $a^n_{q/p} = 33.2$ to a_m and raise perpendicular hq. These perpendiculars locate q; $a_q = 145$ fps^2. Known vectors are drawn solid; vectors defined by the solution are dotted (as an aid to visualizing the polygon).

perpendicular to MQ and pq perpendicular to PQ in accordance with the equations

$$v_q = v_m \mathbin{\text{+\!\!+}} v_{q/m} \qquad \text{and} \qquad v_q = v_p \mathbin{\text{+\!\!+}} v_{q/p}.$$

The various velocities as scaled are shown on the velocity polygon. The angular velocity $\omega_C = v_{m/p}/(MP) = 19.4/5.62 = 3.45$ rad./sec.

For accelerations, the first thing to do is to write the acceleration equation and find as many terms as possible;

$$a^n_m \mathbin{\text{+\!\!+}} a^t_m = a^n_p \mathbin{\text{+\!\!+}} a^t_p \mathbin{\text{+\!\!+}} a^n_{m/p} \mathbin{\text{+\!\!+}} a^t_{m/p}.$$

Taking the terms of the equation in order, we can compute a^n_m because $v_m = 16.7$ fps is known in the velocity polygon;

$$a^n_m = \frac{v_m{}^2}{r_m} = \frac{(16.7)^2}{3} = 93 \text{ fps}^2.$$

We know the direction of a^t_m (perpendicular to D), but not its numerical value.

The components of a_p are found to be

$$a^n{}_p = \frac{v_p^2}{r_p} = \frac{(14.66)^2}{2} = 107.3 \text{ fps}^2 \qquad \text{and} \qquad a^t{}_p = r_p\alpha_B = (2)(10) = 20 \text{ fps}^2.$$

The value of $a^n{}_{m/p}$ can be found from $v_{m/p} = 19.4$ fps in the velocity polygon;

$$a^n{}_{m/p} = \frac{(v_{m/p})^2}{MP} = \frac{(19.4)^2}{5.62} = 67 \text{ fps}^2,$$

pointing in the sense MP. Finally, we know the direction of $a^t{}_{m/p}$ but not its magnitude. Thus, there are only two unknowns in the acceleration equation (magnitudes of $a^t{}_m$ and $a^t{}_{p/m}$) and therefore the equation can be solved. Under the acceleration equation, write all that is known (it will save time to be orderly and complete on this);

(*i*)
$$a^n{}_m \nrightarrow a^t{}_m = a^n{}_p \nrightarrow a^t{}_p \nrightarrow a^n{}_{m/p} \nrightarrow a^t{}_{m/p},$$
$$93 \nrightarrow a^t{}_m = 107.3 \nrightarrow 20 \nrightarrow 67 \nrightarrow a^t{}_{m/p},$$

where the two-headed arrows indicate that the particular senses in which these vectors point are unknown. As you are aware, vectors may be added and subtracted in any sequence without change in the result. Nevertheless, it will definitely be advantageous to keep related vectors together in this work, $a^n{}_p$ with $a^t{}_p$ for example. Then the desired total accelerations are easily found from the acceleration polygon. Be sure to arrange that all absolute acceleration vectors originate at O_a; thus, draw $a^n{}_m = O_a e$ from O_a and then *em* of indefinite extent perpendicular to $a^n{}_m$. Point m is somewhere along *em*. The acceleration equation says that the sum of these two vectors is equal to the sum of those shown on the right-hand side of equation (*i*). Add the right-hand-side vectors in order and reach point f, Fig. 100. Then a perpendicular to $a^n{}_{m/p}$ at f intersects *em* at m and defines $a_m = O_a m$ ($a^t{}_{m/p}$ is perpendicular to $a^n{}_{m/p}$).

The manner of finding a_q is a process similar to finding v_q in the velocity polygon. We may use both a_p and a_m and simultaneous equations:

(*j*)
$$a_q = a_p \nrightarrow a^n{}_{q/p} \nrightarrow a^t{}_{q/p},$$
$$a_q = 109 \nrightarrow 47.6 \nrightarrow a^t{}_{q/p}.$$

(*k*)
$$a_q = a_m \nrightarrow a^n{}_{q/m} \nrightarrow a^t{}_{q/m},$$
$$a_q = 134 \nrightarrow 33.2 \nrightarrow a^t{}_{q/m}.$$

In these equations,

$$a^n{}_{q/p} = \frac{(v_{q/p})^2}{QP} = \frac{(13.8)^2}{4} = 47.6 \text{ fps}^2 \qquad \text{and} \qquad a^n{}_{q/m} = \frac{(v_{q/m})^2}{QM} = \frac{(9.65)^2}{2.8}$$
$$= 33.2 \text{ fps}^2.$$

The direction of $a^n{}_{q/p}$ is parallel to QP; of $a^n{}_{q/m}$, parallel to QM; of $a^t{}_{q/p}$, perpendicular to QP; of $a^t{}_{q/m}$, perpendicular to QM. Since the right-hand sides of equations (*j*) and (*k*) are equal to the same thing, they are equal to each other, which makes one vector equation and two unknowns, the magnitudes of $a^t{}_{q/p}$ and

$a^t_{q/m}$. Add $a^n_{q/p}$ to a_p to get point g and draw a perpendicular to $a^n_{q/p}$ (for $a^t_{q/p}$).
Add $a^n_{q/m}$, which is parallel to QM (not a continuation of em), to a_m to get point h
and draw hq perpendicular to mh. The intersection of gq and hq locates q and
defines a_q.

The angular acceleration of link C may now be found from the relative tangential acceleration of any two points in C, for example, M and P, by equation
(c);

$$a^t_{m/p} = r\alpha = (MP)\alpha_C = 54 \text{ fps}^2$$
$$\alpha_C = \frac{a^t_{m/p}}{MP} = \frac{54}{5.62} = 9.6 \text{ rad./sec.}^2 \text{ CL.}$$

The value 54 fps² is scaled from the polygon, and comparison with the acceleration
equation (i) shows that it points downward ($a^t_{p/m}$ is equal but in the opposite

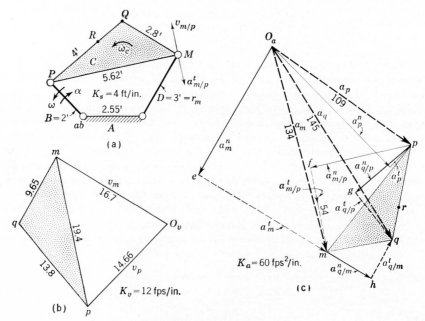

Fig. 100. Repeated.

sense); this sense of the tangential acceleration of M with respect to P shows that
the angular acceleration of C is clockwise, Fig. 100(a). A check on the accuracy
of the acceleration polygon may be obtained by computing α_c also from $a^t_{q/m}$ and
$a^t_{q/p}$; the results should agree with 9.6 rad./sec.² already found.

For better understanding, it is worth noting that if α_C is found immediately
after having found $a^t_{m/p}$, then a_q can be found from *either* equation (j) or (k)—
both are not needed—because both tangential relative accelerations can be
computed;

$$a^t_{q/p} = (QP)\alpha_C \searrow \qquad \text{and} \qquad a^t_{q/m} = (QM)\alpha_C \nearrow.$$

107. Relative Angular Accelerations. If two bodies have angular motion in the same plane or in parallel planes, Fig. 92, the angular-acceleration vectors are parallel and the relation of the angular accelerations can be obtained from § 94 by dividing each term of $\Delta\omega_B = \Delta\omega_A + \Delta\omega_{B/A}$ by Δt and letting $\Delta\omega/\Delta t = \alpha$, the average acceleration during the time interval Δt or an instantaneous value. This operation gives

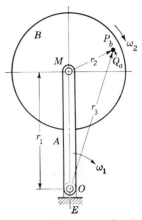

Fig. 92. Repeated.

$$(l) \qquad \alpha_B = \alpha_A + \alpha_{B/A},$$

in which the signs are algebraic; that is, for example, if CL is positive, CC is negative.

108. Example—Planetary Gears. In Fig. 101, the arm B ($WR = 10$ in.) rotates about a fixed center R and carries with it a 4-in. gear D which rolls on a fixed internal gear A (14 in. diameter). The arm has an angular velocity $\omega_B = 2$ rad./sec. CL, an angular acceleration $\alpha_B = 8$ rad./sec.² CC. Determine the absolute acceleration of Q on gear D. (See Chapter 9 for additional discussion of planetary gears.)

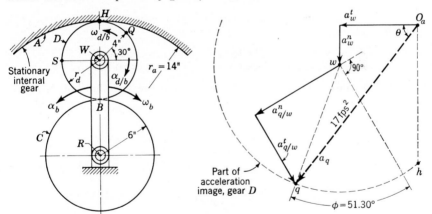

Fig. 101. Planetary Gears. The instantaneous acceleration image of gear D *for this position of B* can be drawn by noting that $a_{q/w}$ has the same magnitude whatever the position of Q on the circumference; thus with center at w and radius wq, draw the circle qh. Then, for example, $O_a h = a_h$, the acceleration of point H. The angle $\phi = \tan^{-1}(\omega^2_D/\alpha_D) = \tan^{-1} 1.25 = 51.3°$.

Solution. Since D rolls on A, the number of turns of D with respect to the arm B as B makes 1 rev. is the ratio of the circumference of A to that of D; or $2\pi r_a/2\pi r_d = r_a/r_d = 14/4 = 3.5$. That is, the ratio of the angular motion of D relative to that of B is 3.5; thus $\omega_{D/B} = 3.5\omega_B = (3.5)(2) = 7$ rad./sec. CC and $\alpha_{D/B} = 3.5\alpha_B = (3.5)(8) = 28$ rad./sec.² CL. The direction of $\omega_{D/B}$ is determined by imagining rotation of B to be CL in Fig. 101 and noting that as D rolls on A it rotates in the opposite sense to B. Evidently, as the arm slows down, so will the gear; hence $\alpha_{D/B}$ is opposite in sense to $\omega_{D/B}$.

Considering the problem, we note that the acceleration of point W can be computed from the given data; then we are concerned with the relative accelerations of two points Q and W in a rigid body. However, it must be noted in $v_{q/w} = r\omega$ and $a^n_{q/w} = r\omega^2$ that ω is the absolute angular velocity of the body, and in

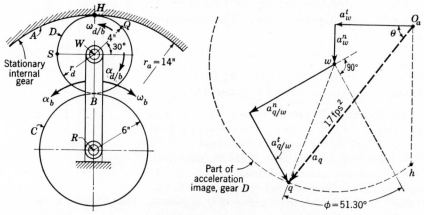

Fig. 101. Repeated.

$a^t_{q/w} = r\alpha$ that α is the absolute angular acceleration (§ 103). Thus, from §§ 94 and 107, we have

$$\omega_D = \omega_B + \omega_{D/B} = (+2 \text{ CL}) + (-7 \text{ CC}) = -5 \text{ rad./sec. CC};$$
$$\alpha_D = \alpha_B + \alpha_{D/B} = (-8 \text{ CC}) + (+28 \text{ CL}) = +20 \text{ rad./sec.}^2 \text{ CL}.$$

Having determined that ω_D is CC, ignore its negative sign (but not its sense). In the relative acceleration equation, we have

$$a_q = a^n_w \mathbin{+\!\!\!+} a^t_w \mathbin{+\!\!\!+} a^n_{q/w} \mathbin{+\!\!\!+} a^t_{q/w},$$
$$a_q = 3.33 \mathbin{+\!\!\!+} 6.67 \mathbin{+\!\!\!+} 8.33 \mathbin{+\!\!\!+} 6.67,$$
$$\quad\;\; \downarrow \qquad \leftarrow \qquad \nearrow \qquad \searrow$$

where the numerical values are computed as follows ($a^n = r\omega^2$, $a^t = r\alpha$):

$$a^n_w = (\text{RW})\omega_b^2 = \frac{10}{12}\,(2)^2 = 3.33 \text{ fps}^2, \quad \downarrow$$

$$a^t_w = (\text{RW})\alpha_b = \frac{10}{12}\,(8) = 6.67 \text{ fps}^2, \quad \leftarrow$$

$$a^n_{q/w} = (\text{QW})\omega_d^2 = \frac{4}{12}\,(5)^2 = 8.33 \text{ fps}^2, \quad \nearrow$$

$$a^t_{q/w} = (\text{QW})\alpha_d = \frac{4}{12}\,(20) = 6.67 \text{ fps}^2. \quad \searrow$$

Adding the vectors on the right side of the equation for a_q, Fig. 101, we get $a_q = 17 \text{ fps}^2$ at $\theta \approx 52°$ as shown.

109. Equivalent Mechanisms. You have observed that certain machine components because of their configurations as built are represented kinematically by four-link mechanisms as a convenience in velocity and acceleration analyses. Examples of this idea are: the eccentric, Fig. 36(c), p. 58, and the drag-link mechanism, Fig. 72, p. 104. Then in Fig. 84, we found that a four-bar mechanism with crossed links is

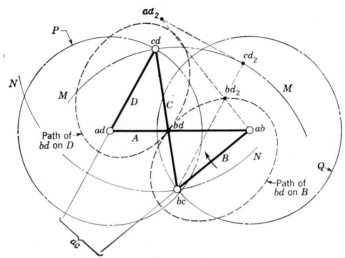

Fig. 84(a). Repeated.

equivalent to elliptical gears. Given the acceleration of a point in one of the gears, the theory of this chapter, relating as it does to two points in a rigid body, does not apply to the situation in going through the coincident points at the point of contact of the gears. However, one could solve this problem by using the material of this chapter in going from one end bc of the equivalent connecting link C (assuming link A is stationary) to the other end cd. Then, knowing the acceleration of one point cd in link D, one can find the acceleration of any other point in this rigid body.

A situation similar to Fig. 84 is the eccentric cam with reciprocating roller follower, Fig. 102. For all positions of the mechanism, the distance between P and Q is constant $(= r_b + r_e)$. It can be seen by inspection of Fig. 102 that the motion of the slider D is the same for either the cam and follower or the equivalent slider-crank mechanism $AB'C'D$. Thus, a_p can be found from a_q; then, since P ($=$ centro de)

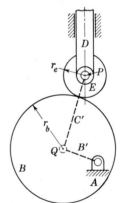

Fig. 102. Slider-Crank for Circular Cam and Roller Follower.

is in both links D and E, the acceleration of any other point in D or E can be found as desired (all points in D have the same acceleration).

In the foregoing examples, there is one set of bar-and-slider links that simulates the entire cycle of motion of the original mechanism. There are a number of other mechanisms which can be replaced by an equivalent bar linkage, but in which the motion is kinematically the same perhaps only for an instant. In this case, such linkages might well be called *instantaneous equivalent linkages* as a reminder that a new position of the real bodies results in a different equivalent linkage. Suggestive illustrations are shown in Fig. 103, in which the equivalent links are shown in

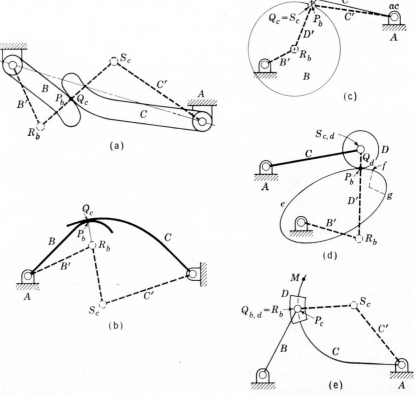

Fig. 103. Equivalent Mechanisms. In (a) and (b), each slight movement of the bodies B and C results in a new equivalent mechanism because the curvature of the contact surfaces is not constant. In (c), the linkage $AB'D'C'$ is equivalent to ABC for the entire cycle of motion; but this would not be true if the follower were a flat-face follower, nor if the curvature of the cam surface changed. In (d), let the radius of curvature of ef be constant, and let the radius from f to g be constant; then one equivalent linkage is obtained while the arc ef is in contact, another while the arc fg is in contact. At point f, where the curvature changes, there will be an instantaneous change of acceleration; so when f is in contact, two analyses should be made, one for arc ef, another for arc fg. In (e), the situation is different from the others, and the equivalent-linkage method fails when the path of R_b on C is a straight line; the center S_c would be at infinity and unavailable.

heavy dots. In each case, R_b is the center of curvature of link B at the point of contact, S_c is the center of curvature of link C at the point of contact. Evidently, if the radius of curvature of the surfaces varies as it often does for bodies in sliding-rolling contact [notice the higher pairing, except in (e)], then the equivalent linkage would be expected to change. In Fig. 103(c), the circular cam with knife-edge follower has a single equivalent mechanism, but this would not be true of a flat-face follower because the point of contact Q_c in the follower would not remain at a constant distance from ac. A method of finding radii of curvatures at the contact point is given in reference **59**.

A more complete theory of acceleration (see reference **141**) explains how to "pass through" such contact points as those in Fig. 103, and for some problems, the more complete theory becomes a necessity; on the other hand, when an equivalent linkage is easily determined and points R and S are at hand for use in the solutions, the equivalent-linkage idea often provides a quick and easy approach.

110. Closure.   The reader is reminded once again that the relative-acceleration equation as developed in this chapter is applicable only to two points in a rigid body.

Since the acceleration of one point is always with respect to some other point, be sure when using v^2/r that the v and the r are corresponding values.   Also in $a^n{}_{p/q} = r\omega^2$ and $a^t{}_{p/q} = r\alpha$, where P and Q are points in a rigid body, remember that ω and α are *absolute* values and r is the distance between the points.

Keep the vector diagrams easy to check and easy to use by putting the numerical value alongside any vector used in computations as well as alongside those asked for in the problem.   We have not always followed our own rule in this matter because the diagrams are naturally small and some would be too crowded and too confusing.

Since jerk is a vector quantity, methods similar to the foregoing of analyzing linkages for jerk can be and have been developed—for example, see reference **77**. However, at this stage, not enough is known of the effect of the magnitude of the jerk to be sure that it is significant.   In a series of pin-connected links, it will never be infinite, and it is the infinite jerk which makes it significant in cams.

PROBLEMS

Note to Students. Unless otherwise stated, lay out all mechanisms full scale on $8\frac{1}{2} \times 11$ paper in the upright position. In some cases, the long dimension of the paper should be horizontal, abbreviated LDH. Dimensions are placed alongside the link when convenient and are in inches. Usually, satisfactory locations of points are indicated as follows: O_v (2, 6), which means that the velocity pole O_v is located 2 in. from the left *edge* of the paper and 6 in. from the bottom *edge*, always actual measure no matter what scale is given.

As an act of courtesy to those who must check your work and *for your own later convenience*, always letter the points in the polygons in the conventional manner. As a regular practice, place the magnitude alongside certain vectors: (a) those given or computed, (b) those used in computing quantities needed in the solution, and (c) those representing the quantities asked for. Be sure the intended length of each vector is clear.

Algebraic Solutions

371. In Fig. 13, p. 20, show that the acceleration of P at the position B is equal to the normal acceleration of Q when it is at B. The point P is moving with harmonic motion. Also show that at any position of P, its acceleration is the horizontal component of the normal acceleration of Q.

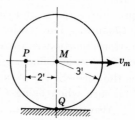

Fig. P103. Problem 373.

373. The center of the rolling wheel in Fig. P103 is moving toward the right with a speed of $v_m = 10$ fps. If the acceleration is $a_m = 5$ fps^2 toward the right, what are a_p and $a_{p/q}$?

Ans. $a_p = 27.4$ fps^2 at 7°, $a_{p/q} = 40.5$ fps^2 at 312°.

Less than Five Links

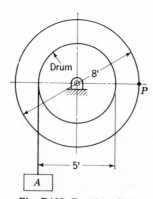

Fig. P102. Problem 372.

372. A mine cage A, Fig. P102, is suspended from a cable wound around a 5-ft. drum and is moving down with a constant acceleration of 2 fps^2. When $t = 0$, its speed is 10 fps. When $t = 3$ sec., what are the normal, tangential, and total accelerations of point P?

Ans. $a^n = 164$ fps^2, $a^t = 3.2$ fps^2.

374. The wheel of Fig. P104 rolls at R with an angular velocity of 4 rad./sec. CL and an angular acceleration of 8 rad./sec.^{2} CC; locate W (3, 9), O_v (6, $7\frac{1}{2}$), O_a (4, 1); use $K_s = 20$ in./in., $K_v = 2$ fps/in., and $K_a = 6$ fps^2/in. (a) Set up the velocity and acceleration polygons with numerical values for points P, Q, and R. (b) Draw the velocity and acceleration images of the circles. (Parts of these images may fall off the paper.)

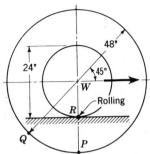

Fig. P104. Problem 374.

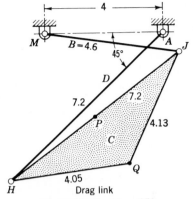

Fig. P105. Problem 378.

375. Use plate No. 18, Wingren. The instantaneous motion of crank B is defined by $\omega_B = 15$ rad./sec. CL and $\alpha_B = 30$ rad./sec.^{2} CL. What are the accelerations of K and L, and the angular acceleration of connecting rod C? Use O_v (2, $10\frac{1}{2}$); O_a (5, 6); $K_v = 3$ fps/in.; $K_a = 50$ fps^2/in.

376. Use plate No. 21, Wingren. Crank D is turning 90 rpm CL with an acceleration of 36 rad./sec. CC; O_v (7, 8); O_a ($2\frac{1}{2}$, 4); $K_v = 2$ fps/in.; $K_a = 30$ fps^2/in. Draw the velocity and acceleration polygons and dot in the acceleration image of C. What are the numerical values of a_l and α_C?

377. Use plate No. 19, Wingren. Draw a perpendicular to C at K and on it 1 in. (actual measure) from K, locate Q in link C. Draw another perpendicular to B at $\frac{1}{3}$ of B downward from J and locate P in B $\frac{1}{2}$ in. toward left. Crank D rotates uniformly at 240 rpm CC; O_v (8, 9), O_a (4, 6); $K_v = 4$ fps/in.; $K_a = 100$ fps^2/in. (a) Determine a_k, a_q, and a_p and the acceleration images of links B and C (including P and Q). (b) What are α_C and α_B?

378. In the drag-link mechanism of Fig. P105, $\omega_B = 10$ rad./sec. CC; $\alpha_B = 0$; $K_s = 2$ in./in.; $K_v = 3$ fps/in., $K_a = 200$ fps^2/in.; M (2, 3) LDH; O_v (8, 4); O_a (0.5, 5). See **297.** (a) Find the acceleration polygon, including the acceleration image of C, and the numerical values of a_h, a_q, and a_p. (b) What is α_C?

379. The crank in the slider-crank mechanism of Fig. P106 has $\omega_B = 20$ rad./sec. CL and $\alpha_B = 2$ rad./sec.^{2} CC when $\theta = 30°$. What are the acceleration of Q and the angular acceleration of link C? Show the velocity and acceleration polygons: O_a (1, 6); $K_a = 100$ fps^2/in.; O_v (4, 6); $K_v = 10$ fps/in.

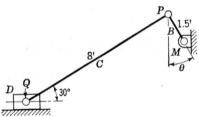

Fig. P106. Problem 379.

380. Use plate No. 17, Wingren. Locate point Q in link C on a perpendicular at K to JD and $\frac{3}{4}$ in. from K, actual measure (but note scale of drawing). The motion of B at this instant is $\omega_B = 10$ rad./sec. CC and $\alpha_B = 30$ rad./sec.^{2} CC; O_v ($8\frac{1}{2}$, 7); O_a (6, 6); $K_v = 1$ fps/in.; $K_a = 10$ fps^2/in. (a) Determine the acceleration polygon. (b) What are a_D, a_q, a_k, and α_C?

381. The same as **380** except that $\omega_B = 11$ rad./sec. CC.

382. In Fig. P107, let v_q be constant at 450 fpm; $\theta = 45°$; find a_p and α_C. Choose pole locations and the scale of the polygons.

383. The same as **382** except that $\theta = 30°$.

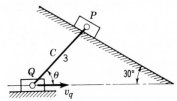

Fig. P107. Problems 382, 383.

384. In Fig. P108, let v_q be constant at 240 fpm and $K_v = 300$ fpm/in.; $\theta = 30°$. Draw the velocity and acceleration polygons, finding v_p, a_q, a_p, and α_C. Choose pole locations and the scale of the acceleration diagram.

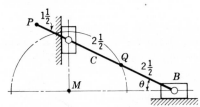

Fig. P108. Problems 384, 385.

385. The same as **384** except that $\theta = 60°$.

386. The wheel M (2, 5) in Fig. P109 rolls toward the right at 10 ips with $\alpha_B = 1$ rad./sec.² CC; $K_v = 3$ ips/in.; O_v (2, 3). Construct the acceleration images of links B and C and give the numerical values of a_q, a_r, a_p, and α_C. If the solution to **299** is at hand, use it as convenient in the solution of this problem. Choose a suitable scale for acceleration.

387. The velocity of J in Fig. P110 is 180 ips with $\alpha_B = 10$ rad./sec.² CC. Construct the acceleration polygon, the image of link C, and note the magnitudes of a_p, a_m, and α_C. If a solution to **301** is at hand, construct this polygon on a separate sheet.

388. Two rolling wheels are connected by link C as shown in Fig. P111. The angular velocity of wheel B is momentarily constant at 5 rad./sec. CC; $K_v = 2$ ips/in.; M $(1\frac{1}{2}, 1)$; O_v (7, 7). Find the acceleration images of links C and D and the numerical values of a_r, a_q, and α_C. If a solution to **298** is at hand, choose a scale and construct this polygon on a separate sheet.

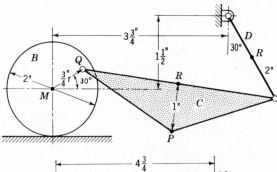

Fig. P109. Problem 386.

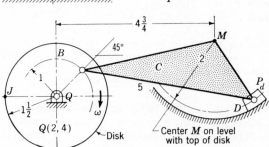

Fig. P110. Problem 387.

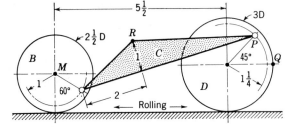

Fig. P111. Problem 388.

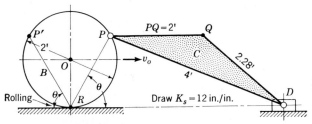

Fig. P112. Problems 389, 390.

389. The center O ($2\frac{1}{2}$, $2\frac{1}{2}$), Fig. P112, of the rolling wheel has a constant velocity of $v_o = 900$ fpm ($K_v = 300$ fpm/in.); $\theta = 60°$; O_v ($1\frac{1}{2}$, 9); O_a (6, 6); $K_a = 2 \times 10^5$ fpm²/in. Find the acceleration images of links B and C and the numerical values of a_p, a_r, a_D, a_q, α_B, and α_C. If a solution to **300** is at hand, use it as convenient in the solution of this problem.

390. The velocity of wheel B in Fig. P112 is $v_o = +10$ fps and its acceleration is $a_o = -30$ fps² (positive rightward); R (5, $\frac{5}{8}$) LDH; O_v ($\frac{1}{2}$, $5\frac{3}{4}$); O_a (9, $7\frac{1}{2}$); $K_v = 3$ fps/in.; $K_a = 20$ fps²/in.; note $K_s = 1$ ft./in. on mechanism. (a) For $\theta = 60°$, determine the acceleration polygon, including the acceleration images of links B and C. (b) What are a_p, a_q, a_r, and a_D? (c) What are α_C and α_B?

391. In Fig. P113, the wheel B rolls so that $v_m = 10$ fps and $a_m = 10$ fps², both toward the left; O_v (6, 8); O_a ($7\frac{1}{2}$, 5); $K_v = 4$ fps/in.; $K_a = 20$ fps²/in. For the position shown, what are the absolute accelerations of the pins P and Q? Compute α_C.

392. The same as **391** except that the wheel is slipping and $\omega_B = 4$ rad./sec. CL, $\alpha_B = 4$ rad./sec.² CL (no change in v_m and a_m).

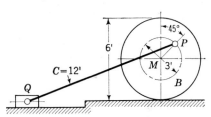

Fig. P113. Problems 391, 392.

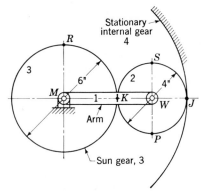

Fig. P114. Problem 393.

393. In the epicyclic gear train of Fig. P114, the motion of arm 1 is defined by $\omega_1 = 3$ rad./sec. CL and $\alpha_1 = 12$ rad./sec.² CC. Since the 4-in. planet gear

2 rolls on the 14-in. internal gear 4, it follows that $\omega_{2/1} = 10.5$ rad./sec. CC and $\alpha_{2/1} = 42$ rad./sec.^{2} CL (§ 108; also Chapter 9). Draw half size on $8\frac{1}{2} \times 11$ paper and find a_p, a_r, and α_2; M (3, 8); O_a (5, 2); $K_a = 3$ fps^2/in.

Ans. $a_p = 16.9$ fps^2 at 121°.

O_v (3, 10) and O_a (6, 6). Let $K_v = 0.4$ fps/in. and $K_a = 2$ fps^2/in. when the angular velocity of link D is 3 rad./sec. CL and its angular acceleration is 6 rad./sec.^{2} CL. Determine the acceleration of point P and the angular acceleration of link B.

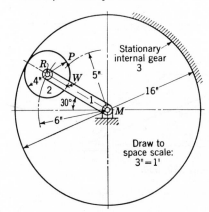

Fig. P115. Problem 394.

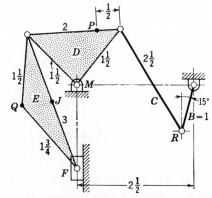

Fig. P116. Problem 396.

394. Epicyclic gear trains are used in gear reduction units of various kinds, for example, the automatic automobile transmission. Figure P115 represents a part of such a train, for which the designer may wish to investigate accelerations. The motion of the arm 1 is defined by $\omega_1 = 10$ rad./sec. CL and $\alpha_1 = 20$ rad./sec.^{2} CL. The motion of planetary gear 2 with respect to the arm is defined by $\omega_{2/1} = 40$ rad./sec. CC and $\alpha_{2/1} = 80$ rad./sec.^{2} CC. Locate P on the circumference of gear 2 at a radius of 5 in. from M; M (4, 8); O_a (8, 6). Let $K_a = 20$ fps^2/in. What is a_p? *Hint.* See § 94.

Ans. 112.7 fps^2.

Six Links

395. Use plate No. 5, Wingren. Name pin joints as follows: centro $df = Q$; centro $bc = R$; centro $bf = M$. Locate

396. The crank B in Fig. P116 turns at 300 rpm CC with $\alpha = 30$ rpm^2 CL; M (2, 6); O_v (2, 3); $K_v = 1000$ fpm/in. The actual length of the crank is 2 ft. (other dimensions in proportion). Construct the acceleration polygon and give numerical values of a_p, a_q, a_F, and α_E. If the solution to **316** is at hand, use it as convenient in solving this problem. Choose a suitable scale for acceleration.

397. In Fig. P117, let $v_j = 180$ ips and $\alpha_B = 0$; $K_v = 60$ ips/in. Construct the acceleration polygon and give numeri-

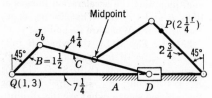

Fig. P117. Problem 397.

cal values of a_p and α_C. If the solution to **315** is at hand, use it as convenient in solving this problem. Choose a suitable scale for acceleration.

398. The wheel B in the toggle mechanism of Fig. P118 turns at a constant speed of 80 rad./min. Lay out with $K_s = 6$ in./in., LDH; M (1, 1); O_v (3, 7); O_a (6, $4\frac{1}{2}$); $K_v = 20$ fpm/in.; $K_a = 500$ fpm²/in. Draw the velocity and acceleration polygons, giving numerical values of all velocities used, a_F, and α_C.

linkage for this position and find a_q, a_r, and α_C. Select suitable scales for the polygons.

400. Using the equivalent linkage for the cam system of Fig. P120, find a_q and α_C if the cam rotates at a constant speed of 50 rad./sec. CC; M (2, 5); O_v ($7\frac{1}{2}$, $3\frac{1}{2}$); O_a (6, 3); $K_v = 30$ ips/in.; $K_a = 500$ ips²/in.

Ans. $\alpha_C \approx 250$ rad./sec.

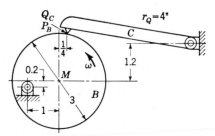

Fig. P120. Problem 400.

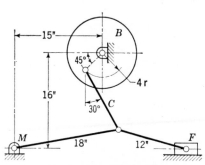

Fig. P118. Problem **398**.

Equivalent Mechanisms

399. The eccentric B, Fig. P119, turns at a constant speed of 30 rad./sec. CL; M (3, 6). Decide upon the equivalent

401. In the linkage of Fig. P121, B turns at the constant speed of 10 rad./sec. CL. What are the acceleration of R and the angular acceleration of C? Use the equivalent-linkage method; M (1, $7\frac{1}{2}$); O_v (5, 9); O_a (7, 4); $K_v = 0.6$ fps/in.; $K_a = 20$ fps²/in.

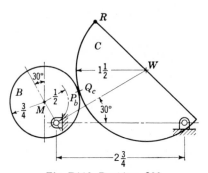

Fig. P119. Problem **399**.

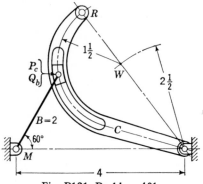

Fig. P121. Problem **401**.

402. In the oscillating-cam arrangement of Fig. P122, the member C is momentarily moving with $\omega = 20$ rad./sec. CC and $\alpha = 50$ rad./sec. CL. Determine a_p and α_B, using an equivalent linkage; M (2, 6); O_v (1, $7\frac{1}{2}$); O_a (3, 1); $K_v = 20$ ips/in., $K_a = 500$ ips²/in.

403–500. These numbers may be used for other problems.

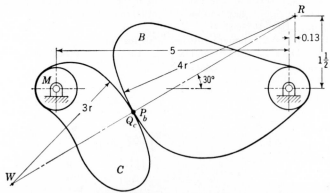

Fig. P122. Problem 402.

7

Spur Gears

111. Introduction. Gears are machine elements whose purpose is to transmit motion and/or power from one shaft to another. Although we speak of *friction gears*, which depend upon a frictional force to transmit motion (and power), the word *gear* used without a modifying adjective means a toothed wheel, Fig. 104. Motion is transmitted by toothed

Courtesy Brown and Sharpe Mfg. Co., Providence, R.I.

Fig. 104. An Assortment of Gears.

wheels because the teeth on one wheel mesh with those on another; the teeth on one push the teeth on the other. There are many types of gears, too many for all of them to be covered in detail in this book. This chapter is concerned with **spur gears,** Fig. 105, in which the teeth are formed on the outside of a right cylinder and the *elements of the teeth* (Fig. 106) are parallel to the axis of the cylinder. From this definition, it can be concluded that spur gears in operation are mounted on parallel shafts. Spur gears are so common that if the word *gear* is used alone without a contrary context, a spur gear is no doubt intended.

179

We shall develop the theory of gearing with reference to spur gears. If this is mastered, one is in a position to understand gear action in other kinds of gears. Since there are many terms new to the reader and related to gearing that must be understood, we shall define some of them first, placing them together for easy reference.*

Courtesy Illinois Gear and Machine Co., Chicago

Fig. 105. Spur Gear. Some gears are large; this one is cast steel, 8 ft. 10 in. in diameter, 12 in. face width, 1 diametral pitch.

112. Definitions. See Fig. 106 while studying these definitions.

When two spur gears are in mesh, there are two imaginary reference cylinders, called the **pitch cylinders,** which are tangent and rolling on each other. The pitch cylinders, whose width is equal to the face width, Fig. 106, are such that if the toothed wheels were replaced by rolling friction cylinders of the same size as the pitch cylinders, the angular velocity ratio would be unchanged. The end projection of a pitch cylinder is the **pitch circle.**

In its most general sense, the **pitch circle** of an involute is that circle whose center coincides with the center of the gear and whose circumference passes through the common centro of two gears in mesh. This definition results in different pitch circles for two meshing involute gears with every change in their center distance (Fig. 108 and § 116). In ordinary usage, *the pitch circle of any gear is an imaginary circle that is the*

---

* Since most failures in college can be attributed to the failing student's deficient knowledge of English, spend some time on these words, all of which are pure English. Engage in conversation with your classmates on the subject of gearing. Perhaps your roommate is having trouble with his dedendum.

basis of measurement of gears, especially of interchangeable gears; let this pitch circle be called the **standard pitch circle.** It is the **standard pitch circle** which will usually be meant when the term *pitch circle* is used, unless the context evidently intends a more general meaning. The size of a gear is the diameter in inches of its standard pitch circle, called the **pitch diameter.** Mating interchangeable gears theoretically have their standard pitch circles tangent, but of course they are not precisely so for any length of time except by chance.

The **pitch point** is the tangent point of the pitch circles and the common centro of two gears (their projection) in mesh. It is also the point of contact between two meshing conjugate teeth when contact is on the line of centers of the gears.

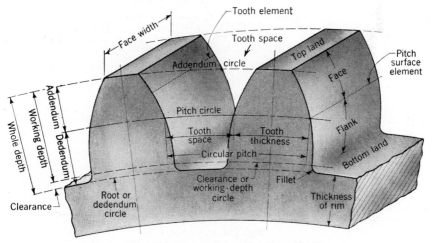

Fig. 106. Nomenclature for Gear Teeth.

The **addendum circle** (or *outside circle*) is a circle which bounds the outer ends of the teeth. The addendum *cylinder* encloses a spur (or helical) gear.

The **addendum** *a* is the radial distance between the pitch circle and the addendum circle (the perpendicular distance from the pitch surface to the addendum surface of a *rack*).

The **top land** is the surface of the top of the tooth.

The **dedendum circle** or **root circle** is a circle that bounds the bottoms of the teeth.

The **dedendum** *d* is the radial distance from the pitch circle to the root circle, that is, to the bottom of the tooth space.

The **bottom land** is the surface of the bottom of the tooth space.

The **working depth** is the radial distance from the addendum circle to the working-depth circle. It is equal to the sum of the addendums of the mating gears. The **working-depth circle** or **clearance circle** marks the distance that the mating tooth projects into the tooth space when it is symmetrically placed in the space, pitch circles tangent.

The **whole depth** is equal to the addendum plus the dedendum.

The **clearance** c is the radial distance between the clearance circle and the dedendum circle; it is the dedendum of one gear minus the addendum of its mating gear.

The **tooth thickness,** also called **circular thickness,** is the width of tooth measured along the arc of the pitch circle.

The **tooth space** is the width of space between teeth measured along the arc of the pitch circle.

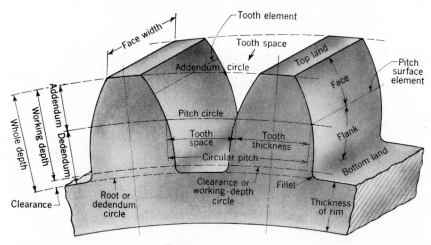

Fig. 106. Repeated.

The **backlash** is the tooth space minus the tooth thickness of the mating tooth (measured on the pitch circle). When backlash exists between two gears, one gear can be turned through a small angle while the mating gear is held stationary. Backlash is necessary to care for inevitable errors and inaccuracies in the spacing and in the form of the tooth, to provide a space between the teeth for a lubricant, and to allow for thermal expansion.

The **face of the tooth** is the surface of the *tooth* between the pitch cylinder and the addendum cylinder.

The **face width** is the length of a spur-gear tooth as measured along an element of the tooth, sometimes roughly spoken of as the *width of the gear*. In general, it is the distance from one end of a tooth to the other end measured along an element of the *pitch surface*.

The **flank** is the surface of the *tooth* between the pitch and root cylinders.

The **pinion** is the smaller of two gears in mesh; the larger of the two is called the *gear;* these words are so used when it is desired to distinguish between meshing gears.

A *rack* is a gear of infinite diameter; its pitch circle is a straight line (Fig. 118).

The *gear ratio* is the number of teeth in the larger gear divided by the number of teeth in the smaller. It is the same as the *velocity ratio* when the pinion drives (§ 116).

Many other technical terms will be defined as we go along.

113. Law of Gearing. If a driven gear has 40 teeth and the driving pinion 20 teeth, you can see without proof that there will be two revolutions of the pinion for each revolution of the gear, a velocity ratio of 2 on the basis of large-scale angular motion. The law of gearing, that is, of gear-tooth action, is not concerned with this large-scale velocity ratio m_ω, which depends on the tooth numbers, but with the velocity ratio while any particular pair of teeth are in driving contact.

In § 64 and Fig. 60 (repeated for convenience), we found the common centro bc of two rotating bodies B and C with sliding contact to be on the

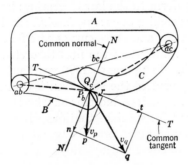

Fig. 60. Repeated.

common normal to the surfaces at the point of contact and also on the line of centers ab-bc-ac. In § 74, we learned that the instantaneous velocity ratio of two bodies, say B and C, Fig. 60, moving as a constrained system with respect to a third body A (probably the frame of the machine), is inversely as the ratio of the distances from the common centro to the centros with respect to the reference body A, that is to say, in relation to Fig. 60 with C driving B,

(a)
$$m_\omega = \frac{\omega_C}{\omega_B} = \frac{ab\text{-}bc}{ac\text{-}bc}.$$

If the curvature of the two contact surfaces in Fig. 60 is such that bc will be at a different point after the bodies have undergone a displacement, then in accordance with equation **(a)**, the velocity ratio will have changed. If the driving body is moving with constant angular velocity, it follows that the speed of the driven body changes; a change in speed results in acceleration, and accompanying the acceleration is a force ($F = ma$). If this sort of force comes into being with the engagement of every succeeding pair of teeth, it is periodic and will result in vibration and noise. This is what happens if gear teeth are made so that the common centro changes position materially, and the tooth surfaces would then be subjected to excessive wear; hence, the law of gearing, which may be stated as follows:

In order for a pair of engaging teeth to maintain a constant velocity ratio, *the shape of the profiles in contact must be such that the normals to the profiles at all points of contact shall all pass through a fixed point on the line of centers.* This fixed point is the pitch point (and common centro) that *must not move.* Two curves that satisfy the law of gearing are called **conjugate curves.**

We see then that the curves used for the profiles of the teeth must be especially chosen to meet this condition. It is possible to choose the curve of one of the contacting surfaces and then design the mating curve of such form as to conform to the law of gearing; but it is not done this way. There are two families of curves, *involutes* and *cycloids,* that properly mated satisfy the law and these curves care for all practical applications. The cycloidal family was the first theoretically correct form to be used for gear teeth. For perhaps 2000 years prior to this use, which was during the latter part of the seventeenth century (**34**), crude projections (for teeth) on cylinders and pins on wheels performed as toothed wheels. Today, except for special situations where the cycloidal curves have some unique advantage, practically all gear-tooth profiles are involutes. The advantages of the involute are given later after the reader has learned more of its properties.

Since it is impossible to manufacture gear teeth exactly as any curve, the actual profile varies in some way from the theoretical. The difference between the theoretical and actual as measured by certain means is called the **profile error.** The greater the profile error, the greater is the violation of the law of gearing and the greater are the noise, vibration, and destructive inertia forces. The amount of error which can be tolerated depends, as would be expected, on the speed. For a given error, the faster the gears turn, the greater the instantaneous accelerations arising from the change of location of the common centro, because each pair of teeth pass through contact quicker; and the greater the accelerations, the greater the vibrating forces. The best of modern gear profiles closely approach the theoretical. Our discussions will be on the basis of theoretically correct profiles.

114. Involute Curve Satisfies the Law of Gearing. The method of drawing an involute is explained in the Appendix, § 179. We know that if a taut string is unwrapped from a stationary cylinder, Fig. 107(a), a point T on this string describes an involute mn on a stationary plane. The circle (projection of the cylinder) from which it is unwrapped (from which the involute is generated) is called the **base circle**. Before proceeding further, note now an important attribute of the involute; at any position of point T, its instant center of rotation (and the instant center of all other points on the string) is at the point where the string is tangent to the circle; that is, at position e, the radius of curvature is ae; and at T, the radius is bT, etc. It follows that *a normal to the involute curve is tangent to the base circle.*

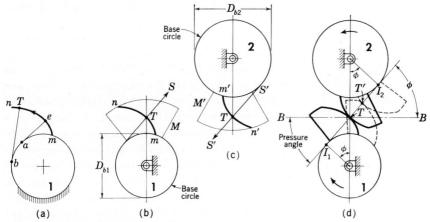

Fig. 107. **Forming the Involute Profile for a Tooth.** In (a), center of curvature at T is b; radius of curvature at e is ae. There are coincident points T_1 and T_2 in (d). Observe that two spur gears in mesh turn in opposite directions.

Another way to generate an involute is an inversion of Fig. 107(a): pull the point T on the taut string in a straight line with respect to a stationary reference, let the cylinder rotate as the string unwinds, and find the path of T on a plane attached to the cylinder, as in Fig. 107(b). The relative motion is the same as before and mn is again an involute.

Choose another base circle 2, any size, and generate an involute $m'n'$, Fig. 107(c). Let these curves mn and $m'n'$ be on steel plates M and M' attached to the cylinders, and cut the plates along the curves so that the involutes can be brought into contact, as in Fig. 107(d). No matter what points of the involutes are in contact, the normal to the curves is tangent to the base circles (a characteristic of an involute, as shown above). For example, let the curves be in contact at any point T', Fig. 107(d). The normal to the curve of base circle 2 is tangent to base circle 2 and the normal to the curve of base circle 1 is tangent to base circle 1; there is one common normal perpendicular to both involutes and tangent to both base circles. To look at it another way, imagine the string to be wrapped about both cylinders 1 and 2, Fig. 107(d), and let the cylinders rotate with the string taut; then some point T on the string *generates both curves simultaneously* on planes attached to the rotating bodies. It is easy to see that as these cylinders rotate, the string always crosses the line of centers at the same place T. Therefore, the common centro is fixed and the law of gearing is satisfied.

115. Line of Action and Pressure Angle. The **line of action** (also called *path of action, path of contact*) is the locus of all points of contact between two gear teeth [the locus is limited by the points of tangency I_1 and I_2, Fig. 107(d); contact should never occur "outside" of these points]. Except for involute teeth, the line of action is curved (see § 130). In the case of involute profiles, the line (path) of action is the common normal I_1I_2, Fig. 107(d), because this single line is normal to the involute profiles at any and all points of contact. Another name for the line of action is the **generating line,** because by its action, a point on the line generates the tooth profile. In involute gearing, the generating line is a straight line tangent to the base circle. The length of that part of the generating line bounded by the actual initial and final points of contact is called the **length of contact.**

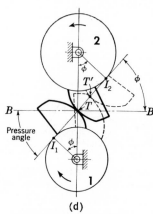

(d)

Fig. 107(d). Repeated.

The **pressure line** is the *straight* line (for any kind of teeth) which is normal to the curves at the point of contact. In involute gearing only, this line is identical with the line of action. The pressure line is so named because the line of action of the force (do not confuse with line of action of the point of contact) between the teeth (friction absent) acts in the direction of the normal to the curves in contact.

When two gear teeth are in contact, the slope of the *pressure line* I_1I_2 with respect to a perpendicular to the line of centers BB, Fig. 107(d), is called the **pressure angle** or the **angle of obliquity,** ϕ in Fig. 107(d). Except by arbitrary definition, § 118, involute gears have no pressure angle until two teeth are brought into contact. At any such point of contact T or T' for involute profiles, Fig. 107(d), the normal I_1I_2 to the curves always makes the same angle ϕ with BB, which is to say that for involutes in contact *the pressure angle is constant*—an important advantage of involute profiles.

116. Properties of Involute Profiles. Suppose the base circles 1 and 2 are mounted a certain distance apart, as shown in Fig. 108. If involutes (mn and $m'n'$) generated from these base circles are brought into contact, the point of contact will lie somewhere along the line I_1I_2. Since this line is normal to both curves, the centro (and pitch point) is at p, where it crosses the line of centers. The slope of the generating line I_1I_2 with the common tangent EE is the pressure angle ϕ. If two gears were to be made for the center distance in Fig. 108, the pitch circles would be HH for 1 and KK for 2, both passing through p. If the centers O and Q are moved closer together, the angle ϕ and the pitch radii r_1 and r_2 become smaller; if the centers are separated farther, ϕ and the r's increase; the

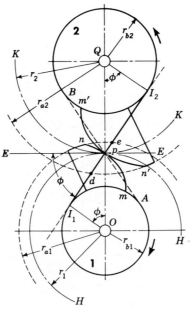

Fig. 108. Center Distance and Pitch Point. As 1 turns CL, these involutes go out of contact at e because the involute 1 does not extend beyond this point. Similarly, because involute 2 extends only as far as d, contact could not begin before point d on line I_1I_2.

velocity ratio remains constant at $r_2/r_1 = Qp/Op = r_{b2}/r_{b1}$. In any position of the centers O and Q, the point p (common centro) is where the line of action I_1I_2 (common normal) crosses the line of centers OQ; Op is the pitch radius r_1 of 1 and Qp is the pitch radius r_2 of 2. The radii of the base circles are, respectively, r_{b1} and r_{b2}; the various diameters would be designated as D_1, D_2 (without letter subscripts for pitch-circle dimensions), D_{b1}, and D_{b2}. Thus, in any particular position, the right triangles OpI_1 and QpI_2 show that $\cos \phi = r_b/r$, or

(b) $r_b = r \cos \phi$ and $D_b = D \cos \phi$.

Since p is the common centro, $v_{p1} = v_{p2} = 2\pi r_1 n_1 = 2\pi r_2 n_2 = r_1\omega_1 =$

$r_2\omega_2$, and we have the **velocity ratio** (and *gear ratio*) for the pinion driving as

(c)
$$m_\omega = \frac{\omega_1}{\omega_2} = \frac{n_1}{n_2} = \frac{r_2}{r_1} = \frac{D_2}{D_1} = \frac{D_{b2}}{D_{b1}} = \frac{N_2}{N_1};$$

that is, the angular velocity ratio is inversely as the pitch-circle diameters and as the base-circle diameters and as the tooth numbers N_1 and N_2. If $m_\omega > 1$, the driver is the smaller; if $m_\omega < 1$, the driver is the larger.

Observe in Fig. 108 that there is no outward limit of the involute; it may continue indefinitely. However, if both sides of the teeth are

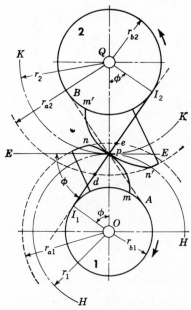

Fig. 108. Repeated.

bounded by involutes, as they generally are, the maximum theoretically usable portion in the outward direction is determined by the teeth becoming pointed (see Fig. 109). In Fig. 108, the profile for the other side of the tooth (not shown) would be generated by a point on the dotted line AB.

It is also evident that the involute starts at the base circle; hence, if a tooth profile should happen to extend inside the base circle, as it often does, then this portion of the tooth profile is not an involute, and since it is not, it should not make contact with an involute. To satisfy the law of gearing, involute teeth should not touch each other except outside of the base circles (between I_1 and I_2 or A and B, Fig. 108).

117. Pitch. We may think of the pitch circles as being defined by the involutes in contact (HH and KK, Fig. 108), or we may define the base circle in terms of a given pitch circle. Thus, let the pitch circle be as shown in Fig. 109; it will theoretically be rolling on another pitch circle when running with another gear. If so and if it is desired that the pressure angle be ϕ, draw a tangent T to the pitch circle, and draw EE through this tangent point at an angle ϕ with line T; then the base circle is tangent to EE with the same center as the pitch circle; $r_b = r \cos \phi$. Now the involute G can be generated.

The amount of the involute used for the tooth profile depends mostly on the tooth height, which in turn is a function of the pitch and type of tooth. The *pitch* of a gear is a measure of the spacing of the teeth, and for a given type of tooth, it is also a measure of the size of tooth. If the tooth is relatively small, as at A, Fig. 109, it lies entirely outside of the

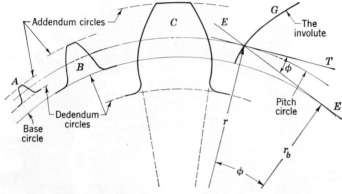

Fig. 109. Adapting the Involute to the Tooth Profile. Note that for the tooth width shown, tooth B, for example, could be very little taller before it would become pointed

base circle. As the tooth size is increased for particular pitch and base circles and for interchangeable teeth (§ 118), a point is reached, say size B, Fig. 109, where the dedendum circle coincides with the base circle. If still fewer (larger) teeth are used, as at C, then with the same proportional dedendum, the tooth must necessarily extend inside of the base circle. That part of the flank which is inside of the base circle is usually *drawn* with radial lines and with a fillet to the dedendum circle; actually it may be manufactured quite differently (§ 119).

There are two different pitches in common use in this country: the circular pitch and the diametral pitch. The *circular pitch* P_c is the distance in inches measured along the arc of the pitch circle between corresponding points on adjacent teeth. Thus, if there are N teeth on a gear of circular pitch P_c in., the circumference of the pitch circle is NP_c; but it is also πD, where D is the pitch diameter. Thus, we have

$$(21) \qquad NP_c = \pi D \qquad \text{or} \qquad P_c = \frac{\pi D}{N}.$$

There must always be a whole number of teeth N. Circular pitch is commonly used for cast teeth because the patternmaker finds this a convenient dimension for a layout.

The **diametral pitch** P_d is the number of teeth per inch of diameter, a ratio. In equation form, we have

$$(22) \qquad P_d = \frac{N}{D} \qquad \text{or} \qquad N = P_d D.$$

The custom is to speak of diametral (pronounced di-am′-e-tral) pitch without units, but it has the unit in.$^{-1}$ For example, a 2-in. gear with an "8 pitch" has 16 teeth.

Courtesy Barber-Colman Co., Rockford, Ill.

Fig. 110. Comparative Sizes of Gear Teeth.

The **base pitch** P_b, applicable to involute teeth, is the distance in inches measured along the arc of the base circle between the center lines of adjacent teeth; this is the circumference of the base circle divided by the number of teeth; that is,

$$(23) \qquad P_b = \frac{\pi D_b}{N} = \frac{\pi D \cos \phi}{N} = P_c \cos \phi,$$

where we have used $P_c = \pi D/N$. Because the base pitch is the distance between adjacent teeth measured along the generating line, it is also sometimes called the *normal pitch* (NQ, Fig. 113).

Since it is often necessary to convert diametral pitch to circular pitch, or vice versa, a relation between these pitches is useful. Multiplying together corresponding sides of equations (21) and (22), we get

$$(24) \qquad P_c P_d = \left(\frac{\pi D}{N}\right)\left(\frac{N}{D}\right) = \pi \qquad \text{or} \qquad P_c P_d = \pi.$$

Thus a 4-pitch gear has a circular pitch of $\pi/4 = 0.7854$ in. Always give the unit of circular pitch. A comparison of tooth sizes for different diametral pitches is shown in Fig. 110.

118. Interchangeable Involute Systems. A group of gears of a particular pitch are interchangeable if any one of the gears will run with any other and have the same pressure angle in each case (normal backlash). In addition to the pitch and the pressure angle being the same, it is also true that, to be interchangeable, every gear has the same addendum, that the dedendum is equal to the addendum plus some clearance, and that the tooth width is virtually equal to the width of space. When gears are made in accordance with the following systems, they have a particular "built-in" pressure angle. That is, two so-called 20° gears when properly mounted will run with a nominal pressure angle of 20°. Owing to inaccuracies in mounting, the actual pressure angle will be slightly different. Moreover, one of these standard gears may be deliberately operated at a distinctly different pressure angle for some reason. To distinguish between operating and "built-in" values, we might call the built-in pressure angle the **degree of involute;** the degree of involute of a cutter is then the pressure angle for which it is designed.

(a) **Full-depth, $14\frac{1}{2}°$ and 20° Involute System.** All full-depth (also called *full-height* and *full-length*) teeth have a working depth of $2/P_d$ in. The clearance and the whole depth may vary, but the usual nominal values as given by the ASA are (**64**):

Addendum $a = 1/P_d$,	Clearance $c = 0.157/P_d$,
Dedendum $d = 1.157/P_d$,	Whole depth $= 2.157/P_d$,

Outside diameter D_a = pitch diameter + (2)(addendum) = $D + 2a$.

The clearance is not a critical dimension and is often made greater.

The difference between the standard $14\frac{1}{2}°$ full-depth and 20° full-depth teeth is in the size of the base circle for a particular gear size, $D_b = D \cos 14\frac{1}{2}°$ versus $D_b = D \cos 20°$; see Fig. 108. The increase in the degree of involute ϕ to 20° results in a somewhat broader and stronger base section, Fig. 111, and it reduces the interference difficulties (§ 124). Draw the tooth thickness equal to the tooth space ($P_c/2$) and let the fillet radius be about 1.35 to 1.5 times the clearance. For most drawing purposes, it is not necessary to develop involutes for every profile; approximations in the form of circular arcs are sufficiently accurate. See Appendix for more detail.

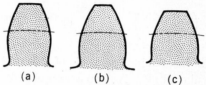

(a)   (b)   (c)

Fig. 111. Comparison of Tooth Profiles. These teeth have the same pitch and are drawn as they would appear on 20-tooth gears; (a) $14\frac{1}{2}°$ full depth, (b) $20°$ full depth, (c) $20°$ ASA stub.

(b) Stub-tooth Systems. The ASA standard proportions are (**64**):

Working depth $= 1.6/P_d$,   Clearance $c = 0.2/P_d$,
Addendum $a = 0.8/P_d$,   Whole depth $= 1.8/P_d$,
Dedendum $d = 1/P_d$,   Degree of involute $= 20°$.

Another system of stub teeth which is often used is that of the Fellows Gear Shaper Company. In this system, the dimensions of the tooth are indicated by *two* diametral pitches, written thus: 5/7 (say *five-seven*, not *five-sevenths*). The first number (numerator) is *the* pitch of the gear and should be used in all calculations involving the pitch *except* in computing the radial dimensions (addendum and dedendum) of the tooth. The second pitch (denominator) is used with the proportions of the full-depth system, paragraph (a) above, to get the radial dimensions of the tooth. This plan results in a shorter (stub) tooth. Thus, if the pitch is 5, the addendum of a full-depth tooth is $\frac{1}{5}$ in.; but a Fellows tooth would have an addendum of only $\frac{1}{7}$ in. in accordance with the Fellows pitch of five-seven. Clearance in the Fellows system is $0.25/P_d$, where the P_d is the second pitch (denominator) and the degree of involute is $20°$. Among the pitches for which standard shaping cutters may be obtained are: 4/5, 5/7, 6/8, 7/9, 8/10, 9/11, 10/12, and 12/14. See § 124 for some advantages and disadvantages of these systems. A comparison of tooth profiles is shown in Fig. 111.

A practical way to provide backlash is to feed the cutter somewhat deeper than standard depth on one or both of the two gears. On accurately cut teeth, the amount of the backlash is of the order of $0.04/P_d$ to $0.025/P_d$.

119. Interference of Involute Teeth. When a relatively large part of the flank of a tooth lies inside of the base circle and is made radial, there is likely to be interference (overlapping of profiles) between this flank and the tip of the mating tooth. This effect is a maximum, if it exists for a particular system of gears, when a rack is engaged with the smallest gear (usually 12 teeth in the $14\frac{1}{2}°$ system), as seen in Fig. 112. The amount of interference decreases as the size of the larger gear decreases (the diameter of a rack is infinite).

Unless the profiles have been modified, interference is indicated whenever the intersection of an addendum circle with the line of action falls "outside" of the tangent point of the line of action and the base circle.

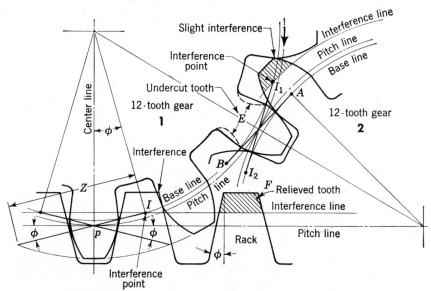

Courtesy Fellows Gear Shaper Co., Springfield, Vt.

Fig. 112. Interference. The "interference points" are where the line of action is tangent to the base circles. The teeth are $14\frac{1}{2}°$ full depth. Notice the interference of the uncorrected rack with an uncorrected 12-tooth pinion. Also observe that the interference of two 12-tooth gears (upper right) is slight.

These tangent points, I, I_1, and I_2, Fig. 112, are called the *interference points*. If the teeth extend beyond these points, contact between an involute tip and a noninvolute flank is indicated. At best, such contact would not satisfy the law of gearing, and since the theoretical profiles actually overlap, interference occurs.

Notice that the fewer the teeth (see Fig. 109), the more flank there is inside of the base circle and the more likely is interference. For a particular system of interchangeable involute gears, the only means of eliminating interference are either (1) to remove some of the metal flank on the pinion, as at E, Fig. 112, which evidently weakens the pinion tooth, or (2) to trim or ***relieve*** the face of the tooth on the gear, as at F, Fig. 112.

Often some of both actions are taken. Cutting the teeth with unrelieved hobs or rack cutters (§ 129) automatically undercuts them.

120. Action of Gear Teeth. Think of the line of action as being limited by the interference points I_1 and I_2, Fig. 113, and master the following facts. The pinion is driving CC in Fig. 113.

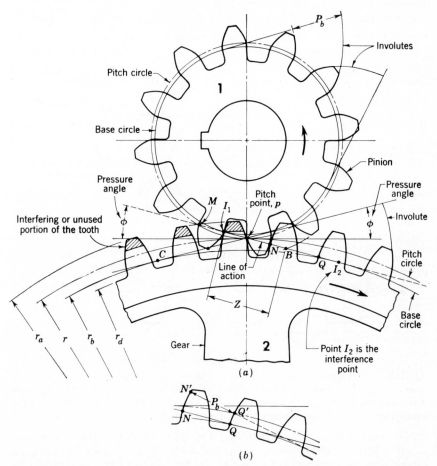

Fig. 113. Action of Gear Teeth. The $\phi = 14\frac{1}{2}°$ pressure angle can be made 20° by increasing the center distance, but if the teeth are left the same, the backlash would be excessive. The base pitch is $P_b = NQ = N'Q'$ [as seen in (b)], measured along the normal to the profiles. Notice that the radius of curvature of the pinion tooth at N is I_1N; of the gear tooth, I_2N.

1. Contact begins between the flank of the tooth on the driver and the face of the tooth on the driven.

2. The initial point of contact in any kind of gearing is where the

addendum circle of the driven gear cuts the generating line—*except, in the case of involute gearing,* when this point of intersection falls "outside" of the corresponding interference point. Since contact should occur only between the interference points, the tooth profiles should be modified, if necessary, so that contact starts at the interference point (assume this situation in the solution of problems).

3. Contact ceases between the face of the driving tooth and the flank of the driven tooth.

4. The final point of contact in any kind of gearing is where the addendum circle of the driver cuts the generating line—*except, in the case of involute gearing,* when this point of intersection falls "outside" of the corresponding interference point. If this point of intersection is not between the interference points, the profiles should be modified so that contact ceases at the interference point (to be assumed in the solution of problems).

In Fig. 113, the intersection point M is not between I_1 and I_2, and thus the shaded portion of the gear teeth may be considered as the interfering portion, and the initial point of contact is taken at I_1. Since the intersection B (see rule 4 above) is inside I_2, the final point of contact is B.

The distance I_1B is the *length of the line of action* or *contact length* or *length of action* Z.

The tangent line through C would be the line of action for the gear driving CC or the pinion driving CL.

The distance $I_1N = NQ$ is equal to the *base pitch.* You can perhaps see this easiest at NQ. Notice the detail in Fig. 113(b). Imagine the string NQ being wrapped onto the gear's base circle. Points N and Q follow the gear-tooth outline and when N and Q lie on the base circle, it is evident that NQ is the distance between teeth as measured on the base circle, or $\pi D_b/N$. See § 117.

121. Angles of Approach, Recess, and Action. The **angle of approach** α is the angle through which a gear turns from the instant a pair of teeth come into contact until the same pair of teeth are in contact *at the pitch point.*

The **angle of recess** β is the angle through which a gear turns from the time a pair of teeth are in contact *at the pitch point* until the instant that the same pair of teeth go out of contact.

The **angle of action** θ is the angle through which a gear turns from the time a pair of teeth come into contact until the same pair of teeth go out of contact. The **arc of action** is the arc on the pitch circle which subtends the angle of action.

(d)   Angle of action $\theta = \alpha + \beta$.

See Fig. 114 for the method of determining these angles graphically. The pinion 1, turning CL, is driving the gear 2. Therefore, the line of action AA slopes as shown; the degree of involute $\phi = 14\frac{1}{2}°$ in this case. Draw pitch, addendum, dedendum, and base circles. Locate the interference points I_1 and I_2 on the line of action. To locate the initial point of contact, find the intersection m of the addendum circle of the driven and AA. Since m is *not* between I_1 and I_2, use I_1 as the initial point of contact. Draw the contacting profiles of both the pinion and gear

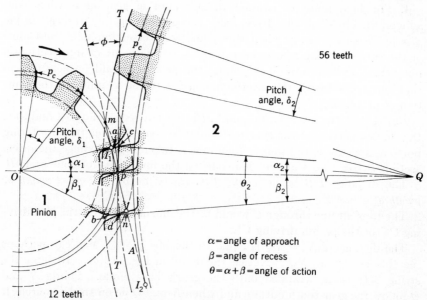

Fig. 114. Angles of Approach, Recess, Action, and Pitch. Notice that point a is on the pitch circle; while contact starts at I_1, it is point a which makes contact when the teeth are in contact at p. The difference between the point of contact and the point whose movement is followed is clearer on recess, points n and d on the gear, n and b on the pinion.

through the initial point of contact, as shown.* Find the intersection n of the addendum circle of the driver and the line of action. Since n is between I_1 and I_2, it is the final point of contact. Draw the contacting tooth profiles for both gears through point n, as shown. (It is not necessary to draw profiles through the pitch point, but it is done here to help visualize the situation. The other sides of the teeth are not necessary for this purpose, but may be drawn if desired.)

Now review the definition of the angles. The profile point a on the pinion's pitch circle is the point in contact at the pitch point p; hence, the pinion's angle of approach α_1 is angle aOp. Similarly, it is the point

* See § 180, Appendix, for a method of drawing gear teeth.

c on the gear which engages point a when contact reaches the pitch point; therefore cQp is the angle of approach α_2 on the gear. At the final point of contact, we see that b is the point on the pinion which was in contact at the pitch point, and angle $pOb = \beta_1$ is the angle of recess on the pinion. Since point d on the gear was in contact with b on the pinion when these teeth engaged at p, the angle $pQd = \beta_2$ is the angle of recess on the gear. Observe that neither the approach angle nor the recess angle is ever greater than the degree of involute ϕ.

Unless the gears are the same size, the angles of approach and recess are not the same, as is evident from Fig. 114. Actually, the ratio of the angles of action, etc., is the same as the velocity ratio; $m_\omega = \omega_1/\omega_2 = \alpha_1/\alpha_2$, etc. This observation is well used as a check on the accuracy of the solution.

122. Contact Ratio. The *pitch angle,* Fig. 114, is the angle subtended by an arc on the pitch circle equal in length to the circular pitch. The ratio of the angle of action divided by the pitch angle is the **contact ratio** m_c:

$$(e) \qquad m_c = \frac{\text{angle of action}}{\text{pitch angle}} = \frac{\alpha + \beta}{\delta} = \frac{\theta}{\delta},$$

where the symbols accord with Fig. 114. This ratio is the same for each of two gears in mesh. The significance of the contact ratio for spur gears lies in the following statements: if the angle of action is equal to the pitch angle, $m_c = 1$, a pair of teeth starts contact at the instant that the preceding pair ceases contact and there is always exactly one pair of teeth in contact; if $\alpha + \beta$ is greater than δ, $m_c > 1$, a pair of teeth starts contact *before* the preceding pair ceases contact and therefore part of the time there are two (or more) pairs of teeth in contact (the average number of pairs of teeth in contact is greater than one); if $\alpha + \beta$ is less than δ, $m_c < 1$, a pair of teeth starts contact *after* the preceding pair ceases contact and therefore there is a period of time when no teeth are in contact (and the drive is not continuous). It follows that to have teeth continuously in contact, and consequently to have a smooth transfer of load to succeeding teeth, the contact ratio should always be greater than unity. A value of m_c a little less than one can sometimes be tolerated when the speed is quite low, but as the speed and power increase, it becomes essential to increase the average number of teeth in contact (which is a significant advantage of helical gears), not because it is kinematically desirable but because it is impossible to make perfect involute teeth. When the form and spacing of the tooth profiles are inaccurate, the transfer of load occurs with shock loading. In general, the greater the average number of teeth in contact for a given tooth error, the less the shock when a pair of teeth make contact.

The contact ratio for involute gears is also given by

$$(f) \qquad m_c = \frac{\text{length of action}}{\text{base pitch}} = \frac{Z}{P_c \cos \phi}$$

in which the length of action (the active length of the line of action) may be calculated (§ 123) or measured if a layout has been made. As previously demonstrated (§ 120), the distance between teeth as measured along the line of action is the base pitch ($P_b = I_1 N = NQ$, Fig. 113). Thus, the total length of action divided by the distance between teeth measured along the contact line is the contact ratio—which justifies equation (f). Observe that the contact ratio can be found from (f) without drawing tooth profiles (see Fig. 115).

123. Computing the Length of Action. If the initial and final points of contact m and n lie between the interference points I_1 and I_2, we get a configuration somewhat as shown in Fig. 115. The length of action is the length mn, which is easiest found if broken into parts, the approach

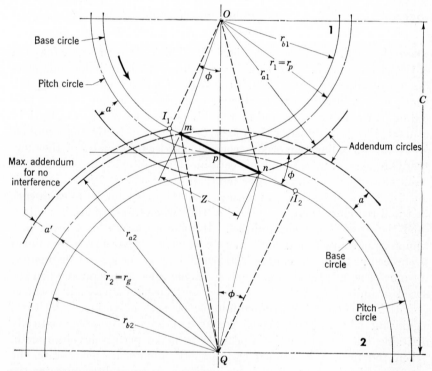

Fig. 115. Computing the Contact Length. The pressure angle and addendum are somewhat exaggerated over usual values for clearness. This illustration also shows the construction for obtaining the contact length mn graphically.

length mp plus the recess length pn; thus

(*g*)   Length of action $Z = mn = mp + pn$.

For the pinion driving, we see that the approach length is

$$mp = I_2m - I_2p;$$
$$I_2m = (r_{a2}^2 - r_{b2}^2)^{1/2};$$ [RIGHT $\triangle QI_2m$]
$$I_2p = r_2 \sin \phi;$$ [RIGHT $\triangle QI_2p$]
(*h*) $$Z_a = mp = (r_{a2}^2 - r_{b2}^2)^{1/2} - r_2 \sin \phi.$$

Similarly the recess length is obtained as follows:

$$pn = I_1n - I_1p;$$
$$I_1n = (r_{a1}^2 - r_{b1}^2)^{1/2};$$ [RIGHT $\triangle OI_1n$]
$$I_1p = r_1 \sin \phi;$$ [RIGHT $\triangle OI_1p$]
(*i*) $$Z_r = pn = (r_{a1}^2 - r_{b1}^2)^{1/2} - r_1 \sin \phi.$$

Substitute the values in (*h*) and (*i*) into (*g*) and get $mn = Z$;

(*j*) $$Z = (r_{a2}^2 - r_{b2}^2)^{1/2} - r_2 \sin \phi + (r_{a1}^2 - r_{b1}^2)^{1/2} - r_1 \sin \phi,$$
[APPROACH LENGTH]   [RECESS LENGTH]
$$= (r_{a2}^2 - r_{b2}^2)^{1/2} + (r_{a1}^2 - r_{b1}^2)^{1/2} - C \sin \phi,$$

the length of action when the intersections of the addendum circles and the line of action *lie between the interference points*; $r_1 + r_2 = C$, the center distance. To use (*j*), recall that $r_b = r \cos \phi$ and that the radius of the addendum circle is $r_a = r + a$ where a is the addendum (§ 118). The value of $Z = mn$ from (*j*) divided by $P_b = P_c \cos \phi$ gives the contact ratio; $m_c = Z/P_b$.

If the point m should fall "outside" of the interference point, then the approach length is I_1p, or

Approach length $Z_a = r_1 \sin \phi$,

provided the interference point is the initial point of contact. If the larger gear is the driver, the words *approach* and *recess* are interchanged; it is expected that the reader can make this transition.

124. Reducing Interference Difficulties. Interference can be eliminated by modifying the parts of tooth profiles which encroach upon one another, as mentioned in § 119. However, this action may reduce the contact ratio below some desired value and in any case it renders useless a part of the tooth's face on the larger gear. This fact suggests using a shorter tooth, as at E, Fig. 116, thus doing away with some or all of the interfering parts—which is why stub teeth, § 118, were introduced. However, shortening the tooth reduces the contact ratio in cases where there would be no interference with the full-depth tooth, and is therefore generally undesirable whenever interference does not exist.

Another basic change that would alleviate interference is to increase the degree of involute. In Fig. 116, for a degree of involute of ϕ_1, draw the generating line AA and find the closest interference point I_1; then draw the circle CC through it. The radial distance between circle CC

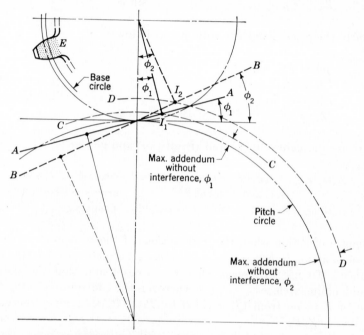

Fig. 116. Pressure Angle and Interference.

and the gear's pitch circle is the maximum addendum which the gear may have without interference. Now draw a generating line BB at a greater degree of involute ϕ_2 and find the interference point I_2. We see immediately that the gear's addendum may be increased significantly without extending beyond interference point I_2. Stated another way, if the actual addendum circle fell between CC and DD, then there would be interference in the ϕ_1 system, but no interference in the ϕ_2 system. This is the reason why the 20° full-depth tooth is often preferred over the $14\frac{1}{2}$° full-depth tooth when interference would occur in the $14\frac{1}{2}$° system. For a given addendum and interference in the $14\frac{1}{2}$° system, a change to the 20° system may well result in a larger contact ratio. However, if the interference condition does not exist, the contact ratio *decreases* with an increase in the degree of involute.

As compared with the original $14\frac{1}{2}°$ full-depth tooth, the stub tooth reduces or eliminates the interference both by shortening the tooth and by increasing the degree of involute. Another means of avoiding interference is to use unequal addendum and dedendum teeth, explained in reference **141**.

The worst interfering condition occurs when the pinion engages a rack. Using an approach similar to that in the following example, it can be shown that there will be no interference with a rack if the minimum number of teeth in the pinion is as follows: for $14\frac{1}{2}°$ full depth, 32 teeth; for 20° full depth, 18 teeth; for 20° stub teeth, 14 teeth.

125. (a) Example—To Find Minimum Number of Teeth on Pinion for No Interference. A gear has 45 teeth of 3 pitch. What minimum number of 20° full-depth interchangeable teeth may the pinion have without interference?

Solution. The diameter of the gear is $D_g = N/P_d = 45/3 = 15$ in.; $r_g = 15/2 = 7.5$ in. For full depth, the addendum is $a = 1/P_d = 0.333$ in. We shall explain a graphical solution because the figure is needed anyway for an algebraic solution. Select center Q, Fig. 117, and draw the pitch circle of the gear, $r_g = 7.5$ in.; draw the gear's addendum circle, $r_{ag} = r_g + a = 7.5 + 0.333 = 7.833$ in.;

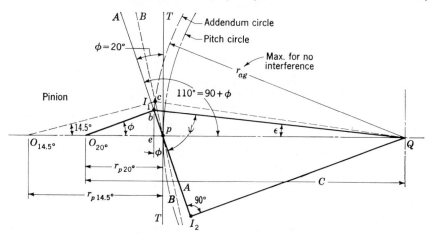

Fig. 117. Conditions for No Interference. The interference point I_1 is point b in the 20° system and point c in the 14.5° system.

draw a tangent TT to the pitch circle; draw the generating line AA (solid) at 20° with TT. Now the intersection b of the addendum circle of the gear and the line AA is the initial point of contact (pinion driving) for no interference. Drop a perpendicular at b and locate O_{20}; then $pO_{20} = r_{p20}$ is the minimum radius of the pinion for no interference; measure $r_{p20} = 2.48$ in. Then

$$N_p = 2r_pP_d = (2)(2.48)(3) = 14.88, \text{ say 15 teeth,}$$

which is the minimum number of teeth on the pinion for no interference, 20° full-depth in mesh with 45-tooth gear. (For good measure, the dotted lines BB and $cO_{14.5}$ show the solution for full-depth 14.5°. Minimum $N_p \approx 26$ teeth.)

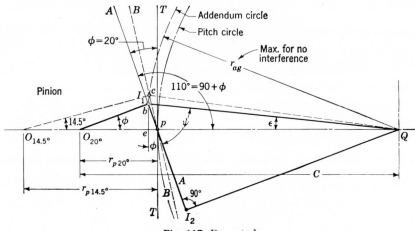

Fig. 117. Repeated.

Usually a graphical solution done to large scale is accurate enough for this purpose. However, if an algebraic solution is desired, one may proceed as follows (refer to Fig. 117):

$$QI_2 = r_g \cos 20 = (7.5)(0.94) = 7.05 \text{ in.,}$$

$$\sin \psi = \frac{QI_2}{Qb} = \frac{7.05}{7.833} = 0.9 \text{ in.,}$$

$$\psi = 64.166°,$$

Angle $Qpb = 90 + \phi = 90 + 20 = 110°,$

$$\epsilon = 180 - 110 - 64.166 = 5.834°.$$

Now apply the law of cosines to triangle Qpb and find

$$(pb)^2 = (Qb)^2 + (Qp)^2 - (2)(Qb)(Qp) \cos \epsilon$$
$$= (7.5)^2 + (7.833)^2 - (2)(7.5)(7.833) \cos 5.834°,$$

from which $pb = 0.854$ in. (Also $be = Qb \sin \epsilon = r_{ag} \sin \epsilon$ and $pb = be/\cos \phi$.) Now since $pb = r_p \sin \phi$, we get

$$r_p = \frac{pb}{\sin \phi} = \frac{0.854}{\sin 20} = 2.495 \text{ in.}$$

The corresponding number of teeth is

$$N_p = 2r_pP_d = (2)(2.495)(3) = 14.97,$$

or, as before, 15 teeth. The actual diameter of the pinion is therefore a little larger than the minimum computed; $D_p = 15/3 = 5$ in. Note that although there will be no interference with 15 teeth on the pinion, it does not follow that the teeth will not be undercut; a rack or hob cutter will result in a small amount of undercut. See § 129.

(b) Example—To Find the Maximum Addendum on Gear for No Interference.
Example (a) looked at the matter of interference from the viewpoint of the
pinion. This example investigates the condition to be met by the gear to avoid
interference. A 15-tooth pinion meshes with a 60-tooth gear. If they are 14.5°
involute and 10-pitch gears, what is the maximum addendum that the gear can
have without interference?

Solution. Refer to Fig. 117 (also Fig. 115). The maximum addendum on the
gear for no interference is that which results in the gear's addendum circle passing
through the interference point I_1. The corresponding addendum-circle radius is
r_{ag}, Fig. 117. The center distance is obtained from the radii.

$$D_p = \frac{15}{10} = 1.5 \text{ in.}, \qquad r_p = 0.75 \text{ in.}, \qquad D_g = \frac{60}{10} = 6 \text{ in.}, \qquad r_g = 3 \text{ in.}$$

$$C = r_p + r_g = 0.75 + 3 = 3.75 \text{ in.}$$

The radius $r_{bp} = OI_1 = r_p \cos \phi = (0.75)(0.968) = 0.726$ in. Now apply the
cosine law to triangle OQI_1 and get

$$r_{ag} = (C^2 + r_{bp}^2 - 2Cr_{bp} \cos 14.5)^{1/2}$$
$$= [3.75^2 + 0.726^2 - (2)(3.75)(0.726)(0.968)]^{1/2} = 3.053 \text{ in.}$$

Therefore, the maximum useful addendum is $r_{ag} - r_g = 3.053 - 3 = 0.053$ in.*
versus the actual standard addendum of $a = 1/P_d = 0.1$ in. The useless excess
of addendum on the gear in this case is $0.1 - 0.053 = 0.047$ in. While this part
of the tooth is never intended to be in contact, we do not actually remove it from
the gear. In interchangeable systems, it is cheaper to leave it as is. However,
in noninterchangeable teeth, we may remove all or part of this useless 0.047 in.
by one means or another. See the next chapter.

There is more than one approach to the trigonometry of this example. Also,
those who are concerned with repeated calculations of this sort would find advantage in a derived equation. Such a derivation follows, but it is hoped that the
beginning student will not use the result, because his aim should be to work with
the configuration until the various relationships are clearly in mind. To take a
different approach, consider the right triangle QI_1I_2, Fig. 117 (or Fig. 115).

$$QI_1 = r_{ag} = [(I_1I_2)^2 + (QI_2)^2]^{1/2}.$$
$$I_1I_2 = I_1P + I_2P = r_p \sin \phi + r_g \sin \phi = C \sin \phi.$$
$$QI_2 = r_g \cos \phi.$$
$$r_{ag} = (C^2 \sin^2 \phi + r_g^2 \cos^2 \phi)^{1/2}.$$

Multiply and divide the terms in the parentheses by P_d^2, use $CP_d = (N_1 + N_2)/2$
and $r_gP_d = N_2/2$, and get the maximum permissible radius of the addendum
circle for no interference as

$$r_{ag} = \frac{1}{2P_d} [(N_1 + N_2)^2 \sin^2 \phi + N_2^2 \cos^2 \phi]^{1/2},$$

in which we note that this radius is dependent only on the tooth numbers for a
given degree of involute and pitch, that is, for a given system of gears. The
maximum addendum then becomes $a_{\max} = r_{ag} - r_g$.

* The slide rule will not produce a useful answer here.

126. The Involute Rack.   The pitch surface of a rack, Fig. 118, is a plane and the pitch line (circle) is a straight line.   Thus, as a rack is driven by a pinion, or vice versa, it moves in a straight line (generally), and hence it must reciprocate with respect to the pinion.   The side (profile) of an involute rack tooth must be a plane because, as it moves

Courtesy Illinois Gear and Machine Co., Chicago

Fig. 118. Rack with Pinion.

through engagement with a pinion, this profile must, to satisfy the law of gearing, be normal at all times to the line of action, which is a straight line in involute gears.   It follows that the inclination (acute angle) of the side of the rack tooth with respect to the pitch line is equal to 90° minus the degree of involute of the system of which it is a member (Fig.

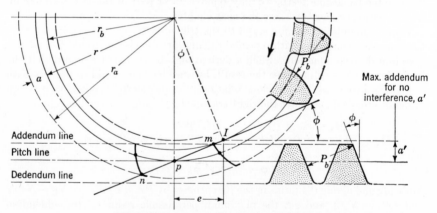

Fig. 119. Geometry of a Rack and Pinion.

119).   Point I, where the line of action is tangent to the base circle of the pinion, is the interference point on the pinion side.   Since the base circle of the rack is at infinity, there is never a problem of interference between the face of the pinion and the flank of the rack; the other interference point is at infinity and the distance of the final point of contact n (pinion driving) from the pitch point p is limited only by the pinion teeth becoming pointed. If the addendum on the rack is greater than a', the intersection of the rack's addendum line and the line of action will occur "outside" of I and interference exists. The initial (m) and final (n) points of contact are found by the same rules as before. The points needed to get the angles of approach and recess on the pinion are shown but not named.

Imagine the degree of involute ϕ decreasing and note that the interference point I moves toward p. If, for example, the addendum remains the same and ϕ is decreased to $14\frac{1}{2}°$ in Fig. 119, interference would exist. Also try to perceive that if ϕ decreases, but not enough to bring about interference conditions, the contact ratio increases. Finally, observe

Courtesy Illinois Gear and Machine Co., Chicago

Fig. 120. Internal Gear on Test.

that there is no change in the pressure angle and no change in the pinion's pitch circle (which marks the point on the profile in contact at the pitch point) as the pinion is moved toward or away from the rack (the pitch line of the rack moves to remain tangent to the pinion's pitch circle).*

The straight side of the involute rack tooth is one of the significant advantages of the involute system, inasmuch as it is easier to manufacture accurate plane surfaces. Such surfaces, in the form of either a rack or a hob (worm-type) cutter, can then be used to *generate* accurate involutes in cutting the teeth (§ 129).

127. Internal Gear and Pinion. An ***internal*** (or *annular*) ***gear*** has teeth cut on the inside of the rim, Fig. 120, with the tooth *space* on the internal gear much the same as the tooth itself on an external gear.

* One does not memorize these statements; they should be thought through and understood.

Study Fig. 121. In the illustration, the addendum circle of the internal gear does not cut the line of action because the end of its tooth extends inside of the base circle where there is no involute. Thus, that part of an internal gear tooth which extends inside of its base circle is useless and subject to interference unless modified. If the pinion is driving, the

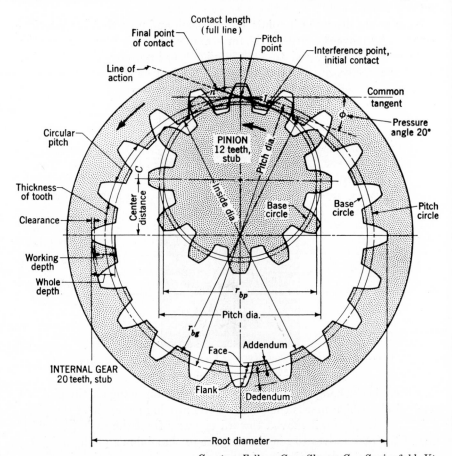

Courtesy Fellows Gear Shaper Co., Springfield, Vt.

Fig. 121. Nomenclature for Internal Gears.

usual situation, the initial point of contact is where the addendum circle of the driven (internal) gear cuts the line of action—except when there is no intersection. In the latter case, contact should not be made before the interference point I (tangent point of *pinion* base circle and line of action) is reached. As usual, the final point of contact is at the intersection of the line of action and the addendum circle of the driver. As in the case of a rack, the final point is only limited by the pinion teeth becoming pointed. Internal gears are in great favor in epicyclic gear trains (Chapter 9) because they are especially adaptable to some situations as in automatic automotive transmissions.

Compared with external spur gears, internal gears have several advantages when they can be used. For example:

1. For a particular velocity ratio, the internal gear arrangement is more compact, center distance less, Fig. 121.

2. Because the teeth travel more nearly in the same direction (§ 128), there is less sliding of the teeth on one another; this tends toward less wear and greater efficiency.

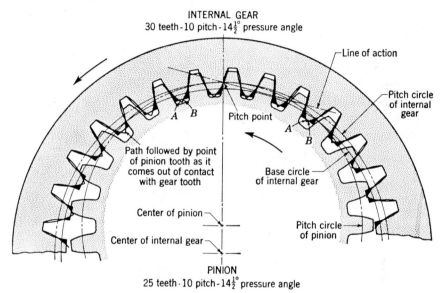

Courtesy Fellows Gear Shaper Co., Springfield, Vt.

Fig. 122. Fouling of Internal Gear Teeth.

3. Because the internal gear "wraps" around the pinion, instead of curving away, the contact ratio is greater than for the same tooth numbers on external gears; this results in a more gradual transfer of load from tooth to tooth and quieter running; also the teeth will have greater life for the same load or the same life for a greater load.

4. Because of the concave *surface* of the internal tooth, the actual surface in contact (*degree of osculation*) is increased; this factor also permits a greater load for the same surface stress and life.

When the pinion is large compared with the internal gear, the teeth may "foul" each other, Fig. 122, a kind of interference peculiar to internal gears. To avoid fouling, let the *minimum* differences in tooth numbers, $N_g - N_p$, be:

For $14\frac{1}{2}°$ full depth, 12 teeth;   for 20° full depth, 10 teeth;
for 20° stub, 8 teeth.

These minimum differences require some modification of the profiles. If no modification is desired, the difference in tooth numbers must be greater.

128. Sliding Velocity of Gear Teeth. At the instant the teeth are in contact at the pitch point, the contact points on the teeth have no relative motion along the line of centers and pure rolling (without slippage) exists. At any other point of contact, there is sliding, and the further the contact point is from the pitch point, the greater is the rate of sliding. Thus, maximum values of the sliding velocity occur at the initial and final points of contact. One could use any of the various principles of finding velocities previously explained, determine the velocity of the contact point in each gear, and obtain the sliding velocity from the components along the common tangent to the profiles at the point of contact. However, there is a much easier way.

If two external gears are in contact, their angular velocities are in opposite senses. From equation (t), § 95, which is $\omega_{B/A} = \omega_B \rightarrow \omega_A$, we see that the relative angular velocity of external gears 1 and 2 is the arithmetic sum of their angular velocities;

$$(k) \qquad \omega_{2/1} = \omega_2 - (-\omega_1) = \omega_2 + \omega_1,$$

[OPPOSITE ROTATION]

where vector signs are not needed because the rotations are coplanar. In Fig. 123, with the pinion driving CL, we find the initial and final points

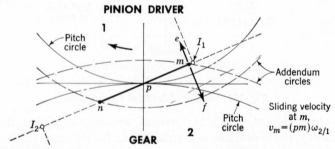

Fig. 123. Sliding Velocities. Notice that at m the radius of curvature of the driving pinion tooth is I_1m; for the driven tooth, it is I_2m. At n, the curvature of the pinion tooth is I_1n; of the gear tooth I_2n.

of contact m and n as usual. Now imagine one of the gears to be stationary, say the pinion 1; the entire relative angular motion of the gear is then $\omega_{2/1}$. Next recall that the relative motion of two points in contact is along the common tangent to the surfaces in contact and that in involute gearing this tangent is normal to the line of action. Thus, at m, Fig. 123, the relative motion is in the sense me or mf, depending on which gear is the reference body.

At the instant shown in Fig. 123, the centro of the gears is p (§ 113);

that is, the gear 2 is rotating about p and all points in it have velocities proportional to their distances from p (velocities with respect to pinion 1). Hence the velocity of m in 2 relative to 1 is $v_{m2/1} = (pm)\omega_{2/1}$, which is the sliding velocity at m. Since the same arguments apply at the final point of contact, we may say

(*l*)   Sliding vel. at initial point, $v_{m2/1} = $ (approach length) $\omega_{2/1}$,

(*m*)   Sliding vel. at final point, $v_{n2/1} = $ (recess length) $\omega_{2/1}$,

in which the approach and recess lengths, *respectively the radii of curvature of the involutes at the end points*, may be measured from a graphical layout, or computed as explained in § 123.

Courtesy Brown and Sharpe Mfg. Co.,   *Courtesy National Broach*
Providence, R.I.   *and Machine Co., Detroit*

Fig. 124. Cutters. (a) Form cutter. (b) Shaving cutter. The shaving cutter is made with many cutting edges on a profile as shown and is an accurate tool for finishing involute profiles.

Suppose the pinion, Fig. 123, turns at 8000 rad./min. with a velocity ratio of 2 and suppose the actual distance $pm = 2.05$ in. Then

$$v_{m2/1} = \frac{2.05}{12}\,(8000 + 4000) = 2050 \text{ fpm.}$$

If the system is an internal gear and pinion, the relative angular velocities subtract arithmetically, equation (*p*), § 94, because they are now in the same sense.

The complete interrelation of all factors which limit the power-transmitting capacity of gear teeth is not yet known. However, a criterion which has been found useful for very high-speed gears is the product of the surface compressive stress at the area of contact and the maximum sliding velocity, which is one reason besides purely kinematic reasons for wanting to know the sliding velocity.

129. Manufacture of Gear Teeth. Before continuing our discussion, it will be worthwhile to consider briefly the methods of cutting teeth. Gears may be cast with the teeth already molded, especially molded plastic gears and some exceptionally large and rough cast-iron gears; or they may be formed by extruding; or if the face width is small, they may be stamped accurately to form; but in most applications, gear teeth

Courtesy Fellows Gear Shaper Co., Springfield, Vt.

Fig. 125. Gear Shaper. The pinion cutter has teeth the same as on a like size gear except that the cutter's addendum is equal to the gear's dedendum in order to cut the clearance. First, the cutter is fed, without turning, into the blank to the proper depth. Then, the gear being cut and the cutter rotate together with pitch circles tangent and rolling (the spindles on which they are mounted are being driven intermittently by gears with the same gear ratio as that of the cutter-gear). Thus the tooth profile on the gear is *generated* as it is being cut by the reciprocating cutter. When the blank has made a complete revolution, all the teeth have been cut.

are cut. One method of cutting is to use a **form cutter** whose profile *abcd*, Fig. 124(a), has the same shape as the space between the teeth on the gear to be cut. At best, this tooth space only *approximates* the theoretical space because the same cutter is likely to be used for cutting gears with several different tooth numbers, and each time the number of teeth changes, the form of the tooth changes. See the basic rack, § 178, for the composite tooth form. These cutters are used for run-of-mill shopwork in milling machines or in special machines designed for milling gear teeth.   Because of the approximations involved, this method does not produce the most accurate gear teeth.

Especially at high speeds, where the dynamical effects are most pronounced, accurately cut teeth are quite advantageous.   Accuracy can be obtained by generating the involute surfaces as they are being cut. Since involutes run together properly, teeth may be generated by using a **pinion cutter** whose profiles are precise involutes, as in the Fellows gear shaper, Fig. 125.   Not only does this type of machine, properly utilized, cut the teeth more accurately than form cutters, but also it cuts more

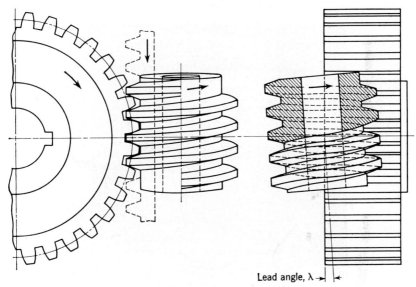

Lead angle, λ →| |←

Fig. 126. Principle of Hobbing Gear Teeth.   For straight teeth, the hob is mounted with the axis at the lead angle as shown so that the threads pass straight across the face of the gear.   Then the hob is fed slowly across the face of the blank while hob and blank are rotated at the proper velocity ratio for the required number of teeth. After the hob has been fed entirely across the face of the blank, all the teeth have been cut.

rapidly.   For best accuracy, rigidity of tool and work supports is important, and finishing cuts are on one side at a time.   Notice that if a standard cutter profile is a true involute (its tip may be somewhat relieved), it automatically undercuts the gears with small tooth numbers, thereby removing the interfering metal.   (Standard cutters can be so used as to avoid undercutting.)   This same shaper method is used to cut helical gears, § 133, by oscillating the proper cutter on a helical guide as it makes its cutting stroke.

Teeth may also be cut and generated with a **hobbing cutter** or **hob,** which is one shaped like a worm, Fig. 126, but with the threads gashed for the purpose of forming cutting edges, Fig. 127. In Fig. 126, the principle is illustrated with an ordinary worm. The profile of an involute worm, that profile cut by a plane normal to a helix, is straight-sided like a rack tooth of the same system. For this reason, the generation of the teeth is analogous to generating them with a rack cutter (see next paragraph). These straight sides can be manufactured accurately, and the accurate cutting tool can generate accurate involutes. A hob of a particular pitch cuts all interchangeable gears of that pitch, and it will undercut the flanks of gears whose working profiles extend inside their base circles, unless the tips of the hob teeth have been relieved in order to

Courtesy Barber-Colman, Rockford, Ill.

Fig. 127. Hobbing a Helical Gear. In this case, the axis of the hob is turned so that the cutting edges pass across the face of the gear in the direction of the teeth.

minimize or avoid this. Hobs can be used on standard milling machines or on special gear-hobbing machines, Fig. 127.

A *rack cutter,* Fig. 128, as the name suggests, is in the form of a rack, say three or more teeth, except that the addendum of the cutter is equal to the dedendum of the gear being cut in order to cut the clearance. Having straight sides, the cutter profiles may be made quite accurately. If unrelieved, except for perhaps a small tip radius, the rack profile undercuts teeth when theoretical interference exists. A single cutter cuts all interchangeable gears of a particular pitch. Maag gear-cutting machines operate on this principle.

There are other types of machines and other ways of manufacturing gears, for example, by extruding, but the foregoing ones serve our purpose. When accuracy is highly important, teeth are often given a finishing operation by broaching, shaving, or grinding. If teeth are to be shaved, a process of removing a thin layer of metal with multiple cutting edges in the form of a tooth, Fig. 124(b), it is recommended

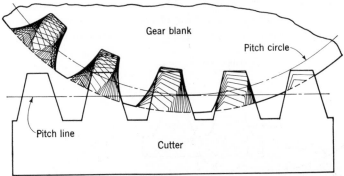

Fig. 128. Generating Action of a Rack Cutter. The cutter is first fed into the blank to the required depth—for interchangeable teeth, until the standard pitch line of the rack is tangent to the standard pitch circle of the gear. Then the pitch circle rolls on the pitch line through a small amount after each cut. Since the length of the rack cutter is limited, the rolling action continues for a short distance, say equal to the circular pitch, after which the cutter and work are repositioned for additional rolling after each cutting stroke.

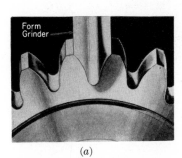

| (a) | (b) |

Courtesy The Gear Grinding Machine Co.,   *Courtesy Pratt & Whitney Co.,*
                Detroit   *West Hartford, Conn.*

Fig. 129. Finish Grinding. In (a) is seen the form-grinding process. The wheel in (b) has straight sides in the manner of a rack. Such a machine can be adjusted to finish-grind any shape of involute tooth, including helical gear teeth as seen.

that the clearance be increased to $0.35/P_d$ (dedendum = $1.35/P_d$—compare with § 118). If a very hard tooth surface, as a casehardened surface, is to be finished, grinding is necessary. It is done by form grinders, Fig. 129(a), or by a straight-sided grinding wheel that is made to generate the tooth involute while it is grinding, in accordance with the tooth action of a rack and pinion, Fig. 129(b).

130. The Cycloidal System. The family of cycloidal curves, since they can be made to satisfy the law of gearing, can be used for gear-tooth profiles. In fact, the cycloidal system was historically the first theoretically correct form to be used. These curves are generated by rolling a circle, called a *generating* or *rolling circle,* on a path, called the *directing* line or path; a point on the circle generates a curve on the plane contain-

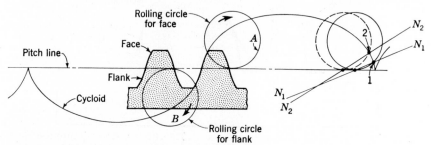

Fig. 130. **Forming a Cycloidal Tooth on a Rack.** Since the path of contact always follows the generating line or curve, the rolling circle in this case, we see immediately that the pressure angle is variable with the location of the point of contact. The normal to the curve at point 1 is N_1N_1; at 2, it is N_2N_2.

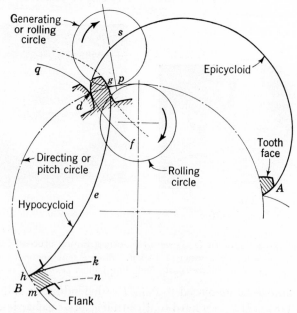

Fig. 131. **Forming Cycloidal Teeth on Gears.** If the outer rolling circle started at d, it would generate the epicycloid ds rolling toward the right and dq rolling toward the left. The face at g is from the epicycloid dq. Similarly, at B, the hypocycloid mn is the same as the one hk. The manner of putting these curves together is evident from the illustration.

ing the path. If the circle is rolling on a straight line, the curve generated is called simply a *cycloid,* Fig. 130. As seen, the rack tooth is
made up of parts of two cycloids, one for the face and one for the flank.
Observe that the profile is of double curvature and recall that the involute
rack tooth has straight sides—a disadvantage of the cycloid.

When the generating circle rolls on the *outside* of another circle, a
point on it generates an *epicycloid;* the profile of the *face* of a cycloidal
tooth is part of an epicycloid, as at *A*, Fig. 131.

When the generating circle rolls on the *inside* of another circle, a point
on it generates a *hypocycloid;* the profile of the *flank* is part of a hypocycloid, as at *B*, Fig. 131.

In order for cycloidal teeth to satisfy the law of gearing, it is essential
that the directing (pitch) circles be tangent and that the *flank and face in
contact* have been generated by the same size rolling circle. In Fig. 132,

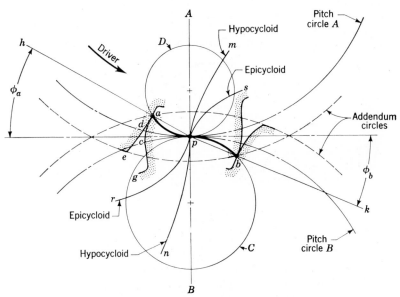

Fig. 132. Cycloidal Tooth Action. Notice that pressure angle ϕ_a is not equal to ϕ_b;
that the direction of the normal force along *hp* is not the same as that along *pk*.

the generating circle *D* simultaneously generates the epicycloid *ps* and
the hypocycloid *pm* on the planes of the pitch circles *B* and *A*, respectively, as these planes move, with the pitch circles and generating circle
all rolling on one another (the center of *D* remains stationary). It
follows that the face *ca* of the tooth on *B* should be taken from *ps* and
the flank *da* on *A* should be taken from *pm*. In a similar manner, the
epicycloid *pr* and the hypocycloid *pn* are generated simultaneously by

rolling circle C on A and B, forming, respectively, the face de on A and the flank cg on B.

As before, the initial point of contact a is at the intersection of the addendum circle B on the driven and the generating line (circle D). The final point of contact b is the intersection of the addendum circle A of the driver and the generating curve (circle C). To see that the law of gearing is satisfied, observe that p is the centro of circle D no matter which pitch circle it rolls on, so that each point in D is moving in a direction perpendicular to the radius of said point with respect to p. Thus at the instant pictured in Fig. 132, the curve ac at point a is normal to the straight line

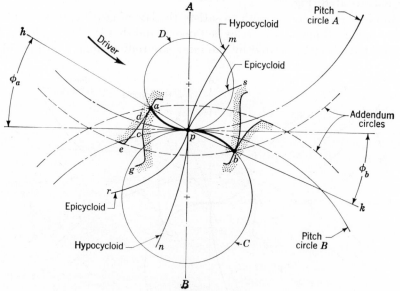

Fig. 132. Repeated.

ap and the curve ad at a is normal to ap; hence, a common normal to the profiles at point a passes through the pitch point p. The same reasoning applies to any other point of contact. Apprehend the following statements concerning cycloidal teeth.

1. The path of contact or line of action apb is curved.

2. Therefore, the pressure angle varies from a maximum ϕ_a at a to zero at p (when contact is at the pitch point), and then to another maximum at the final point of contact b. This is one of the significant disadvantages of the cycloidal system as compared with involute teeth, in which the path is a straight line and the pressure angle is constant. The consequence of the curved path is a varying separating force (the component of the tooth force in a vertical direction, Fig. 132), which is periodic and therefore conducive to vibration, noise, and wear.

3. The directing (pitch) circles must be exactly tangent in order to satisfy the law of gearing; otherwise there is a different normal to each curve at the point of contact. This characteristic constitutes a second impelling disadvantage of cycloidal teeth as compared to involute. The center distance of involute gears can be changed without affecting the velocity ratio (without destroying conjugate action), so that inaccuracies in mounting, or adjustments to provide more or less backlash, have no effect on the smoothness of operation of involute gears.

4. The faces and flanks in contact must have been generated by the same size rolling circle. It follows that for an interchangeable system, *all* faces and flanks must have been generated by the same size rolling circle.

5. If the *diameter* of the generating circle is equal to the *radius* of the directing (pitch) circle, the hypocycloid (flank of tooth) is radial. (Satisfy yourself by demonstration of this fact.)

6. If the *diameter* of the generating circle is *larger* than the radius of the directing circle, the corresponding tooth flanks appear undercut, but contact remains theoretically correct.

The outstanding advantage of cycloidal teeth is that there is no interference problem, no matter how few teeth are used. Thus cycloidal profiles are often used for gears with, say, three or four teeth.

Standard interchangeable systems (see item 4 above) have been based on the size of generating (rolling) circle which would result in radial flanks on a gear of 12 or 15 teeth. Let the standard diameter of generating circle be equal to the radius of the 12-tooth gear. The addendum and dedendum are the same as in the full-depth involute systems (§ 118). Methods of constructing cycloidal curves are given in the Appendix, § 181. It is believed that the reader can now decide upon the mating curves for a cycloidal internal gear and pinion.

131. Closure. Some points that might be emphasized in closing (see references **65, 66**) are: (1) The involute is completely defined by its base circle. (2) The pressure angle of two involutes in contact is a function of the center distance of the involutes' base circles. 3. An involute does not have a pitch diameter. When two involutes move in contact, the contact point on the line of centers (where the path of contact intersects the line of centers) is the centro and pitch point; the tangent circles at this point, with centers at the centers of the base circles, are called the pitch circles. (4) The velocity ratio of two involutes in contact is inversely as the radii of their base circles; it is independent of the center distance. Many other significant points have been made, but all cannot be repeated here.

The predominant advantages of involute teeth were mentioned in items 2 and 3, § 130. From a practical manufacturing point of view, the straight sides and plane surfaces of the involute rack tooth and also the single curvature of involute gear teeth (no change in the profile curve as in cycloidal teeth) are decisive. However, not only do actual gear-tooth profiles vary from true involutes because of inescapable manufacturing errors; but even if the profiles were perfect to start with, they would depart from perfection as soon as a load is applied because of the deflection of the tooth. In extreme cases, the effect of deflection is corrected for to some extent.

If a layout of teeth or of gears is made for a 1 diametral pitch, the various dimensions for other pitches and similar gears is obtained by dividing 1-diametral-pitch dimensions by the given P_d.

Since some engineers have spent years studying and developing gearing, the general textbook can only touch some of the more important points. Changes and developments are still taking place, and situations arise where standard interchangeable gears do not seem to be a good answer. Thus not only is the engineer often faced with difficulties in providing gears with enough capacity to transmit the required power, another phase of the design, but also he is confronted with intriguing situations which call for difficult decisions on the kinematics phase of the design involving the use of special tooth profiles (**141**).

PROBLEMS

Dimensions and Nomenclature

501. Two spur gears with $14\frac{1}{2}°$ full-depth involute teeth have a velocity ratio of 0.75 and the driver turns at 600 rpm. The driven gear has 36 teeth and $P_d = 3$. Determine (a) the number of teeth in the driver, (b) the center distance, (c) the circular pitch of each gear, (d) the tooth width on the pitch circle, (e) the outside (addendum) diameters, (f) the diameters of the base circles, (g) the rpm of the driven gear and the pitch-line speed of each, (h) the base pitch.

502. Two spur gears with 20° full-depth involute teeth have a velocity ratio of 3.2. The driven gear has 64 teeth and $P_c = 1\frac{1}{4}$ in. Determine (a) the number of teeth in the driver, (b) the center distance, (c) the diametral pitch of each gear, (d) the tooth width on the pitch circle, (e) the outside (addendum) diameters, (f) the diameters of the base circles, (g) the rpm of the driven gear and the pitch-line speed of each, (h) the base pitch.

503. The center distance of two 20° ASA stub-tooth gears is 9 in. and the velocity ratio is 0.5; $P_d = 2$; the driver turns 240 rpm. Determine (a) the diameter and tooth number for each gear, (b) the circular pitch and tooth width on the pitch circle, (c) the diameters of the addendum, dedendum, and base circles, (d) the rpm of the driven gear and the pitch-line speed of each, (e) the base pitch.

504. It is necessary to connect two shafts about 7.36 in. on centers with a velocity ratio of about 3.2. Using gears of 7 pitch, list 4 pairs of tooth numbers which approximate the desired conditions. If the center distance were the most important, which pair would you recommend; which pair if the velocity ratio is the more important?

505. The outside (addendum) diameter of a gear is measured to be closely 10.67 in. The gear has 30 teeth, 20° fulldepth. Compute (a) the diametral and circular pitches, (b) the diameters of the pitch, dedendum, and base circles, (c) the clearance and the base pitch. (d) The backlash should probably be between what two values?

506. The outside diameter of a gear is 7.5 in. It has 20° full-depth teeth, 4 pitch. Compute (a) the circular pitch and the number of teeth, (b) the diameters of the pitch, dedendum, and base circles, (c) the clearance and base pitch. (d) The backlash should probably be between what two values?

507. For a 4-pitch gear, calculate the addendum, dedendum, and clearance for the following interchangeable systems: (a) $14\frac{1}{2}°$ full depth; (b) 20° full depth; (c) ASA stub tooth; (d) Fellows stub tooth. Arrange answers in tabular form.

508. The same as **507** except that $P_d = 7$.

509. If the dedendum and base circles coincide in the $14\frac{1}{2}°$ involute full-depth system, how many teeth are in the gear?

510. The same as **509** except that $\phi = 20°$.

511. A gear has a base circle of 9.5 in. diameter and it is operating with a pitch circle 10.1 in. in diameter. What is the pressure angle?

512. If a $14\frac{1}{2}°$ involute full-depth tooth has an addendum of 0.1667 in., what is its base pitch?

513. The same as **512** except that $\phi = 20°$.

514. Given a 25-tooth $14\frac{1}{2}°$ full-depth involute gear; $P_d = 4$. (a) What is its base-circle diameter? Compute this and also solve by a graphical construction. (b) Now suppose that the teeth are 20° full depth. What is D_b (algebraic and graphical—same drawing)? (c) Compute the addendum and draw in the working-depth circle. Is it the same for each? (d) Is any part of the tooth profile outside of the working-depth circle not involute for the $14\frac{1}{2}°$ profile; for the 20° profile?

515. The same as **514** except that the number of teeth is 35.

Gear-tooth Action

516. Sketch the basic geometry of two gears in mesh, as in Fig. 117, and prove that the maximum radius of the addendum circle r_{ag} for no interference is

$$r_{ag} = [r_{bg}^2 + (C \sin \phi)^2]^{1/2},$$

where r_{bg} is the radius of the base circle on the gear.

517. Considering the geometry of two gears in mesh, say Fig. 115, show that the maximum addendum a_g on the gear that could be used without interference is given by

$$a_g = \frac{D_2}{2} \left\{ \left[1 + \frac{2N_1}{N_2} \sin^2 \phi \left(1 + \frac{N_1}{2N_2} \right) \right]^{1/2} - 1 \right\},$$

where the subscripts 1 and 2 refer to pinion and gear, respectively.

518. Suppose that the gear teeth are designed so that the contact begins at one interference point I_1, Fig. 115, and ends at the other I_2. Show that the distance $I_1 I_2 = [(D_1 + D_2)/2] \sin \phi$ and then prove that for continuous drive ($m_c \gtrless 1$), the sum of the tooth numbers must be

$$N_1 + N_2 \gtrless \frac{2\pi}{\tan \phi}.$$

519. Use plate No. 36, Wingren. The pinion on the upper shaft turns at 1000 rpm CC; $P_d = 3$; $m_\omega = 2$; $C = 6$ in.; the teeth are $14\frac{1}{2}°$ full depth. (a) Construct the pitch, addendum, dedendum, and base circles for each gear. Name the interference points I_1 and I_2. (b) Draw two teeth on each gear (away from the pitch point) by generating involute curves and approximating these curves with circular arcs. Locate pinion teeth close to the top of the sheet on the left. (c) Locate the initial and final points of contact and find the angles of approach, recess, and action. Indicate by crosshatching the interfering parts of the teeth. (d) Determine the contact ratio m_c in two ways, from measurements only. (e) What is the maximum addendum possible on each gear for no interfer-

ence, by measurement? (f) Draw a radial line from the axis of the pinion 30° above the horizontal toward the right (1st quadrant) and then perpendicular to this line, draw the pitch line of a rack. Construct 3 rack teeth and crosshatch the parts of the teeth that would be involved in interference. (g) On a separate sheet, find the following by computation only: radii of base circles; base pitch; length of action; contact ratio; the maximum addendum on the gear for no interference; the maximum addendum on the rack for no interference; the sliding velocity at the initial and at the final points of contact.

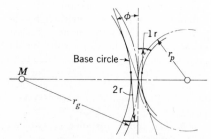

Fig. P123. Problems 520-531.

520. A 4.333-in. pinion with 20° full-depth teeth and with $\omega = 1000$ rad./min. drives a 10-in. gear; $P_d = 3$. In Fig. P123, locate M (2, 4) LDH and solve the problem by a graphical layout. (a) With the pinion rotating clockwise, locate the initial and final points of contact and state the length of action. Name the interference points. (b) Using a single arc for the profile of each gear, 1 in. radius on the pinion and 2 in. on the gear, determine the angles of approach, recess, and action. (c) Determine the contact ratio by two methods: from the angle of action and from the measured length of action. (d) Draw one tooth on the gear and indicate by crosshatching that part which is subject to interference. What maximum addendum (as measured on the drawing) could the gear have without causing interference? If the pitch were 9 instead of

3, would there be any interference? Explain. (e) What is the radius of curvature of the tooth profile on the pitch circle?

521. The same as **520** except that $\phi = 14\frac{1}{2}°$. See the Appendix for radii for these tooth profiles.

522. A 4.333-in. pinion with 20° full-depth teeth drives a 10-in. gear; $P_d = 3$ (same gears as in **520**). Without using any graphical answers, compute (a) the length of action and contact ratio, (b) the maximum addendum that the gear could have without interference, (c) the maximum sliding velocity between the teeth during approach and during recess.

523. The same as **522** except that $\phi = 14\frac{1}{2}°$.

524. A 4.333-in. pinion with full-depth teeth drives a 10-in. gear; $P_d = 3$. What minimum pressure angle should be used if interference is to be absent? (Compare data with **520**.)

525. The same as **524** except that $D_g = 20$ in.

526. A 4-in. 3-pitch pinion with full-depth $14\frac{1}{2}°$ involute teeth drives an 8-in. gear. These gears may be located, Fig. P123, by M (3, 4) LDH. Solve by graphical means, answering questions (a) to (e) in **520**. See Appendix for reasonable radii of profiles.

527. The same as **526** except that $\phi = 20°$. From your knowledge of the properties of an involute decide upon reasonable radii for drawing the profiles.

528. A 4-in. 3-pitch pinion with full-depth $14\frac{1}{2}°$ involute teeth drives an 8-in. gear (same gears as in **526**). Without using any graphical answers compute (a) the length of action and contact ratio, (b) the maximum addendum that the gear could have without interference, (c) the maximum sliding velocity between the teeth during approach and during recess.

529. The same as **528** except that $\phi = 20°$.

530. A 4-in. pinion with full-depth teeth drives an 8-in. gear; $P_d = 3$. What minimum pressure angle should be used if interference is to be absent? (Compare data with **526**.)

531. The same as **530** except that $P_d = 5$.

532. A 25-tooth 5-pitch $14\frac{1}{2}°$ involute pinion engages a 40-tooth gear. If the standard pitch circles are separated 0.05 in., what is the actual pressure angle?

533. The same as **532** except that the separation is 0.1 in.

534. A gear has 30 teeth of 3 pitch. What minimum number of 20° full-depth interchangeable teeth may the pinion have without interference? (a) Solve graphically. (b) Solve mathematically.

535. The same as **534** except that the minimum number of $14\frac{1}{2}°$ full-depth teeth is desired.

536. What is the smallest number of full-depth teeth that two equal gears may have without interference if the degree of involute is 20°?

537. The same as **536** except that $\phi = 14\frac{1}{2}°$.

538. The same as **536** except that the teeth are ASA stub.

539. It is desired to have a contact ratio of 2 for the case of a velocity ratio of 1. Let the design be such that the teeth are just long enough to reach the interference points, which then become the initial and final points of contact. (a) Lay out these gears and derive an equation which gives the minimum number of teeth required. (b) Compute this number for $14\frac{1}{2}°$ and for 20° involute teeth. What diameters would you use for $P_d = 8$?

540. It is desired to have a contact ratio of nearly 2 and it is therefore decided to have an approach length $Z_a = P_c$. Assume that contact begins at the interference point I_1 (and the gear tooth does not extend beyond I_1). Let $P_d = 5$. If the velocity ratio is $2\frac{1}{3}$ and the teeth are full-depth interchangeable-type $14\frac{1}{2}°$ involute, how many teeth may be used on each gear? Lay out this problem for a graphical solution and also, setting up an appropriate equation, compute N_p. What is the contact ratio?

541. The same as **540** except that $\phi = 20°$.

542. A 45-tooth gear is in mesh with a 15-tooth pinion. (a) If $\phi = 14.5°$ and $P_d = 3$, what maximum addendum may the gear have without extending past the interference point? (b) How long is the interfering part? (Do not substitute in an equation, but show detailed solution.)

Ans. (b) 0.052 in.

543. A 12-tooth full-depth $14\frac{1}{2}°$ involute pinion is driving a 30-tooth gear; $P_d = 10$. Set up the necessary equations and compute (a) the maximum outside diameter of the gear for no interference, (b) the standard interchangeable outside diameter, (c) the radial length of tooth that is interfering.

Ans. (c) 0.056 in.

Rack and Pinion

544. (a) Using Fig. P124, determine the minimum pressure angle ϕ that may be used if a 15-tooth pinion meshes with a rack without interference. (b) The same as (a) except the pinion has 25 teeth. (c) Are either of these angles standard for interchangeable gears?

Ans. (a) 21.4°.

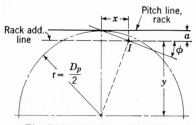

Fig. P124. Problems 544–549.

545. (a) Show that the minimum number of interchangeable full-depth teeth on a pinion to mesh with a rack without interference is given by $N_p = 2/\sin^2 \phi$. See Fig. P124. (b) The same as (a) except that the teeth are ASA stub and the minimum $N_p = 1.6/\sin^2 \phi$.

546. Show that the minimum number of interchangeable teeth on a pinion to mesh with a rack without interference is

given by $N_p = 2P_d a/\sin^2 \phi$, where a is the addendum on the rack.

547. For a 2.5-in. 20° involute pinion driving a rack, compute the maximum addendum a for the rack if its teeth do not extend beyond the interference point. To what full-depth pitch does this a correspond? What is the corresponding number of teeth on the pinion? See Fig. P124.

548. The same as **547** except that the degree of involute is $14\frac{1}{2}°$.

549. The same as **547** except that the teeth are ASA stub.

550. An 18-tooth pinion of 2 pitch is driven by a rack. The teeth are $14\frac{1}{2}°$ full depth. (a) For the pinion, find the size of the base circle, the angles of approach, recess, and action, the base pitch, and the contact ratio. (b) What maximum addendum can be used on the unmodified rack profile without interference? (c) For a check, compute the diameter of the base circle, the contact ratio, and item (b). (d) What is the maximum sliding velocity between the teeth when the pinion has an angular velocity of 150 rad./min.?

551. The same as **550** except that $\phi = 20°$.

552. The same as **550** except that the teeth are ASA stub.

Internal Gears

553. A 4-in. ($= 2r_p$) 4-pitch pinion with full-depth 20° involute teeth drives a 10-in. *internal* gear, Fig. P125. For each gear, find by a graphical solution (a) the size of the base circle, (b) the angles of approach, recess, and action, and (c) the contact ratio. (d) Is the interference condition present? If so, how much of the gear tooth is involved? If the pitch were 10 instead of 4, would interference occur? Explain. (e) For a check, compute the base-circle diameters and the contact ratio. (f) Compute the maximum sliding velocity during both approach and recess when the angular velocity of the pinion is 400 rad./min.

554. The same as **553** except that $\phi = 14\frac{1}{2}°$.

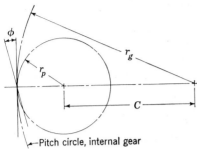

Fig. P125. Problems 553-557.

555. The same as **553** except that the teeth are ASA stub.

556. The same as **553** except that the gear is 14 in. in diameter.

557. If a 4-in. 4-pitch pinion with full-depth teeth drives a 10-in. internal gear (see **553**), Fig. P125, what minimum pressure angle is needed to avoid interference? Solve graphically and algebraically.

558. A 6-in. 5-pitch pinion drives a 20-in. gear. The teeth are full depth, 20° involute. The pinion turns at 4000 rad./min. Without using answers obtained graphically, compute the maximum sliding velocity between teeth (a) if the gear is an internal gear, (b) if the gear is a spur gear. Use sketches to make your solution clear.

Cycloidal Gears

559. What should be the diameter of the generating circle in the standard 12-tooth cycloidal system for each of the following gears: (a) $N = 30$, $P_d = 4$; (b) $N = 80$, $P_d = 20$; (c) $D = 20$ in., $P_d = 5$ in.; (d) $N = 25$, $P_c = \frac{1}{2}$ in.; (e) $P_c = 1.25$ in.?

560. The same as **559** except that the generating circle is sized according to the 15-tooth system.

561. An 18-tooth 6-in. pinion drives a 9-in. gear. If the cycloidal teeth are based on the 15-tooth system, determine the initial and final points of contact, the line of action, and the pressure angles at the initial and final points.

562. The same as **561** except that the

cycloidal teeth are based on the 12-tooth system.

563. A 40-tooth gear of $\frac{3}{8}$ in. pitch is in mesh with a rack. If the cycloidal teeth are based on the 15-tooth system, determine the initial and final points of contact, the line of action, and the pressure angles at the initial and final points.

564. Show that if the diameter of the generating circle is equal to the radius of the directing circle, the flank of a cycloidal tooth will be radial.

565. The flanks and faces of 3 cycloidal gears have been generated by rolling circles as follows: gear A, face by a 2-in. circle, flank by a 3-in. circle; gear B, face by a 3-in. circle, flank by a 2-in. circle; gear C, face by a 2-in. circle, flank by a 3-in. circle. Which gears will run together in conjugate fashion?

Larger Paper Needed

566. A 14-in. pinion with $14\frac{1}{2}°$ full-depth teeth of 1 pitch turns 450 rad./min. while driving a gear with a velocity ratio of 15/7, Fig. P126; M (9, 3) for 11 × 17 paper. Use scale of 3 in. = 1 ft. (a) Draw at least 2 teeth on the gear by any convenient method. (Do not draw any teeth in the vicinity of the pitch point. If sectional view is omitted, put teeth entirely on left side of gear.) (b)

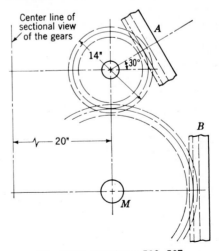

Fig. P126. Problems 566, 567.

Draw 2 teeth on pinion by approximating an involute of your own construction with a circular arc. (c) Locate the initial and final points of contact. Determine and record the values of the base pitch, length of action, angles of approach, recess, and action. Compute the contact ratio in two ways. (d) Draw a portion of a rack engaging the gear and a portion engaging the pinion. Is there interference between the rack and gear? (e) Indicate all interfering tips of teeth by crosshatching. (f) If it is desired to complete a working drawing with sectional view, use the following dimensions: face width = 5 in.; pinion shaft = 3 in.; pinion key = $\frac{1}{2} \times \frac{3}{4}$ in.; pinion teeth cut from solid bar; gear shaft = 4 in.; hub diameter = $7\frac{1}{2}$ in.; gear key = $\frac{3}{4} \times 1$ in.; gear rim = 1 in.; gear bead = 1 in.; gear hub length = $5\frac{3}{4}$ in.; 6 arms in gear, elliptical section, $1\frac{1}{2} \times 3$ in. at center of shaft, $1\frac{1}{8} \times 2\frac{1}{4}$ at the pitch circle. (g) Without the use of graphical answers, compute (on separate paper) the contact ratio for the pinion and its rack and the maximum sliding velocity during approach and during recess for the two gears.

567. The same as **566** except that $\phi = 20°$.

568. The gears to be drawn, Fig. P127, have $P_d = 3$, $14\frac{1}{2}°$ full-depth teeth, $m_\omega = 2$, $D_p = 4$ in.; M (5, 6) for 11 × 17 paper; space scale, full size. Answer items (a) to (g) in **566** except part (f).

569. The same as **568** except that $\phi = 20°$.

570–620. These numbers may be used for other problems.

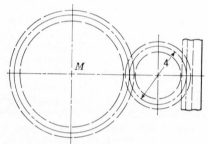

Fig. P127. Problems 568, 569.

8

Helical, Bevel, and Worm Gearing

132. Introduction. There are many forms of gears other than spur gears, too many for all of them to be described here. The other principal types of gears are helical, bevel, and worm gears, which we shall cover briefly in this chapter.

133. Helical Gears. All elements that form the face and flank of a spur gear tooth are parallel to the axis of the gear. The elements which form the surface of a helical gear tooth are cylindrical helices; one end of the tooth is advanced circumferentially over the other end, Fig. 133. As a result, the leading end comes into contact first, the trailing end last, with the tooth thus picking up the load gradually. Contrast this action with that in spur gears, in which the entire face width theoretically makes contact at the same instant. If the leading end of a helical tooth is still in contact when the trailing end comes into contact, the line of contact across the tooth's surface is diagonal, reaching from a point high on the face at one end to some point low on the flank at the other end of the tooth. As a consequence of the sloping contact line, the maximum bending moment on a helical gear tooth is only a little greater than half that on the same size spur gear tooth—assuming the same load on each tooth. It follows that on the basis of strength, a helical tooth has a greater power-transmitting capacity than a spur gear tooth of the same pitch and face width. Another important advantage of helical gears is that there is *always more than one* pair of teeth engaged—not true of straight tooth gears.

Because of the greater strength and smoother engagement, as explained above, helical (and herringbone) gears are commonly used for heavy-duty gear boxes with parallel shafts, Figs. 133 and 134.

225

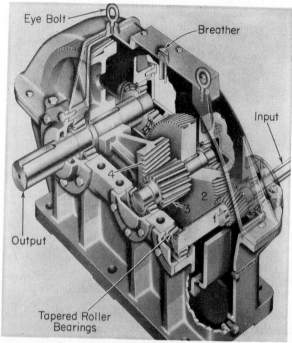

Eye Bolt

Breather

Input

4

2

3

1

Output

Tapered Roller
Bearings

Courtesy Westinghouse Electric Corp., Pittsburgh

Fig. 133. Helical Gears. These teeth are hobbed after heat-treatment; full depth, unequal addendum and dedendum teeth. The small gears are cut from wrought-steel blanks; the large ones are of heat-treated cast steel.

Input

Courtesy Farrel-Birmingham Co., Inc., Ansonia, Conn.

Fig. 134. Gearbox. This gearbox has double helical or herringbone gears (§ 135). It is the reduction unit between an electric motor and a heavy-duty mixing machine. Notice the spray lubrication near the meshing teeth.

The **helix angle** ψ, Fig. 135, is the angle between a tangent to the pitch helix and an element of the pitch cylinder. (In full, this is the *pitch helix angle*, but without a modifying adjective, pitch helix angle is intended.) The helix angle for a particular *lead*, § 142, varies with the size of the cylinder; for example, the helix on the pitch cylinder is different from the one on the base cylinder. *The* pitch of a helical gear is the pitch in a plane of rotation or a diametral plane, which is not the same as the pitch in a plane normal to the pitch helix. The diametral plane is

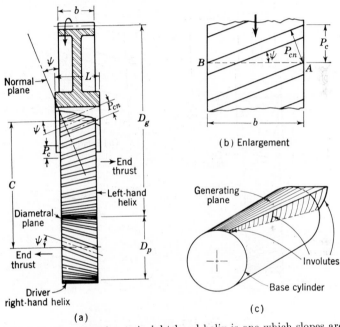

Fig. 135. Helical Gear Notation. A right-hand helix is one which slopes around the cylinder in the manner of a right-hand screw thread; that is, the tooth on the front side of the gear slopes upward toward the right with the *axis vertical*. A plane whose edge is cut at an angle with an element of the base cylinder generates the helical tooth profile as suggested in (c).

also called the transverse plane. To distinguish between the pitches, let P_c and P_d be the circular and diametral pitches, respectively, in the transverse plane, P_{cn} and P_{dn} be the pitches in the normal plane. Notice in Fig. 135 that the normal circular pitch is smaller than the transverse circular pitch and that the relation between them is

(a)
$$P_{cn} = P_c \cos \psi,$$

from which it follows that $(P_c = \pi/P_d, \ P_{cn} = \pi/P_{dn})$

(b)
$$P_{dn} = \frac{P_d}{\cos \psi},$$

where ψ is the helix angle. If N is the number of teeth and D is the pitch diameter, we have the following relations:

(**c**) $\quad P_{cn} = P_c \cos \psi = \dfrac{\pi D \cos \psi}{N}$ $\quad$ and $\quad P_{dn} = \dfrac{P_d}{\cos \psi} = \dfrac{N}{D \cos \psi}.$

Shaper cutters, Fig. 125, for helical gears result in the pitch being standard in the transverse plane. However, if helical gears are manufactured with standard hobs, the *normal pitch* is standard. In the latter event,

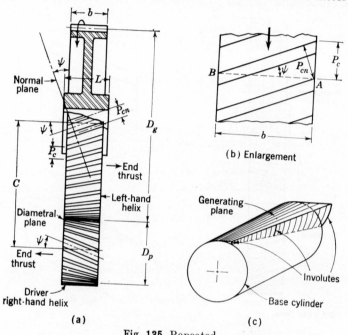

(b) Enlargement

(a)

(c)

Fig. 135. Repeated.

the pitch in the transverse plane P_d contains a decimal fraction. For example, for $N = 20$ teeth, $P_{dn} = 6$, and $\psi = 23°$, the pitch and pitch diameter are

$$P_d = P_{dn} \cos \psi = (6)(\cos 23°) = (6)(0.9205) = 5.523,$$
$$D = \frac{N}{P_d} = \frac{20}{5.523} = 3.621 \text{ in.}$$

As seen in Fig. 135(b), if the leading end B of one tooth is initiating contact at the same instant that the trailing end A of the preceding tooth is ending contact, then $P_c/b = \tan \psi$. However, in practice, some additional overlapping of contact is provided and b is increased a minimum of about 15%, or

(**d**) $$b_{\min} = \frac{1.15 P_c}{\tan \psi} = \frac{1.15 P_{cn}}{\sin \psi}.$$

Notice that for the shafts of helical gears to be parallel, the helix on one gear must be of opposite hand to that of the other and the helix angles ψ must be the same in magnitude. Apply the screw-thread rule to determine whether the helix is right- or left-handed; see Fig. 135. Helical gears are inherently noninterchangeable; hence there is no reason to use standard interchangeable tooth proportions except as they are convenient in cutting. There is no standard helix angle but it is usually between 15° and 25°. The larger the helix angle, the greater the axial thrust, Fig. 135, and caring for this force involves the bearing design, which may present difficulties for large power transmission and too large ψ.

The velocity ratio of two gears *of any kind* is inversely as the number of teeth, and for helical gears on parallel shafts, inversely as the diameters.

134. Normal Pressure Angle for Helical Gears. The relation between the pressure angle in the normal plane and that in the plane of rotation is obtained from the geometry of the teeth. Figure 136 is intended to help

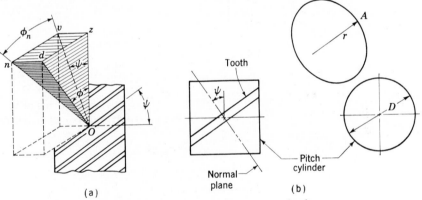

Fig. 136. Relation between Pressure Angles.

in visualizing the space relations involved; zOv is in a vertical plane; zOd is in a vertical plane at right angles to zOv; On is normal to the profile at the mid-pitch point; ϕ is the pressure angle in the diametral plane; ϕ_n is the pressure angle in the normal plane. Thus (angle $nvO = 90°$)

$$(e) \qquad \tan \phi_n = \frac{nv}{vO} = \frac{nv}{zO} \times \frac{zO}{vO} = \frac{dz}{zO} \times \frac{zO}{vO} = \tan \phi \cos \psi,$$

from which the normal pressure angle can be computed from the transverse pressure angle, or vice versa, if the helix angle is known. Note that the normal pressure angle is always less than the transverse pressure angle.

The shape of the tooth in the normal plane depends upon the radius of curvature of the pitch surface in that plane. A plane oblique to the axis of a cylinder cuts across an elliptical area, Fig. 136(b), and the radius of curvature of the ellipse at A is the one desired. The form of the tooth is then the same as that of a tooth on a pitch circle of radius r, Fig. 136(b), with the pitch and degree of involute as in the normal plane. This imaginary gear of radius r is called an equivalent gear (for form purposes) and the corresponding number of teeth N_v is called the **virtual** or **forma-tive** (sometimes *equivalent*) number of teeth. See also § 139. From the mathematics of an ellipse (or from a handbook), we find the radius of

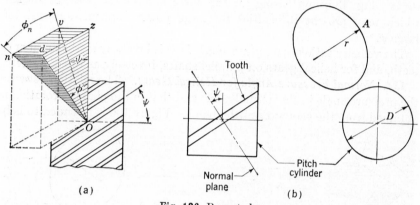

Fig. 136. Repeated.

curvature at the end of a minor axis to be $r = D/(2 \cos^2 \psi)$ in terms of the symbols in Fig. 136(b). The corresponding (virtual) number of teeth is $P_d(2r)$, or

$$(28) \qquad N_v = P_{dn} \frac{D}{\cos^2 \psi} = \frac{P_d}{\cos \psi} \frac{D}{\cos^2 \psi} = \frac{N}{\cos^3 \psi},$$

where $N = P_d D$ is the actual number of teeth in the gear. It is N_v which would suggest whether or not interference effects were involved. Think of a short cylinder with a helix of small helix angle, say Fig. 135. Then imagine the helix angle increasing until it is closer to 90° than zero. At some value, the helix will make a complete turn on the cylinder, which is the situation in the case of a worm, § 142. At any rate, if a helical gear has a large helix angle, as often happens for nonparallel drives, § 136, the actual tooth numbers may be small without undercut or radial flanks. For example, let $N = 5$ and $\psi = 60°$ ($\cos 60 = 0.5$); then $N_v = 5/0.5^3 = 40$ teeth. The actual tooth action (**66**) is such that this computation does not reveal for certain whether or not interference conditions exist; yet it is somewhat reassuring that the form of the tooth in the normal plane is the same as that of a 40-tooth gear.

135. Herringbone Gears. In order to avoid the disadvantages of the axial thrust which accompanies a single helix, gears with both right- and left-hand helices, called *herringbone gears,* are quite common. See Figs. 134, 137, and 138.   In order to divide the load equally between the two parts of the herringbone gear, one shaft is mounted so that it floats in an axial direction.   Since the thrust is approximately equalized, the helix angles used for herringbone gears are larger than for helical gears— 30° to 45°.   This larger angle permits more overlapping of tooth contact

Courtesy Farrel-Birmingham Co., Inc.,
Ansonia, Conn.

Fig. 137. Sikes Herringbone Gear. This gear is over 10.5 ft. in diameter. The cutting edge of a shaper type of cutter, seen in inset, moves up to the center of the gear blank, leaving no unused width of blank. The standard pressure angle for these cutters is 20° in the transverse plane.

Courtesy Lukenweld, Inc.,
Coatesville, Pa.

Fig. 138. Herringbone Gear. This is a welded gear blank, wrought-steel parts. Except for the Sikes machines for cutting herringbone gears, Fig. 137, a space around the center of the gear as seen here is necessary for clearance of the cutting tool. These teeth may be shaper-cut or hobbed.

and hence quieter and smoother operation.   The minimum face width is $b_{\min} = 2.3P_c/\tan \psi$; compare with equation (*d*).

136. Crossed Helicals.   Helical gears on nonparallel shafts are called *spiral gears* or *crossed helical gears,* Fig. 139.   The tooth action of these gears is fundamentally different from that when the helical gears are on parallel shafts.   Contact occurs at a point, theoretically, versus a line for the other gears studied so far.   Moreover, if one visualizes the gears running together, one will note that the teeth slide *across* one another (the only sliding previously discussed for gear teeth is up and

down on the profile). Because of the small contact area, the capacity of
these gears is small and hence they are most frequently used where the
power transmitted is inconsequential, as the drive for the distributor for
an internal-combustion engine. Because of the friction of the cross
sliding, their efficiency is relatively low.

An essential condition to be met in meshing spiral gears is that the
gears have *the same normal pitch*. The shaft angle Σ depends upon the

Courtesy Fellows Gear Shaper Co., Springfield, Vt.

Fig. 139. Crossed Helical Gears.

helix angles and whether or not the hands of the meshing gears are the
same or opposite:

$$(f) \qquad \Sigma = \psi_1 + \psi_2 \qquad \text{or} \qquad \Sigma = \psi_1 - \psi_2,$$
$$\text{[SAME HANDS]} \qquad\qquad \text{[OPPOSITE HANDS]}$$

where ψ_1 and ψ_2 are the pitch helix angles of gears 1 and 2 to be meshed.
Equation (*c*) applied to gears 1 and 2 ($P_{cn1} = P_{cn2} = P_{cn}$) gives

$$P_{cn} = \frac{\pi D_1 \cos \psi_1}{N_1} = \frac{\pi D_2 \cos \psi_2}{N_2}.$$

Using $P_{cn} = \pi/P_{dn}$ and solving for D_1 and D_2, we get

$$(g) \qquad D_1 = \frac{N_1}{P_{dn} \cos \psi_1} \qquad \text{and} \qquad D_2 = \frac{N_2}{P_{dn} \cos \psi_2},$$

from which the center distance $C = (D_1 + D_2)/2$ is

$$(h) \qquad C = \frac{1}{2P_{dn}} \left(\frac{N_1}{\cos \psi_1} + \frac{N_2}{\cos \psi_2} \right).$$

If it is desired to design a pair of helical gears for a particular velocity
ratio $m_\omega = N_2/N_1$, use $N_2 = m_\omega N_1$ in equation (*h*) and get

$$(28a) \qquad C = \frac{N_1}{2P_{dn}} \left(\frac{1}{\cos \psi_1} + \frac{m_\omega}{\cos \psi_2} \right).$$

From this equation, we see that for a particular center distance C and
velocity ratio m_ω, the tooth numbers and the helix angles are interde-
pendent. One of the helix angles can be eliminated by equation (*f*), say
$\psi_2 = \Sigma - \psi_1$, so that if the shaft angle is fixed, the unknowns are N_1 and

ψ_1. Since there must be a whole number of teeth, one could assume a reasonable helix angle ψ_1, say in the vicinity of $\Sigma/2$ for same hands, and solve for N_1. (If the helix angles happen to be equal, the crosswise sliding is minimized.) Then if this number is reasonable, take a nearby whole number and solve for ψ_1 by trial. If repeated designs are necessary, a quicker solution has been devised by Saari (**79**) for $\Sigma = 90°$. Extreme accuracy in meeting the conditions of equation (**g**) is seldom essential. Another plan to meet the requirements of a prescribed center distance is to make the final adjustment by having one blank oversize or undersize and then using noninterchangeable teeth (**80, 141**). If an exact center distance is not required, the solution is relatively simple.

It is important to note that the velocity ratio $m_\omega = N_2/N_1$ is *not* inversely proportional to the diameters unless $\psi_1 = \psi_2$. From equations (**g**), we have

(**i**)
$$m_\omega = \frac{N_2}{N_1} = \frac{D_2 \cos \psi_2}{D_1 \cos \psi_1}.$$

137. Sliding Velocity, Crossed Helicals. The absolute motion of the point of contact P on the pitch cylinder of each gear, Fig. 140, is of course peripheral for each gear (tangent to a pitch circle); $v_1 = \pi D_1 n_1$ and $v_2 = \pi D_2 n_2$. The vector difference of those velocities (relative velocity) is the sliding velocity,

(**j**)
$$v_{1/2} = v_1 \rightarrow v_2.$$

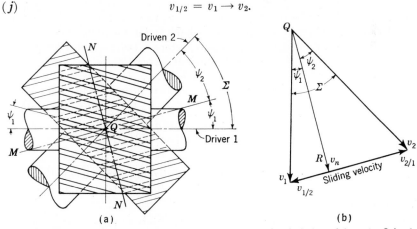

(a)   (b)

Fig. 140. **Sliding in Crossed Helical Gears.** The gear 2 is below driver 1; Q is the point of contact on the underneath side of 1. The shaft angle Σ is measured on the plane projection as shown. Visualize the teeth on top of 1 passing to the bottom side and note that the line MM is tangent to the helix on the bottom of gear 1; line NN is normal to the helices at Q. The velocities at Q are represented in (b), wherein the condition to be satisfied is that v_1 and v_2 have the same component $v_n = RQ$ in the normal direction. (The top lands of the teeth are represented only in a conventional manner by parallel straight lines.)

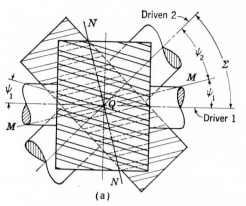
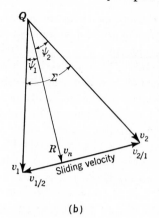

Fig. 140. Repeated.

As seen in Fig. 140, the components of the absolute velocities in a direction normal to the helices at Q, namely, v_n, must be the same; otherwise the teeth would overlap or separate. This condition may be used to find v_1 or v_2 if one of them is known. Let v_1 be known completely; v_2 is known to be normal to the shaft of 2; and $v_{1/2} = -v_{2/1}$ is known to be in a direction tangent to the pitch helix at Q; these conditions define the sense and magnitude of v_2.

138. Example—Crossed Helicals. Two spiral gears with a shaft angle of $\Sigma = 90°$ are to have a velocity ratio $m_\omega = 2$ and a center distance of $C = 4$ in.; $n_1 = 1800$ rpm. The gears are of the same hand and it is expected that a normal pitch of $P_{dn} = 8$, 20° full depth, will be satisfactory. Determine (a) N_1, N_2, and the helix angles, (b) the pitches and the pitch diameters, (c) the pressure angles in the transverse planes, and (d) the sliding velocity.

Solution. (a) First assume values of ψ_1 and ψ_2, say $\psi_1 = \psi_2 = 45°$, and solve for N_1 from equation (28a), using $N_2 = m_\omega N_1$ and $\cos \psi_2 = \cos (90 - \psi_1) = \sin \psi_1$:

$$N_1 = \frac{2P_{dn}C}{\dfrac{1}{\cos \psi_1} + \dfrac{m_\omega}{\cos \psi_2}} = \frac{(2)(8)(4)}{1.414 + (2)(1.414)} = 15.1;$$

try $N_1 = 15$ teeth. Then $N_2 = (2)(15) = 30$ teeth. Arrange equation (28a) as follows and solve by assuming values of ψ_1 until the left-hand side equals the right-hand side.

$$\frac{1}{\cos \psi_1} + \frac{m_\omega}{\sin \psi_1} = \frac{2P_{dn}C}{N_1} = \frac{(2)(8)(4)}{15} = 4.267.$$

Assume $\psi_1 = 58.7°$; $\cos 58.7° = 0.52$; $\sin 58.7° = 0.855$.

$$\frac{1}{\cos \psi_1} + \frac{2}{\sin \psi_1} = \frac{1}{0.52} + \frac{2}{0.855} = 4.264,$$

which is suspiciously close for slide-rule answers; therefore $\psi_1 = 58.7°$; $\psi_2 = 31.3°$; N_1 and N_2 are suitable as assumed.

(b) The diametral pitches in the plane of rotation are

$$P_{d1} = P_{dn} \cos \psi_1 = (8)(0.52) = 4.16, \qquad P_{d2} = (8)(0.855) = 6.84.$$

The pitch diameters are

$$D_1 = \frac{N_1}{P_{dn} \cos \psi_1} = \frac{15}{(8)(0.52)} = 3.61 \text{ in.}$$

$$D_2 = \frac{N_2}{P_{dn} \cos \psi_2} = \frac{30}{(8)(0.855)} = 4.39 \text{ in.}$$

These values check the center distance $C = 4$ which is given.

(c) The pressure angles are obtained from Fig. 136,

$$\tan \phi_1 = \frac{\tan \phi_n}{\cos \psi_1} = \frac{\tan 20°}{\cos 58.7°} = \frac{0.3645}{0.52} = 0.701, \qquad \phi_1 \approx 35°.$$

$$\tan \phi_2 = \frac{\tan 20°}{\cos 31.3} = \frac{0.3645}{0.855} = 0.426, \qquad \phi_2 \approx 23°.$$

(d) From Fig. 140, we see that the sliding speed is

$$v_{1/2} = v_1 \sin \psi_1 + v_2 \sin \psi_2$$
$$= \frac{\pi(3.61)(1800)}{12} (0.855) + \frac{\pi(4.39)(900)}{12} (0.52) = 1987 \text{ fpm.}$$

The student should check this answer by a graphical layout similar to Fig. 140 because it will be helpful in visualizing the relationships involved.

139. Bevel Gears. *Bevel gears* are ones in which the pitch surface is a cone, called the *pitch cone,* Fig. 142. They are used to connect non-

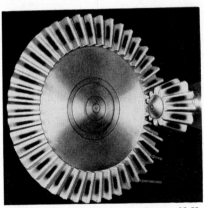

Courtesy The Gleason Works, Rochester, N.Y.

Fig. 141. Straight Bevel Gears. A straight bevel gear is one whose tooth surfaces consist of straight elements which converge to a point at the cone center. This pair, given the trade-marked name of *Coniflex gears* by the Gleason Works, has been manufactured so that the elements of the tooth surfaces are slightly curved with the high part near the center of the face width. The purpose of this plan is to guard against a slight misalignment shifting the contact to an end of the tooth with a resulting possible breakage. The dark spots show the contact area.

parallel shafts that are either intersecting or nonintersecting. The shaft angle Σ is most commonly 90°, as in Fig. 141, but other angles are necessary on occasion. The apex of the cone where the elements of the pitch cone intersect is called the *cone center.* When two bevel gears operate

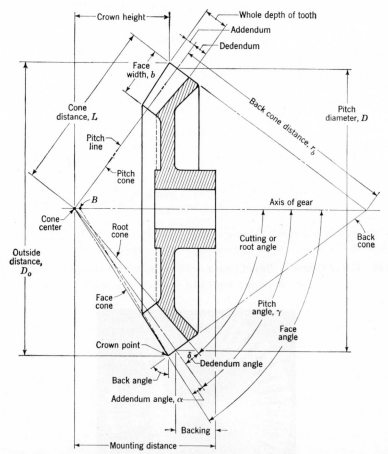

Fig. 142. Nomenclature and Dimensions of a Bevel Gear. In order to eliminate fillet interference at the small ends of the teeth, the more recent practice is to turn *the face cone parallel to the root cone of the mating gear;* hence, its elements do not intersect at the cone center, but at some point B where the dotted line intersects the axis.

on *intersecting* axes, their cone centers are coincident. The length of a pitch-cone element is called the *cone distance.* Top lands and bottom lands mean the same as before. The cone formed by the elements of the top lands is called the *face cone;* the cone formed by the elements of the bottom lands is called the *root cone.* The cone formed by the elements tangent to the base circles is called the *base cone.* The angle γ

between a pitch element and the axis is called the **pitch angle** (not to be confused with the meaning of pitch angle for spur gears, § 122). See Fig. 142 for the meanings of **root** (or **cutting**) **angle, face angle, addendum angle,** and **dedendum angle.**

Since a section of the tooth decreases in size as the cone center is approached, there is not a single value of the pitch, addendum, etc. Unless otherwise indicated by an adjective, these names refer to the largest tooth section, Fig. 142. *The* pitch of a bevel gear is the pitch at the large end; *the* diameter is the diameter of the largest pitch circle. A series of elements perpendicular to the pitch-cone elements at the large end forms a cone in the back called the **back cone.** The length of a back-cone element is called the **back-cone distance,** r_b in Fig. 142.

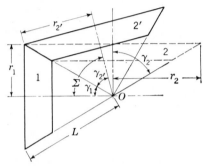

Fig. 143. Angular Gears. Bevel gears whose axes make other than a 90° angle are called *angular gears,* as gears 1 and 2′.

If the shaft angle is 90°, dotted gear 2, Fig. 143, it is seen that

$$(k) \qquad \gamma_1 = \sin^{-1}\frac{r_1}{L} = \tan^{-1}\frac{r_1}{r_2} = \tan^{-1}\frac{N_1}{N_2}, \qquad [\Sigma = 90°]$$

$$(l) \qquad \gamma_2 = \sin^{-1}\frac{r_2}{L} = \tan^{-1}\frac{r_2}{r_1} = \tan^{-1}\frac{N_2}{N_1}. \qquad [\Sigma = 90°]$$

If the shaft angle $\Sigma < 90°$, Fig. 143, the pitch angle may be computed as follows: Start with the definition of the velocity ratio $m_\omega = \omega_1/\omega_{2'} = N_{2'}/N_1$ (solid gear 2′, Fig. 143). Since the cones roll on one another $m_\omega = r_{2'}/r_1$ also.

$$m_\omega = \frac{\omega_1}{\omega_{2'}} = \frac{N_{2'}}{N_1} = \frac{r_{2'}}{r_1} = \frac{r_{2'}/L}{r_1/L} = \frac{\sin \gamma_{2'}}{\sin \gamma_1} = \frac{\sin (\Sigma - \gamma_1)}{\sin \gamma_1}.$$

Expand $\sin (\Sigma - \gamma_1)$, solve for $\tan \gamma_1$, and get

$$(m) \qquad \tan \gamma_1 = \frac{\sin \Sigma}{(N_{2'}/N_1) + \cos \Sigma},$$

which gives the pitch angle on gear 1 when $\Sigma < 90°$. By a similar procedure, eliminating γ_1 by using $\gamma_1 = \Sigma - \gamma_{2'}$, we get

$$(n) \qquad \tan \gamma_{2'} = \frac{\sin \Sigma}{(N_1/N_{2'}) + \cos \Sigma}.$$

A graphical method of finding the pitch cones for *any velocity ratio* of two bevel gears with intersecting axes is given in the Appendix, § 182.

If a generating plane is rolled on the base cone, Fig. 144, any point on the plane remains at a constant distance from the cone center; that is, the point remains in the surface of a sphere and is said to have spherical motion. Thus, a point on this plane does not generate an involute on a plane surface, as in the case of a generating plane rolling on a cylindrical surface, but it generates a curve on a spherical surface which is called a

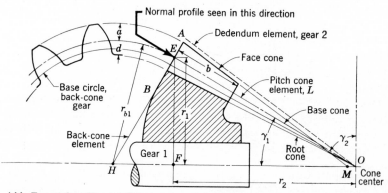

Fig. 144. Formative Gear. The gear-tooth action is taken as equivalent to that of the back-cone (formative) gears. If a circle with O as a center and radius OE is drawn, representing the trace of a sphere on this plane, it is seen that the plane AB is a good approximation of the spherical surface. In the Gleason system, the elements of the face cone AM are parallel to the elements AO of the mating root cone.

spherical involute. Since there is no convenient and economic way of cutting such a curve, the tooth profiles on straight bevel gears are not spherical involutes. The back end of the tooth AB, Fig. 144, is a plane, instead of being spherical. The profiles used on straight bevel gears are those which are conjugate with straight-sided teeth on a crown gear, Fig. 150. The path of contact for such gears is not exactly a straight line but the utilized path is close enough to be so considered for most purposes. The actual profile is closely approximated by the involute teeth on a back-cone gear, Fig. 144. These teeth are seen in their true profile when viewed in line with a pitch-cone element, perpendicular to AB. The radius of the corresponding spur gear is the back-cone radius r_b, and the number of teeth on this imaginary gear is $N_{v1} = 2r_{b1}P_d$, where P_d is *the pitch* of the straight bevel teeth and N_v is called the **virtual** or **formative number of teeth.** From the similar triangles OEF and HEF (sides mutually perpendicular), we have $r_{b1}/L = r_1/r_2$ or $r_{b1} = L(r_1/r_2)$. Using this value in the definition of the virtual number of teeth, we get

(o) $$N_{v1} = 2r_{b1}P_d = \frac{(2r_1P_d)L}{r_2} = N_1\frac{L}{r_2} = \frac{N_1}{\cos \gamma_1},$$

where $\cos \gamma_1 = r_2/L$ as seen in Fig. 144. Considering gear 2 in the same manner, we find the virtual number of teeth as $N_{v2} = N_2L/r_1 = N_2/\cos \gamma_2$.

The bevel-gear-tooth action is taken as that between the two back-cone gears whose pitch circles are tangent at E, that is, contact ratio, undercut, etc. It may be worthwhile to note the *relative* motion of two bevel gears. Imagine one gear stationary and the other rolling around the stationary one. Inasmuch as any point on any circle of the rolling gear remains at a constant distance from the cone center, the motion is spherical motion.

140. Gleason System of Straight Bevels. The industry (AGMA) has adopted as standard the tooth form developed by the Gleason Works, in which the addendum on the pinion is longer than that on the gear except for a velocity ratio of unity (miter gears). The teeth are full depth, with a working depth of $2/P_d$ and a pressure angle of $\phi = 20°$. (Other pressure angles are available.) As the velocity ratio increases, the addendum on the pinion is increased with the results that, as compared with equal addendum gears, interference is reduced or eliminated, the contact ratio is increased, the capacity is increased, and the gears are quieter in operation. The addendum of the gear a_g is obtained from A in Table 2 (p. 240); $a_g = A/P_d$; then the addendum on the pinion is

$$a_p = \frac{2}{P_d} - a_g.$$

For tooth thicknesses, see Table 2.

The face cone of one gear is parallel to the root cone of the mating gear, Fig. 144, instead of converging to the cone center as do other elements. The consequences are that there is a constant clearance along the tooth and that it becomes possible to use a larger edge radius on the cutting tooth without fillet interference at the small end.

141. Types of Bevel Gears. Straight bevel gears have been discussed above. The Gleason Works has developed straight bevel gears, called **Coniflex bevel gears,** whose teeth are so relieved as to localize the pressure between teeth, as suggested by the dark spots on the teeth of Fig. 141. This plan of curving the tooth element in order to avoid damage to the tooth by having the load concentrated on the corner of the tooth is also used for spur gears, but it is especially advantageous for bevels because of the greater difficulty of aligning these gears (and maintaining alignment).

Table 2. Tooth Dimensions for Straight Bevel Gears

(For the gear, $a_g = A/P_d$; for the pinion, $a_p = 2/P_d - a_g$; working depth $= 2/P_d$; whole depth $= 2.188/P_d + 0.002$; d = whole depth $- a$; a stands for addendum, d for dedendum. The circular thicknesses t of the teeth on the pitch circles are: for the gear, $t_g = P_c/2 - (a_p - a_g) \tan \phi$; for the pinion, $t_p = P_c - t_g$.)

Velocity ratios			Velocity ratios			Velocity ratios		
From	To	A	From	To	A	From	To	A
1.00	1.00	1.000	1.23	1.25	0.840	1.76	1.82	0.680
1.00	1.02	0.990	1.25	1.27	0.830	1.82	1.89	0.670
1.02	1.03	0.980	1.27	1.29	0.820	1.89	1.97	0.660
1.03	1.04	0.970	1.29	1.31	0.810	1.97	2.06	0.650
1.04	1.05	0.960	1.31	1.33	0.800	2.06	2.16	0.640
1.05	1.06	0.950	1.33	1.36	0.790	2.16	2.27	0.630
1.06	1.08	0.940	1.36	1.39	0.780	2.27	2.41	0.620
1.08	1.09	0.930	1.39	1.42	0.770	2.41	2.58	0.610
1.09	1.11	0.920	1.42	1.45	0.760	2.58	2.78	0.600
1.11	1.12	0.910	1.45	1.48	0.750	2.78	3.05	0.590
1.12	1.14	0.900	1.48	1.52	0.740	3.05	3.41	0.580
1.14	1.15	0.890	1.52	1.56	0.730	3.41	3.94	0.570
1.15	1.17	0.880	1.56	1.60	0.720	3.94	4.82	0.560
1.17	1.19	0.870	1.60	1.65	0.710	4.82	6.81	0.550
1.19	1.21	0.860	1.65	1.70	0.700	6.81	∞	0.540
1.21	1.23	0.850	1.70	1.76	0.690			

Zerol bevel gears, Fig. 145, also developed by Gleason, have curved teeth, as in spiral bevels, Fig. 146, but with a zero spiral angle ψ. These teeth are generated as are spiral bevels, Fig. 147. They are used in situations where straight bevels would be used and they have localized tooth pressure as do the Coniflex gears. Also, since Zerol teeth can be accurately finish-ground, they are suitable where a very hard tooth surface must be accurately finished.

Spiral bevel gears, Fig. 146, which are to straight bevel gears as helical gears are to straight-tooth spur gears, operate on intersecting axes and have much the same advantages as helical gears: progressive and smoother tooth engagement, quietness of operation, greater strength, and higher permissible speeds. The curvatures of the mating profiles are slightly different in order to obtain the advantages of localized contact. Except for $m_\omega = 1$, the addendum on the pinion is longer than that on the gear, after the manner described for Gleason straight teeth, but the teeth are somewhat stubbed, the *working depth* being $1.7/P_d$. Although full-depth teeth are better from a kinematic viewpoint, the shortened tooth height is desirable in order to keep the top lands from being so narrow that the tips of the teeth become brittle during the hardening process. For a particular pitch in the diametral plane, the tops naturally become narrower as the spiral angle is increased. The standard spiral angle is

Courtesy The Gleason Works, Rochester, N.Y.

Fig. 145. Zerol Bevel Gears.

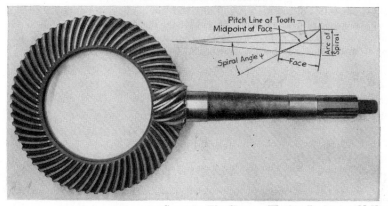

Courtesy The Gleason Works, Rochester, N.Y.

Fig. 146. Spiral Bevel Gears. The inset in the upper right-hand corner shows the meaning of the spiral angle ψ. The pinion has a left-hand spiral, the gear a right-hand spiral. Compare with screw threads where, if the axis is held in a vertical position, the right-hand outside thread slopes upward toward the right, the left-hand thread upward toward the left.

$\psi = 35°$, but other values are used. In general, ψ should be large enough that the arc of spiral, Fig. 146, may be some $1.25P_c$ to $1.4P_c$. The resulting overlapping of tooth contact on the pitch cone permits the use of smaller tooth numbers than for straight bevels with the maintenance of smooth operation. The basic pressure angle on spiral bevels

Courtesy The Gleason Works, Rochester, N.Y.

Fig. 147. Cutting Spiral-bevel Gear Teeth. The cutting blades on the cutter have straight sides. The Gleason Works describes the action as follows: "The tooth surfaces are generated by rolling the gear blank and revolving cutter together as though the blank were meshing with a crown gear of which the cutter represents a tooth . . . . The various ratios of roll for gears of different pitch angles are obtained through the use of change gears. The feed movement for the depth of cut is imparted to the work head by a cam which swings the work alternately into the cutter for the feed and away from the cutter to permit indexing."

is 20°, but certain ratios with $14\frac{1}{2}°$ and 16° are obtainable. When the pitch-line speed is over 1000 fpm, spiral bevels (or hypoid gears) are generally preferred over the tooth types mentioned above. Spiral-bevel gear teeth are cut with a rotary cutter, Fig. 147.

Hypoid gears are quite similar to spiral bevels in appearance, except that their axes do not intersect. See Fig. 148. These gears are widely used for various reasons, an example of which is in the automotive rear-axle drive, where their use makes it possible to lower the center of gravity of the car. Because of the offset of the axes, the pitch surfaces are hyperboloids, whence their name. When two (pitch) hyperboloids are in contact along a common element, two tangent circles rotate in different planes, each in a plane normal to its axis. Moreover, a pair of mating pitch circles with their axes offset make contact in such a way that they

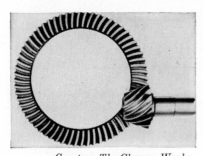

Courtesy The Gleason Works, Rochester, N.Y.

Fig. 148. Hypoid Gears.

do not roll on one another. The outcome of the various characteristics of hypoid gears is that there is considerable sliding lengthwise along the tooth. This additional sliding results in a greater energy loss, in higher localized temperatures, and generally it requires the use of so-called extreme-pressure lubricants whose function is to inhibit welding at the hot spots. The shafts of hypoid gear sets may be offset enough that the shafts can continue past one another; this idea is used in arranging several hypoid pinions on a single shaft with drives to several machines from this one input shaft. The elements of the pitch surface of hypoid gears do not intersect.

Courtesy The Gleason Works,
Rochester, N.Y.

Fig. 149. Angular Gears.

Miter gears are bevel gears of the same size mounted on shafts at 90°; $m_\omega = 1$; $\gamma = 45°$.

The name *angular gears* is sometimes given to bevel gears whose axes intersect at an angle either greater or less than 90°; Figs. 143 and 149.

A *crown gear* is one in which the pitch angle is 90°; that is, the pitch surface (cone) has become a plane, Fig. 150; $\gamma_2 = 90°$. Notice that for a

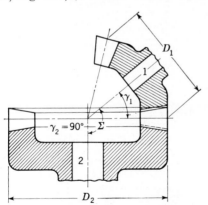

Fig. 150. Crown Gear and Pinion. The size of the teeth is somewhat exaggerated.

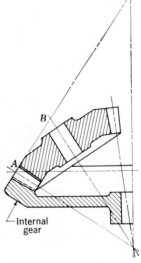

Fig. 151. Internal Bevel Gear and Pinion.

particular size D_2 of crown gear, the shaft angle depends upon the velocity ratio, or vice versa; $\sin \Sigma = D_1/D_2 = N_1/N_2 = 1/m_\omega$. The cutting edge for generating straight bevels simulates a straight-sided tooth on a crown gear blank.

Fig. 152. High-ratio Hypoid. Fundamentally a worm drive, § 142; applicable to high-ratio gears.

If the teeth are cut so that they project toward the inside of the pitch cone, instead of outside, the gear is called an *internal bevel gear*, Fig. 151.

There are a number of other more or less special forms of gears; examples, Gleason's *high-ratio hypoids,* Fig. 152, Fellow's *face gear* (see Fellow's catalogs), and a *spiroid pinion and gear* (81).

142. Worm Gearing. Worm gearing, Fig. 153, is used to transmit power between nonintersecting shafts (usually, but not necessarily) at right angles to each other. (Note that helical gears on nonparallel shafts can do the same thing, but the usual worm gearing has a much greater power-transmitting capacity. A helical gear in a crossed drive is essentially a worm, but on a worm, in contrast with helical gears, a tooth helix makes at least one complete turn about the cylinder.) Compara-

Fig. 153. Worm and Worm Gear. This is a double-reduction worm gear box. The power enters on worm *A* and is delivered by gear *D*. Notice the large lead angle on worm *C*. The size of the teeth on gear *D* must be considerably larger than those on gear *B* for the same power transmission because of its slower speed.

tively high velocity ratios may be obtained satisfactorily in a minimum space, but sometimes at a sacrifice of efficiency. Since the worm threads slide into contact with the gear teeth and slide across the teeth, these gears properly mounted are quiet. The profiles of involute teeth on the worm are straight-sided (§ 129).

The teeth on the worm are called *threads*. A single-threaded worm, Fig. 154, is one in which the pitch helix advances a distance P_c in one

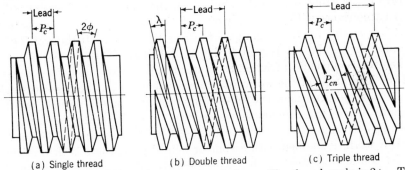

(a) Single thread   (b) Double thread   (c) Triple thread

Fig. 154. Pitch and Lead. The pressure angle is ϕ. The thread angle is 2ϕ. The lead angle is λ; $\psi = 90 - \lambda$. The circular pitch P_c is commonly called *linear pitch* when applied to worms.

turn. The pitch helix of a double thread advances $2P_c$ in one turn; of a triple thread $3P_c$; etc. Five or six threads (or more) are not uncommon on worms because, for a particular worm diameter, the lead angle can be increased by increasing the number of threads (teeth) and the efficiency of the drive increases as the lead angle increases, up to about 40° **(5)**. The axial distance that a thread advances in one turn is called the **lead** L; for a single thread, $L = P_c$; for a double thread, $L = 2P_c$; etc. If a pitch helix is unwrapped from the cylinder onto a plane, Fig. 155, its

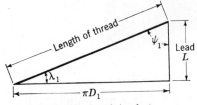

Fig. 155. Lead Angle λ.

angle λ with respect to a diametral plane is the lead angle; $\tan \lambda_1 = L/(\pi D_1) = N_1 P_c/(\pi D_1)$, in which D_1 is the pitch diameter of the worm. Recall that the helix angle ψ is the slope of the helix with respect to a plane perpendicular to the axis; $\lambda_1 + \psi_1 = 90°$, Fig. 155. Consideration of Fig. 154 shows that the circular pitch in the normal plane is $P_{cn} = P_c \cos \lambda$. As usual, the velocity ratio is $m_\omega = \omega_1/\omega_2 = N_2/N_1$, where there are N_2 teeth in the gear and N_1 threads (teeth) on the worm. Since

(p) $$N_2 = \frac{\pi D_2}{P_c} \quad \text{and} \quad N_1 = \frac{\pi D_1 \tan \lambda_1}{P_c},$$

(q) $$m_\omega = \frac{\omega_1}{\omega_2} = \frac{N_2}{N_1} = \frac{D_2}{D_1 \tan \lambda_1} = \frac{\pi D_2}{L},$$

where L is the lead of the worm. Notice that the lead angle λ_1 of the worm is the same as the helix angle ψ_2 of the worm gear, shafts at 90°.

In accordance with your study of analytic mechanics (2), it may or may not be possible for the worm gear to drive the worm. If the worm gear cannot drive, the set is said to be self-locking. This condition arises when the coefficient of friction is greater than approximately the tangent of the lead angle of the worm (it is also dependent on the pressure angle ϕ). In some situations, the self-locking property may be desired, but not usually where the primary purpose is to transmit power, because self-locking corresponds to low lead angles ($<6°$) and therefore low efficiencies.

If a worm is meshed with a spur or helical gear, there is theoretically *point* contact between teeth. This plan would be used only when the transmitted power is not significant. To obtain line contact, the teeth on the gear are cut by a hob which is the same size as the worm to be used, and the cutter is *not* fed *across* the face of the gear. Rather the hob is centered over the face and fed radially inward toward the gear center. In this case the gear partially envelops the worm.

Courtesy Michigan Tool Co., Detroit

Fig. 156. Hourglass Worm Set.

Pressure angles ϕ of $14\frac{1}{2}°$ and 20° are widely used, but as the lead angle is increased, a point is reached where it becomes impossible to get the tool to cut the correct flank and bottom lands. The remedy for this situation is to increase the pressure angle; $\phi = 25°$ for $25° < \lambda < 35°$ and $\phi = 30°$ for $35° < \lambda < 45°$.

Special forms of worm drives are found. The hourglass or Hindley worm drive is featured by the worm wrapping somewhat around the gear (as well as the gear wrapping about the worm). See Fig. 156. With more teeth in contact, the pressure between the teeth is reduced with the expectation of a longer gear life for the same power transmitted.

143. Closure. The types of gears discussed in this chapter are in general highly specialized. If unusual problems of a kinematic nature are involved, manufacturers' specialists can give valuable assistance.

PROBLEMS

Helical Gears, Parallel Axes

621. A helical gear with a helix angle of 23° has 26 teeth which were cut with a standard $14\frac{1}{2}°$ full-depth hob of diametral pitch 7. Compute (a) the circular pitches in the normal and transverse planes, (b) the pitch diameter, (c) the transverse pressure angle, and (d) the face width if the leading end of one tooth is just opposite the trailing end of the preceding tooth.

622. The same as **621** except that the pitch of the hob is 2.5.

623. If a 14-in. helical gear has 56 teeth and a helix angle of 23°, find (a) the normal diametral and circular pitches, (b) the normal pressure angle if a 20° shaper cutter is used, and (c) the face width if the leading end of one tooth is just opposite the trailing end of the preceding tooth.

624. A helical gear, with 28 teeth cut with a standard $14\frac{1}{2}°$ hob of 7 pitch, has a face width of 1.875 in. The tooth advance is 0.539 in. Compute (a) the normal circular pitch, (b) the pitch helix angle, (c) the pitch diameter, (d) the lead of the helix (see Fig. 155), (e) the diameter of the base circle in the diametral plane, (f) the helix angle of the base helix (noting that the lead is the same for all helices of a particular tooth).

625. The same as **624** except that the tooth advance is 0.74 in.

626. A pair of helical gears with a velocity ratio of 1.6 is desired. Neither gear should have fewer than 15 teeth cut with a standard 20° full-depth hob of 5 pitch. If the center distance must be 4.25 in., determine the helix angle, the diametral pitch, the transverse pressure angle, and the minimum face width.

627. A pair of helical gears with a velocity ratio of about 1.6 is desired. Neither gear should have fewer than 15 teeth and the center distance should not be less than 4 in. nor more than 4.5 in. If the teeth are cut with a standard 20° full-depth shaper cutter of 5 pitch, determine suitable values of the helix angle, the diameters, the normal pitch and pressure angle, and the minimum face width.

628. In a 6-ft. ball mill, 75 hp. pass through a pair of $14\frac{1}{2}°$ full-depth steel spur gears: $P_d = 2$, $N_p = 16$, $N_g = 176$, face width $b = 5$ in. These gears wear out, as does a subsequent replacement. Then it is decided to replace them with helical gears, naturally maintaining the same center distance and velocity ratio. A small change in the face width can be tolerated. Let the gears be cut with a standard $14\frac{1}{2}°$ hob. Decide upon at least 3 pairs of gears which can be used (3 different helix angles), choose one pair as the desired answer, and specify the tooth numbers, the pitch in the plane of rotation, the transverse pressure angle, the pitch diameters, and the face width.

629. The same as **628** except that a 20° full-depth shaper cutter of 2 pitch is used.

630. The same as **628** except that it has been decided to use herringbone gears cut by a standard $14\frac{1}{2}°$ hob of 4 pitch.

Crossed Helical Gears

631. Two shafts, $6\frac{7}{8}$ in. apart, making a shaft angle of 90°, are to be connected with crossed helical gears. Gear 1 drives; $N_2 = 40$; $P_{cn} = 0.3928$ in.; $\phi_n = 14\frac{1}{2}°$; $\psi_1 = 60°$; $\omega_1 = 12$ rad./sec. Determine (a) the diameter of the pinion, (b) the number of teeth on gear 1 and the velocity ratio, (c) the circular pitches of each gear, (d) the transverse pressure angle, (e) the crosswise sliding speed of the teeth.

632. In **631**, it is decided to equalize the helix angles as nearly as possible, with a small change of the center distance permissible. Design a set of such gears, specifying any changes in given data and the answers to the questions in **631**.

633. Two shafts in different planes are at 90° and are to be connected with crossed helicals which will be hobbed by 4 pitch, $14\frac{1}{2}°$, full depth. The center distance is to be 14 in. and the desired velocity ratio is 1.5; $\omega_1 = 8$ rad./sec. For the trial calculation, assume equal helix angles and determine the following results: (a) tooth numbers, (b) helix angles, (c) diameters (check C), (d) transverse pressure angles, (e) sliding speed (graphically and algebraically). See Fig. P128.

634. The same as **633** except that ψ_1 is to be assumed as 20° for the trial calculation.

635. The same as **633** except that ψ_1 is to be assumed as 70° for the trial calculation. If solutions to **633** and **634** are available, compare answers.

636. Two helical gears are as follows: $N_p = 24$, $N_g = 40$, $n_p = 800$ rpm, $P_{dn} = 10$, $\phi_n = 14\frac{1}{2}°$, full depth, $\psi_p = 23°$ left hand, $\psi_g = 48°$ right hand. Compute (a) the shaft angle, (b) the circular and diametral pitches of each gear, (c) the diameters and velocity ratio in terms of the diameters, (d) the sliding speed at the pitch point. Solve (d) algebraically and graphically. Show a sketch of your results similar to Fig. P128.

Ans. (a) 25°, (c) 2.61 in. 5.97 in., 1.667, (d) 345 fpm.

637. The same as **636** except that $\psi_g = 48°$ left hand.

638. Two nonintersecting shafts at an angle of 60° (projected) are to be connected by crossed helical gears at the point of their nearest approach, where the center distance is 3 in. The teeth are cut with a standard $14\frac{1}{2}°$ hob of 4 pitch. The velocity ratio is to be 2; $\omega_1 = 10$ rad./sec. Specify the following items: tooth numbers, helix angles, diametral pitches, diameters, pressure angles, sliding speed (graphically and algebraically).

639. Crossed helical gears are to be used to provide a velocity ratio of 3.5 between shafts at $\Sigma = 70°$ with each other. Let the gears be cut with a standard 20° full-depth hob of 6 pitch and keep as close to a minimum 2-in. diameter as possible without going below it. The driver turns at 3000 rad./min. (In general, the larger the gear, the greater the cost.) Let the helix angles be 35° each, same hand, and compute (a) the tooth numbers, (b) the pitch diameters, (c) the diametral pitches, (d) the transverse pressure angles, (e) the sliding speed (graphically and algebraically). Sketches as suggested by Fig. P128 are helpful, especially when ψ_1 and ψ_2 are

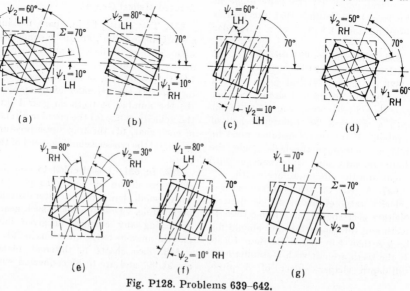

Fig. P128. Problems **639–642.**

different. For clarity, the entire bottom gear is dotted. Of course, there are an infinite number of combinations of ψ_1 and ψ_2 which will give a shaft angle of 70°. This illustration ignores variations of diameters.

Suggestion to Instructor. Divide the class into three parts and let each part work one of the problems **639**, **640**, or **641**. It will then be instructive to assemble all answers on one data sheet for comparison.

640. The same as **639** except that $\psi_1 = 10°$ right hand, Fig. P128(c).

641. The same as **639** except that $\psi_2 = 10°$ left hand, Fig. P128(d).

642. The same as **639** except that $\phi_n = 14\frac{1}{2}°$.

Bevel Gears

643. Use plate No. 38, Wingren. A 16-tooth 4-pitch bevel pinion on shaft *A* drives a 24-tooth gear on shaft *B*. The face width is 1.25 in. The teeth are standard (Gleason). (a) Draw the pitch and back cones. Draw the pitch circles for the formative teeth tangent at the point common to all 4 cones. See § 183, Appendix. Draw two teeth on each formative circle by Tredgold's approximation. Measure and dimension the following: pitch angles, root (cutting) angles, cone distance, dedendum angles, back-cone distances. Dimension the addendums and dedendums on the drawing. (b) On a separate sheet, compute the pitch angles, the outside diameters, and dedendum angles for a check.

644. Two straight bevel gears with a velocity ratio of 4 are mounted on 90° shafts. The pinion has 15 teeth of 8 pitch. Find (a) the gear's pitch and outside diameters, (b) the pitch angles, (c) the back-cone distances, (d) the virtual number of teeth on each, (e) the cutting angles. (f) Considering the formative gear for the pinion, decide whether or not there will be interference conditions. Could you say offhand whether or not interference conditions would exist if the teeth were the interchangeable type, full depth, $14\frac{1}{2}°$? Solve graphically as far as possible.

645. The same as **644** except that the solution is to be algebraic as far as possible.

646. The same as **644** except that the shaft angle is 75°.

647. The same as **646** except that the solution is to be algebraic as far as possible.

648. A 15-tooth bevel pinion of 5 pitch drives a 35-tooth gear with shafts at 90°. Using a graphical solution as far as possible, determine the pitch angles, the outside diameters, the addendum angles, the cutting angles, back-cone distances, and formative number of teeth. Considering the formative gear for the pinion (draw the back cone circles in mesh), decide whether or not there will be interference conditions. Could you say offhand whether or not interference conditions would exist if the teeth were of the full-depth $14\frac{1}{2}°$ interchangeable type?

649. The same as **648** except that the solution is to be algebraic as far as possible.

650. The same as **648** except that the shaft angle is 60°.

651. The same as **650** except that the solution is to be algebraic as far as possible.

652. After the manner of obtaining equations (*m*) and (*n*), derive equations from which the pitch angle of the gear may be computed when $\Sigma > 90°$, showing that

$$\tan \gamma_g = \frac{\sin (180 - \Sigma)}{(N_p/N_g) - \cos (180 - \Sigma)}$$

when $N_g \sin (\Sigma - 90) < N_p$.

653. (a) What is the acute shaft angle when the velocity ratio of a pinion and crown gear is 1? (b) If the acute angle between shafts of a crown and bevel gear is 30°, what is the velocity ratio?

654. A crown gear with 30 teeth of 4 pitch is driven by a 12-tooth pinion. (a) Determine the acute shaft angle, the cutting angle on the pinion, its back-cone distance, and outside diameter. Compute the addendum, dedendum, etc., and draw two formative teeth for the pinion (Gleason proportions) and indicate the amount of radial flank, if any, outside of the working-depth circle. Use a graphical approach. (b) Check the values from the graphical solution by algebraic computation.

655. A worm-gear drive with a velocity ratio of 22 is desired. The teeth are to have a diametral pitch of 3 in the normal plane; the pitch diameter of the worm is to be 1.88 in. Compute the following: the lead angle and the center distance (a) for a triple-threaded worm, (b) for a quadruple-threaded worm, (c) for a quintuple-threaded worm. (d) If this drive is primarily for the transmission of power, would you use one of the foregoing drives? Base your decision on these calculations and the fact that the efficiency of a worm drive is not approaching its best value until the lead angle is greater than 35°.

656. A triple-threaded worm drives a 45-tooth gear; $\Sigma = 90°$. The lead of the worm thread is $1\frac{7}{8}$ in. and the center distance is 5.6 in. The tooth profiles are standard 20° full-depth in the plane of rotation. Compute (a) the velocity ratio and the pitch and outside diameters of the gear and worm (for the gear, the outside diameter at the mid-point of

the face width), (b) the lead angle of the worm and the helix angle of the gear, (c) the normal circular pitch and the normal pressure angle.

657. A worm-gear drive with a velocity ratio of 13 is desired. Let the circular pitch be 1.5163 in. and $\phi = 26°$; the lead angle is 41.5°; and the center distance is to be about but not exceed 16 in. Look into the use of triple-, quadruple-, and quintuple-threaded worms, and on the merits of the proportions, decide which to use. Then compute the normal circular and diametral pitches and the normal pressure angle. What is the helix angle on the gear?

658. The lead angle of a double-threaded worm is 13.32° and the circular pitch is 1.3125 in.; $\phi = 27.5°$; the velocity ratio = 20. Compute the center distance, the normal circular and diametral pitches, and the normal pressure angle.

Larger Paper Needed

659. The worm gear of Fig. P129 has 32 teeth of 5 pitch, full depth, 20° involute (plane of rotation). (a) For a single-threaded worm, compute the lead

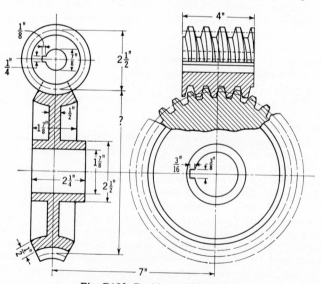

Fig. P129. Problems 659, 660.

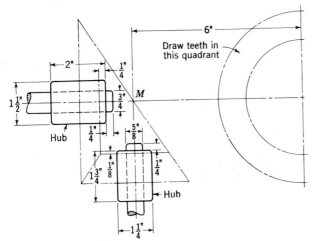

Fig. P130. Problem 661.

angle, addendum, and dedendum. (b) Complete the layout and dimension the drawing. Scale full size on 11 × 17 paper.

660. The same as **659** except that the worm is double-threaded.

661. A 4-in. 5-pitch bevel pinion drives a gear on 90° shafts with a velocity ratio of 1.5, Fig. P130; M (7, 7); face width = 1 in. (a) Lay out full size (11 × 17 paper) the pitch cones with the projection of the Gleason teeth thereon. Show at least two formative teeth for each gear, indicating on the drawing the radii used. If any part of the tip of a tooth may cause interference, crosshatch it. Compute the outside diameters. (b) Complete the drawing of the gears using the dimensions shown and: pinion key = $\frac{1}{8} \times \frac{1}{8}$ in.; gear key = $\frac{3}{16} \times \frac{3}{16}$ in.; web thickness on pinion = $\frac{1}{2}$ in., on gear = $\frac{5}{8}$ in.; rim thickness on pinion = $\frac{3}{8}$ in., on gear = $\frac{3}{8}$ in. (large end). Construct the front view of the gear with teeth projected to one quadrant only. See § 183, Appendix. Let the side view of the gear be half projection and half section.

662. Bevel gears at right angles have a pitch of 1; $D_p = 16$ in.; $D_g = 24$ in.; face width $b = 4.5$ in.; M (3, 5) on

11 × 17 paper; lay out to scale 3 in. = 1 ft. Draw the pitch cones and erect the projections of the teeth as shown in Fig. P131. On a separate sheet, show the necessary computations, including the pitch angles, addendums, dedendums, outside diameters, cutting angles (Gleason system), back-cone distances. To make a study of the kinematic action, draw the pitch circles of the formative gears, locating the center of the large one at Q, Fig. P131. Using graphical methods, determine the angle of action, the length of action, the base pitch, and the contact ratio. Construct at least two formative teeth for each gear. See § 183.

663–670. These numbers may be used for other problems.

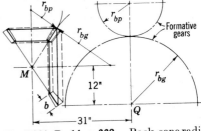

Fig. P131. Problem 662. Back-cone radii are r_{bp} and r_{bg}.

9

Gear Trains

144. Introduction. A *gear train* consists of two or more gears arranged to transmit power from a driving shaft to some driven shaft. Such a train is used when the distance between shafts is not too great, when a certain particular velocity ratio is necessary or desirable, and often when a large velocity change or large mechanical advantage is desired. In general, a *train* may include friction wheels, belt drives, etc. (Chapter 10); but if the principles of this chapter are understood, there will be no difficulty in extending them to other elements in rotation.

There are many reasons why trains are necessary. Perhaps it is desired to have several trains so arranged in a "box" that one has a choice of trains by shifting gears, as in an automotive transmission, Figs. 159 and 168. Of course, the main reason is that the source of power (called the prime mover, the machine that does the work, such as a turbine or electric motor) does not turn at the speed that is satisfactory for the driven device. When this is so, it is generally more economical to let the machine doing the work turn at its most advantageous speed and introduce a train to provide the proper speed of the driven elements. Since the faster a prime mover turns, the smaller it may be for a particular amount of power delivered, the tendency is for their speeds to be high and get higher. Hence, most often a step-down in speed is desired, but sometimes a step-up is required.

145. Train Value. We already know well the definition of velocity ratio and we have repeatedly applied the term to pairs of rotating elements. In dealing with longer trains, however, there will be some advantage in using the *train value* e, which, for the case of all shaft axes fixed in space, is the reciprocal of the velocity ratio:

(*a*)
$$e = \frac{1}{m_\omega}.$$

[AXES STATIONARY]

The velocity ratio of a train is the angular velocity of the first member divided by the angular velocity of the last member in the train. To shorten the presentation, consider only that part of the train of Fig. 157 including shafts C, D, E, and F. Since the angular velocities are

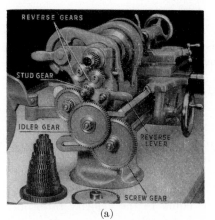

(a)

Courtesy South Bend Lathe Works, South Bend, Ind.

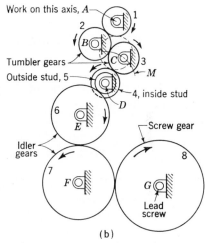

(b)

Fig. 157. Thread-cutting Gear Train. The top gear 1 turns with the work. The screw gear is attached to the lead screw and they turn together. The lead screw drives the tool carriage in the direction of the axis of the work. The number of threads cut depends on the relative speeds of the work and lead screw and on the lead of the thread on the lead screw. A thread-cutting lathe has attached to it a plate which specifies the proper gears to use for a particular number of threads to be cut. But see Fig. 158. The reverse lever shifts gear 2 out of the train and places gear 3 in mesh with 1 and 4, Fig. 157(b), thus reversing the direction of rotation of the lead screw. Gears 2 and 3 are called *tumbler* or *reverse* gears.

inversely proportional to the tooth numbers, we have the angular velocity ratios for succeeding pairs of gears, Fig. 157, as

$$\frac{\omega_C}{\omega_D} = \frac{N_4}{N_3}, \qquad \frac{\omega_D}{\omega_E} = \frac{N_6}{N_5}, \qquad \frac{\omega_E}{\omega_F} = \frac{N_7}{N_6}.$$

If the product of the left-hand sides of these equations is equated to the product of the right-hand sides, we find the velocity ratio between shafts C and F as

$$(b) \qquad m_{\omega C/F} = \frac{\omega_C}{\omega_F} = \left(\frac{N_4}{N_3}\right)\left(\frac{N_6}{N_5}\right)\left(\frac{N_7}{N_6}\right) = \left(\frac{D_4}{D_3}\right)\left(\frac{D_6}{D_5}\right)\left(\frac{D_7}{D_6}\right).$$

From this development, we see that the over-all velocity ratio of a gear train with axes fixed in space is the product of the velocity ratios between each successive pair of shafts. In accordance with equation (*a*), the train value including shafts *C*, *D*, *E*, and *F*, Fig. 157, is

(*c*) $$e = \frac{1}{m_\omega} = \frac{N_3 \times N_5 \times N_6}{N_4 \times N_6 \times N_7}.$$

Compare equation (*c*) with Fig. 157 and classify the gears as *drivers* or *drivens* (or *followers*): 3 is a driver, 4 is a driven, 5 is a driver, 6 is a driven and a driver, 7 is a driven. *Gears which are both driver and follower are called **idler gears;*** that is, for the part of the train being considered, 6 is an idler. With this knowledge of the gears and equation (*c*), we can define the train value as

(29) $$e = \frac{\text{product of tooth numbers of } driving \text{ gears}}{\text{product of tooth numbers of } driven \text{ gears}}.$$

Equation (29) is the basic definition of train value *whether or not the axes are fixed in space*, but its reciprocal is equal to the velocity ratio only when the axes are fixed. A ratio of pitch diameters can be used in place of a ratio of tooth numbers, but one must be careful to keep diameters and tooth numbers properly paired, as in equation (*b*).

146. Idler Gears. An **idler** or *idler gear* is defined above. In the train *C-F*, Fig. 157, previously discussed, gear 6 is an idler. For the *whole train*, gears 2, 3, 6, and 7 are idlers. The tooth numbers (N_6) of an idler appear in both the numerator and the denominator of equation (29) and consequently cancel. An idler has no effect on the magnitude of the train value. Such gears are placed in an ordinary gear train for the purpose of changing the direction of rotation of some driven shaft or to help fill up space between shafts when the center distance is too great for just two gears.

The most satisfactory way of determining the direction of rotation of any shaft in an ordinary train is by inspection of a sketch of the train. A rule for *spur gears* is: When there is an *odd* number of shafts, the first and last shafts rotate in the *same* direction; and in opposite directions when there is an even number. If *one* of the gears is an internal gear, the foregoing rule is reversed: odd-opposite; even-same. When bevels, worms, or crossed helicals are involved, no simple rule is possible because the direction of rotation depends upon the way the bevel gears are mounted, the hand of the helices, and besides, the first and last shafts might not be parallel.

147. Example. Sometimes a gear train can be conveniently represented by designating gears on horizontal lines, each representing a shaft. For example,

1. $N = 20$
2. $N = 12$ _____ $N = 30$
3. $6'', P_d = 5$ _____ $N = 15$
4. $N = 60$(internal) _____ $N = 20$(bevel)
5. $N = 40$(bevel) _____ 2-th. worm
6. $N = 100$

Let $n_1 = 3600$ rpm and find the speed of shaft 6.

Solution. The diagram indicates that a 20-tooth gear drives a 12-tooth gear. On shaft 2, a 30-tooth gear is shown, but it is presumably a part of another train. At any rate it does not mesh with another gear in this train. The 12-tooth gear on 2 drives the 5-pitch 6-in. gear ($= 30$ teeth) on shaft 3. Also on 3 is the 15-tooth gear driving the internal gear. Then two bevels transmit the motion to shaft 5, thence to shaft 6 via a double-threaded worm and a 100-tooth gear. We know that

$$n_6 = \frac{n_1}{m_\omega} = en_1.$$

Using the definition of e in equation (29), we get

$$n_6 = \frac{20 \times 12 \times 15 \times 20 \times 2}{12 \times 30 \times 60 \times 40 \times 100}(3600) = 6 \text{ rpm.}$$

148. Illustrations of Gear Trains. We have already used the thread-cutting train of Fig. 157(b) for explanation purposes. These trains vary in detail in different lathes. With the fewest gears, there is only one idler instead of gears 6 and 7, Fig. 157(b). On the other hand, as in the photograph, Fig. 157(a), compound gears may be involved where idlers might have been. (*Compound gears* are two gears in a particular train that must turn together on a single axis.) In the thread-cutting train, at least one idler will be mounted on an adjustable bracket so that it can be moved to accommodate different diameters of lead-screw gears and outside stud gears. However, since it takes time and costs money to change gears manually, quick-change gearboxes, Fig. 158, are common in situations where the thread-cutting train is used often. By moving the levers seen in the inset, upper left, to match holes in accordance with instructions on the machine, standard screw threads from 8 to 224 per inch can be cut with this set of gears.

A *reverted gear train* is one in which the axis of the driven shaft is along the same line as the axis of the driving shaft. The gears of Fig. 158 are reverted with a set of idlers between the shafts. The automotive transmission is a simpler illustration of reverted trains. The operation of a 3-speed transmission is explained in Fig. 159.

149. Gear Trains for a Given Velocity Ratio. Sometimes, designing a train to produce a certain exact velocity ratio is troublesome. If a velocity ratio need only be approximated, as in most ordinary power-transmitting gearboxes, one tooth more or less on the large gear is of no significance. If the exact velocity ratio is such a number as 2, $2\frac{1}{8}$, etc., that it can be easily transformed into a whole number (to represent a whole number of teeth), the problem again is easy. For example, if 2.125 were the desired value, one could use a pinion of 16 teeth (because it is divisible by 8) and a gear of 34 teeth. The same would be true for any velocity ratio that can be factored into numbers such as to produce convenient tooth numbers. There are, however, limiting values of tooth numbers which would be considered practicable. There is no sharp dividing line, because the practicable limits would depend upon various circumstances, including as usual the economic aspects. If a velocity

Courtesy South Bend Lathe Works, South Bend, Ind.
Fig. 158. Quick-change Gearbox. As seen from rear.

ratio of 2.43 is a necessity, one could use gears with 100 and 243 teeth. These gears have large tooth numbers; the top limit is often placed at 100. If such a limit exists here, then something else must be done. On the low side, 12 teeth are often taken as the minimum.

Given a particular velocity ratio, say $m_\omega = 2.473$, one cannot say off-hand whether an exact solution is possible with gears of reasonable tooth numbers. Maybe the situation is such that a variation of 1% or 0.1% is permissible; this much allowance would make the problem easy. In some cases, a variation of less than one part in a million may be specified in a situation where an exact solution is practically impossible. Given money, time, and space, one can approach such a ratio to any desired degree of accuracy except 100%.

There are several ways to approach such problems, and if one is repeatedly concerned, the subject should be studied in more detail, especially with the help of ratio tables (**32, 86, 87, 134**). The student can see what the nature of the problem is from a short discussion.

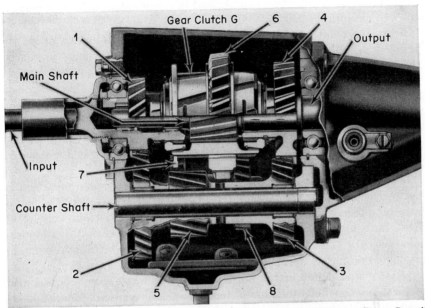

Courtesy General Motors Corp., Detroit

Fig. 159. Three-speed Automobile Transmission. The transmission is shown in neutral. The engine is directly connected to the input shaft via a clutch (not shown). When this clutch is *in*, gear 1 turns with the engine; hence gears 2, 3, 4 also turn because they always mesh as seen. The gear clutch G is splined to (and turns with) the main shaft, and it has internal teeth that engage external teeth (not visible) when it is moved right or left. For high gear (third speed), G is moved toward the left until the left gear clutch (integral with 1) engages and the drive is direct, 1-G-main shaft. For second speed, move G toward the right until the right gear clutch (integral with 4) engages and the drive is 1-2-3-4-G-main shaft. For low gear (first speed), G remains in its neutral position and gear 6 is moved into engagement with 5; the drive is 1-2-5-6-G-main shaft. For reverse, G remains in its neutral position and gear 6 is moved into engagement with 8; gears 7 (barely visible) and 8 are integral with a small reverse countershaft and the drive is 1-2-5-7-8-6-G-main shaft.

Suppose the velocity ratio of $m_\omega = 2.473$ is desired in a pair of gears; maximum and minimum tooth numbers are to be 100 and 12, respectively. While slide-rule answers are not significant where the ratio must be exact, it does permit a quick series of approximations. For example, set this ratio as closely as possible over the left index and start with 12 teeth. Running the eye along the D scale, note that the following pairs look fairly close: 13 and 33; 15 and 37; 17 and 42; 19 and 47;

21 and 52; 23 and 57; 40 and 99; and multiples of these. If a calculating machine is not available, the only alternative is long-hand division to find out which ratio comes closest. The velocity ratio of the 47/19 pair is closest, an error of 0.00068, or about 0.0275% high. If this is not close enough, a double reduction can be used (a triple reduction would permit an even closer approach).

From the point of view of the arithmetic, note that if two ratios are handled as follows, the resulting ratio will be between the two. Given the pairs 47/19 and 42/17:

$$\frac{47 + 42}{19 + 17} = \frac{89}{36},$$

which will be found to be between 47/19 and 42/17, but not necessarily closer to the desired value than 47/19 even though these ratios straddle 2.473.

If the solution is to be carried further, some orderly plan of action is highly desirable. In the method set forth below, the first step is to find a ratio of integers which closely approaches the desired train value $e = 1/m_\omega$, for example, 19/47 (it does not have to be the closest of the foregoing ratios). Then it can be said that the exact ratio is this approximate ratio plus or minus a small amount. However, since, in the end, the numerator and denominator must not only be whole numbers but also factorable into whole numbers of teeth for a multiple-reduction gear set, use the following procedure: Let the approximate ratio be P/Q. The ratio remains the same if both numerator and denominator are multiplied by the same factor K; $P/Q = PK/QK$. Then we can say that the train value

(*d*) $$\epsilon = \frac{PK}{QK + I},$$

where I is an integer. To use (*d*), put in the exact value of e and the approximate numbers P and Q, assume that I is equal to 1 (or 2, etc.), and solve for K, proceeding as illustrated below. Let

$$\frac{1}{2.473} = \frac{17K}{42K + 1},$$

from which $K = 24.4$. Try $K = 24$ and evaluate the right-hand side of (*d*):

$$\frac{17 \times 24}{42 \times 24 + 1} = \frac{408}{1009}.$$

Unfortunately, the denominator of this ratio is unfactorable. We could try $K = 25$, but it, too, results in an unfactorable number. The remedy in a general sense is to get larger numbers in the ratio, and in particular, to assume a larger value of the integer I. Using $I = 2$ results in one factor being greater than 100; so it is not used because the maximum number of teeth is to be 100. Assuming $I = 3$, we get $K = 73.2$; try the closest value $K = 73$ and get

$$\frac{17K}{42K + I} = \frac{17 \times 73}{42 \times 73 + 3} = \frac{1241}{3069} = \frac{17 \times 73}{31 \times 99} = \frac{17 \times 73}{33 \times 93},$$

where the numerator has only one pair of factors and the most appropriate ones are shown for the denominator. Any grouping can be used, 17/31 and 73/99, 17/99 and 73/31, etc. There would be no point in one which had both a step-down and a step-up in speed. Perhaps the preferred group is 17/33 and 73/93. The error for this combination is less than 2.5 parts in a million. It might be possible to get the ratio closer with a 4-gear train, and almost certainly with a 6-gear train.

150. Planetary Gear Trains. In a *planetary train,* also called an *epicyclic train,* one or more of the rotating members moves about a central axis after the manner of planets revolving about a sun; these gears are called the *planet gears* or just *planets.* Proceeding by degrees, attach a wheel B to the carrier or arm A, Fig. 160, and place a mark with

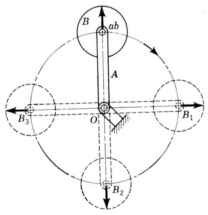

Fig. 160.

an arrowhead on the wheel. Now be assured that B will have made a revolution when the arrow has pointed successively in all directions around the clock. For example, let B be attached to the arm A and then let A make a complete rotation; as a consequence B has made a complete turn, Fig. 160, in the same sense as the arm.

Suppose by some means, gear B is made to turn in constrained motion relative to a moving arm or carrier, say by letting B roll on a stationary gear. This is the situation in Fig. 161, where gear 2 rolls on 1. If the arm A rotates CL, gear 2 turns CL owing to its rolling on 1. If the train value $e = D_1/D_2 = N_1/N_2$, considering the stationary gear as the first gear in the train, the number of times 2 turns due to its rolling is seen to be e for each turn of the arm or en_a CL for n_a CL turns of the arm. But as demonstrated above, gear 2 (B in Fig. 160) turns once for each turn of the arm by virtue of the fact alone that it is moving about center ac with the arm. Thus the total turns of 2 is the algebraic sum of these motions, or

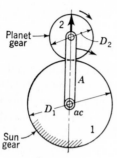

Planet gear

Sun gear

Fig. 161.

$$(e) \qquad n_2 = en_a + n_a,$$

the arithmetic sum of the two since en_a and n_a are in the same sense. At this point, the reader has decided correctly that planetary systems are problems in relative motion and has perhaps thought of the demonstration in § 94. Using the results of § 94 and the symbols of Fig. 161, we write

$$(f) \qquad \omega_{2/f} = \omega_{a/f} + \omega_{2/a} \qquad \text{or} \qquad n_{2/f} = n_{a/f} + n_{2/a},$$

which are general equations, applicable to the angular motions of bodies in the same or parallel planes. In words, equations (f) say that the angular velocity of gear 2 with respect to some frame of reference (the frame of the machine or the ground) is equal to the angular velocity of the arm with respect to the same frame algebraically plus the angular velocity of gear 2 with respect to the arm. [A revolution (n) rather than a radian (ω) is the more convenient angular measure in this phase of our study.]

To get a feel for equation (f) apply it to the situation in Fig. 161. The way to determine the turns of 2 with respect to the arm is to imagine the arm stationary with gear 1 turning in the opposite sense CC to n_a CL; that is, if clockwise is positive,

$$n_{2/a} = +\frac{N_1}{N_2} n_{1/a} = en_{1/a} \qquad \text{or} \qquad n_{2/a} = +en_{a/f} \quad \text{CL},$$

[ARM STATIONARY]   [1 STATIONARY]

in which we have used the train value $e = N_1/N_2$, considering the stationary gear as the first gear in the train, and we have noted that with 1 rotating CC, gear 2 rotates CL. Substituting this value of $n_{2/a}$ into equation (f), we get $n_{2/f} = n_{a/f} + en_{a/f}$ CL, which is seen to be the same as equation (e).

If a planet gear rolls on an internal gear, Fig. 162, the same approach may be used with due regard for directions of rotation. Equation (f) applies and $n_{2/a}$ is found as before by holding the arm stationary and rotating the internal gear 1 in the opposite sense. With the arm (carrier) rotating n_a CL, then gear 1 would rotate CC; and 2 also rotates CC with respect to its axis. Hence, if clockwise is positive, $n_{2/a}$ is negative, or $n_{2/a} = -(N_1/N_2)n_{1/a} = -en_{1/a} = -en_{a/f}$ CC. Substituting this value in (f), we get

$$n_{2/f} = n_{a/f} + (-en_{a/f}) = n_{a/f} - en_{a/f}.$$

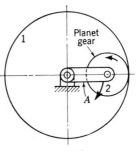

Fig. 162.

In the general case, the sun gear (or internal gear), as well as the arm, may rotate. If this is so, there are two input motions, via two gears or an arm and a gear. The analysis will be shorter if we start with the relative velocity equation in the form $n_{b/c} = n_b - n_c$. Using Fig. 161 with the sun gear in motion, we can write

$$n_{2/a} = n_{2/f} - n_{a/f},$$
$$n_{1/a} = n_{1/f} - n_{a/f},$$

the motion equations for 2 and A and for 1 and A, where the f suggests the frame of reference. A cross multiplication of the right-left sides of these equations gives

$$n_{2/a}(n_{1/f} - n_{a/f}) = n_{1/a}n_{2/f} - n_{1/a}n_{a/f}.$$

Solve for $n_{2/f}$, the "absolute" angular velocity of gear 2, and get

(g) $$n_{2/f} = n_{1/f}\frac{n_{2/a}}{n_{1/a}} + n_{a/f}\left(1 - \frac{n_{2/a}}{n_{1/a}}\right).$$

Since this is a general equation for any two-gear planetary system, Figs. 161 and 162, a consistent convention of signs must be employed; for clockwise $+$, counterclockwise $-$. Now note in equation (g) that $n_{2/a}/n_{1/a}$ is the speed ratio of 2 and 1 *with respect to the arm* and that this ratio is found by considering the arm stationary. If the gears are external gears, the angular velocities are necessarily in the opposite directions; hence the ratio would be a negative number. Also we can say that $n_{2/a}/n_{1/a} = N_1/N_2 = e$, the train value, considering gear 1 as the driver. Then we can drop the subscript f and simply remember that the velocities are absolute or relative to some convenient frame. Taking these steps, we write equation (g) as follows:

$$n_2 = en_1 + n_a(1 - e).$$

In this form, this equation applies to any planetary gear train if the correct interpretation is given the symbols. To help the interpretation, we write the equation as follows:

(30) $$n_L = en_F + n_a(1 - e) \qquad \text{or} \qquad e = \frac{n_L - n_a}{n_F - n_a},$$

where n_a = absolute turns of arm, CL +, CC —.

n_F = absolute turns of first gear in train, CL +, CC —.

n_L = absolute turns of last gear in train, CL +, CC —.

e = train value as defined by equation (29), considering first gear in train as a driver. If first and last gears turn in *same* direction, considering the train as an ordinary gear train (arm stationary), make sign of e *positive*. If as an ordinary gear train first and last gears turn in *opposite* directions, make e negative.

It does not matter which end of the train is taken as the first gear, but once a decision has been made, consistency is essential. The advantage of equation (30) is that, for better or worse, it avoids the detailed and repetitious thinking in applying the relative-velocity principle. However, an engineer, as opposed to a technician, would be expected to be able to think through such problems from the fundamentals; hence, the serious student will desire to solve at least a few problems in a basic manner. It is so easy to have a mix-up in signs that the reader is urged *to write in the sign of each term, including the positive sign.*

Study equation (30) and note that it divides the absolute motion of the last gear into three parts:

1. The turns that the gear has because it moves with the arm, n_a.

2. The turns that the the gear has because it rolls on a stationary sun gear, either directly or indirectly via a train with a train value of e (with gear 1 stationary), en_a.

3. The turns that the gear has because of the motion imparted by the rotation of the sun gear, en_F.

Since it often happens that a planetary gear train is adjoined to an ordinary gear train, it is imperative to note that equation (30) *applies only to the planetary train* and the ordinary part or parts of the whole train should be handled separately.

The velocity ratio remains as previously defined: the angular velocity of the input member (which may be either a gear or the arm) divided by the angular velocity of the output member. Quite high velocity ratios are attainable in planetary trains with only four gears. See problem **713** and others. If there is a double input, then the velocity ratios must be defined.

151. Example. In the planetary train of Fig. 163, the sun gear 1 turns 400 rpm CL and the internal gear 4 turns 50 rpm CC. The tooth numbers are shown in parentheses, Fig. 163. Gears 2 and 3 are attached together. (a) What is the angular velocity of the arm A and the angular velocity ratio 1 to A? (b) What is n_2?

Solution No. 1. (a) Considering gear 1 as the first gear in the train and therefore the driver in computing the train value, we have from equation (29)

$$e_{1\text{-}4} = -\frac{15 \times 15}{25 \times 55} = -\frac{9}{55},$$

negative because the first and last gears turn in opposite directions with the arm stationary (trace it out on the illustration). Using equation (30) with $n_F = n_1 = +400$ and $n_L = n_4 = -50$, we get

$$n_L = en_F + n_a(1 - e)$$

$$-50 = \left(-\frac{55}{9}\right)(+400) + n_a\left[1 - \left(-\frac{9}{55}\right)\right],$$

from which $n_a = +13.28$ rpm CL (the positive sign tells us that the sense is CL according to the assumed convention); $m_{\omega 1/a} = \omega_1/\omega_A = 400/13.28 = 30.1$.

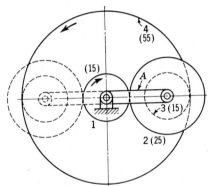

Fig. 163.

Solution No. 2. (a) Let gear number 4 be the first gear in the train. Then

$$e = -\frac{55 \times 25}{15 \times 15} = -\frac{55}{9}; \qquad n_F = -50; \qquad n_L = +400.$$

$$+400 = \left(-\frac{55}{9}\right)(-50) + n_a\left[1 - \left(-\frac{55}{9}\right)\right],$$

from which $n_a = +13.28$ rpm as before.

(b) Knowing the speed of the arm from the previous example, we may compute $n_2 = n_3$ starting at either end of the train. With 1 as the driver, the train value is $e_{1\text{-}2} = -15/25 = -0.6$; $n_F = n_1 = +400$; $n_a = +13.28$.

$$n_L = n_2 = (-0.6)(+400) + 13.28[1 - (-0.6)],$$

from which $n_2 = -218.7$ rpm, the negative sign indicating that it is turning opposite to the direction which has been assumed as positive, to wit, CC. From the internal gear as the first gear in the train, $e = +55/15 = +11/3$; $n_F = n_4 = -50$; $n_a = +13.28$.

$$n_L = n_3 = n_2 = \left(+\frac{11}{3}\right)(-50) + 1.328\left(1 - \frac{11}{3}\right),$$

from which $n_2 = -218.7$ rpm CC, as before.

Table 3

	Gear 1	Gears 2 and 3		Gear 4	
Turns of gear because it turns with arm	n_a	n_a		n_a	
Turns of gear 1 added to above line to give known rotations of 1 (arm stationary)	$400 - n_a$	Corresponding turns of gears 2 and 3	$\Big\} e_{1-2}(400 - n_a)$	Corresponding turns of gear 4	$\Big\} e_{1-4}(400 - n_a)$
Total turns by adding columns	$n_1 = 400$ (known)	$n_2 = n_a + e_{1-2}(400 - n_a)$		$n_a + e_{1-4}(400 - n_a) = -50 = n_4$	

Solution No. 3. A method of solving epicyclic gear problems by a separation of motions is illustrated in Table 3. On the first line is inserted the turns of the arm, because if the arms and gears are locked as a unit and the arm is turned n_a times, each gear turns n_a about its own axis. If there is no stationary gear in the train, choose a gear whose net motion is known, as gear 1 (or 4), and in the next line, insert the number of turns, positive or negative, which added to n_a gives the

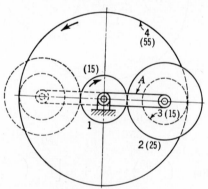

Fig. 163. Repeated.

correct number of turns for that gear. In Table 3, the known gear is taken as 1 and $400 - n_a$ added to n_a gives $+400$, the known turns of 1. Then in the column for the next gear in the train (the intermediate gears may be skipped if there is no interest in them), in this case gears 2 and 3, insert the number of turns of gear 2 which corresponds to the turns given gear 1 (arm stationary). This is the product of the train value and the turns of 1; similarly for all other gears in the train whose angular velocities are desired. Examination of the sums of these motions in the third line reveals that the results are the same as previously found. The use of different solutions as a check is highly recommended.

152. Reverted Planetary Trains and Balancing Centrifugal Forces.

In using epicyclic gear trains, we find it most convenient to have the

initial and final shafts with their axes collinear. When an internal gear
is involved as in Fig. 163, the axes of the arm, gear 1, and the internal
gear are naturally collinear. When only external spur gears are used,
somewhere the train must be reverted, as in Fig. 164, for example.
Here, the sun gear 1 may be stationary or it may be mounted on a con-
centric shaft and driven separately by an outside means. See some of
the problems. The shaft M is connected to (indicated by the dotted
lines of a key) and drives the arm. The
spindle for the planet gears 2 and 3, which are
attached together but free to turn on the
spindle, is fixed in the arm. Gear 4 is attached
to shaft Q. Observe that the center distance
must be the same for each pair (1-2 and 3-4)
of meshing gears. If standard interchangeable
gear teeth all have the same pitch, it follows
that the sum of the tooth numbers for each
pair is the same; $N_1 + N_2 = N_3 + N_4$. If
unequal addendum and dedendum gears are
used, this condition may not be met. Moreover, it does not necessarily follow that the
teeth are the same size. Gear 4, for example,
would actually need larger teeth than gear 1 if
it had a lower pitch-line speed, because in this

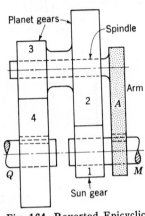

Fig. 164. **Reverted Epicyclic
Train.**

event, the force on the tooth for the same power transmitted would
be greater.

If the planet gears revolve other than slowly, it is desirable and usually
necessary to balance their centrifugal (inertia) force. The most obvious
way would be to have two sets of planets, 2 and 3, Fig. 164, the second set
being at 180° from the set shown. This same
idea is indicated by the dotted planet gears of
Fig. 163. More likely, three sets are used,
Fig. 165, sometimes four and more. Not
only is this a means of balancing centrifugal
forces, but with the power passing through
several sets of meshing teeth, the entire unit
may be smaller for the same power-transmitting capacity.

In the internal-gear type with several equally
spaced planets, one must be sure that the teeth
fit together properly on assembly. In the
simplest train, Fig. 165, a necessary condition

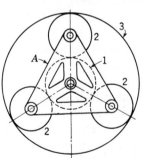

Fig. 165. **Multiple Planet
Gears.**

is met when the number of teeth in the internal gear 3 and the number
in the sun gear 1 are each divisible by the number of planets 2; in this case

$N_1/3$ and $N_3/3$ must be integers. To meet this condition, extended-center-distance gears are sometimes necessary.

153. Example—Planetary Train with Bevel Gears.

Four bevel gears arranged as shown in Fig. 166, with tooth numbers in parentheses, constitute a planetary

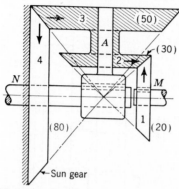

train. Gear 4 is stationary; gear 1 turns with shaft M at 50 rpm, its top surface moving upward (take this as the positive sense). Find the angular velocity of the arm A.

Solution. With 1 as a driver, the train value is

$$e = -\frac{20 \times 50}{30 \times 80} = -\frac{5}{12}.$$

The sign is negative because 1 and 4 turn in opposite directions as an ordinary train (arm stationary), as indicated by the arrows. This system of determining the directions of rotations of bevel gears, that is, by an arrow showing the direction of

Fig. 166. Planetary Train of Bevel Gears.

motion of the top surface, is advisable. Substituting known values in equation (30) ($n_L = n_4 = 0$, $n_F = n_1 = +50$ rpm), we get

$$n_4 = 0 = en_1 + n_a(1 - e)$$

$$= \left(-\frac{5}{12}\right)(+50) + n_a\left[1 - \left(-\frac{5}{12}\right)\right],$$

from which $n_a = +14.7$ rpm, the positive sign indicating that the arm A turns in the same sense as gear 1.

154. Bevel-gear Differential.

When an automobile moves in a curved path, the outside rear wheel must go farther and therefore rotate faster than the inside rear wheel. When the rear wheels are driving wheels receiving power from a single source, the dilemma is resolved by means of a differential. In Fig. 167(a) is depicted in diagrammatic outline the usual arrangement in automotive differentials. Gear 1 is driven directly by the drive shaft from the transmission. While the vehicle is moving in a straight line, there is no epicyclic action. The driving effort is transmitted to gear 2, through the pin A and the meshing teeth on P, 3, 4, through the splines between 3, 4 and the shafts to the rear wheels. The gears P, 3, 4 turn as a unit, that is, without relative motion. When the car goes around a curve, there is epicyclic action, as explained in the caption to Fig. 167.

When there is no resistance to the rotation of one wheel, with a differential like Fig. 167, no effort is transmitted to the other wheel—a situation equivalent to having one wheel on an ice slick and the other not.

Differentials (not illustrated) are now designed so as to transfer automatically some of the power to the wheel which has not lost its traction, greatly minimizing the chances of "getting stuck." In Fig. 167(b) is shown a planetary train 5 (internal gear), 6 (planet), 7 which can be connected into (or disconnected from) the line of power flow in order to provide a greater speed reduction (for increased torque on the rear wheels). The illustration is not clear enough for a detailed explanation of the operation, but the arrangement is similar to that in Fig. 165. There is no change in the function of the bevel differential itself.

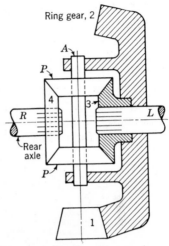

(a) Usual arrangement

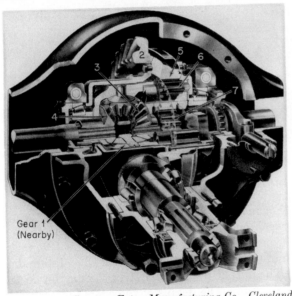

Courtesy Eaton Manufacturing Co., Cleveland

(b) Two-speed differential

Fig. 167. Bevel-gear Differentials. In (a), gear 4 always turns with the right axle R, gear 3 always turns with the left axle L. To understand the differential action, assume the extreme case in which the left wheel is jacked up while the right wheel remains on the ground. If power is now delivered to gear 1, gear 2 and the arm A turn, and gear 4 is the stationary sun gear with the planets P rolling around it. The result is a planetary drive A-P-3 to the left axle. If the automobile makes a turn toward the right, the action is similar except that gear 4, instead of being stationary, moves slower than 3. In (b) is shown a two-speed differential. Extra speed reduction results in extra torque available for the rear wheels.

155. Automatic Transmission. A hydromatic transmission is shown in Fig. 168. Since some of the detail is difficult to follow in this picture, the arrangement is shown diagrammatically in Fig. 169, the reverse gear train omitted. The parts of the transmission are named alike in these two illustrations. The main fluid coupling (toruses) are named M and C. The part B is permanently connected to the engine and always turns at engine speed. The small torus B, D operates sometimes (see caption). The sprags E and G are sometimes on, sometimes not. See Fig. 203, Chapter 11, for an illustration of a sprag connection. Also the clutches

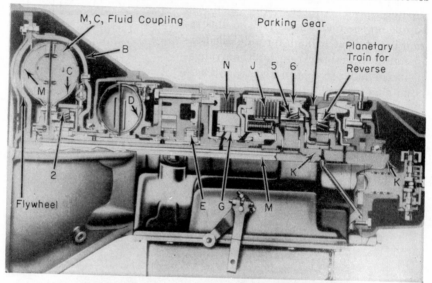

Courtesy General Motors Corp., Detroit

Fig. 168. Automatic Transmission.

N and J may or may not be engaged. At the left, the gear 3 is stopped when the sprag E engages and gears 1, 2, 3 then constitute an epicyclic train of the type in Fig. 165. At the right, gears 4, 5, and 6 constitute another train of the same type that operates when internal gear 6 is held stationary. The output motion is always the motion of the arm which carries the planets 5. The details of the events in each drive are given in the caption to Fig. 169.

156. Closure. The solution of actual gear-train problems is not always obvious and direct. These mechanisms, especially epicyclic trains, are often too complex to use the space for their description in a general textbook of this kind. The ingenuity apparent in some cases is wondrous to behold, a worthy monument to engineering accomplishment. A good acquaintance with the foregoing ideas should form a reasonable background for further study.

One should be warned against the low efficiency (large power loss) that may well accompany very large gear reductions in planetary trains. In general, the efficiency decreases as the velocity ratio increases. However, given the need for a high ratio, as 200, it might be achieved with several 4-gear planetary trains of widely different efficiencies (**91**). If a low-efficiency train is used in practice, it should be done knowingly.

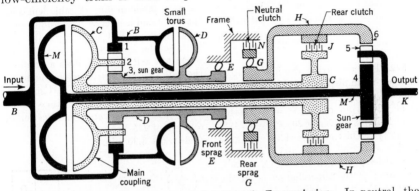

Fig. 169. Diagrammatic Arrangement of Automatic Transmission. In neutral, the neutral clutch N is open, sprag E is on and gear 3 is stationary, sprag G off, clutch J open, torus D, B is empty; B turns with the engine, causing C to idle because J is open; C drives M, so with K stationary, M, 4 drives 6, H, which turns idly.

In first speed (low-low), sprag E is on and 3 is stationary; B, D is empty; clutch N is engaged, sprag G is on, clutch J is open, and therefore part H (with internal gear 6) is stationary; the drive is B-1-2-C-M-4-5-K, through both gear trains.

In second speed, sprag E is off and 3 is not stationary; clutch N is engaged, sprag G is on, clutch J is open, and therefore part H (gear 6) is stationary; B, D is full and B is driving D, both at engine speed so that 3 and 1 have the same angular velocity and train 1-2-3 is turning as a unit; thus with no reduction in train 1-2-3, the drive is B-C-M-4-5-K.

In third speed, sprag E is on and gear 3 is stationary; clutch N is engaged, sprag G is off, clutch J is engaged so that part H is connected to and rotates with C; M also turns at the same speed as C via the coupling, so that the train 4-5-6 is turning without relative motion (not operating); the drive is B-1-2-C-M-K.

In fourth speed, the drive is direct, $\omega_K = \omega_B$; sprag E is off, coupling B, D is operating, so that $\omega_D = \omega_B = \omega_1 = \omega_3$ and 1-2-3 have no relative motion; clutch N is engaged, clutch J is engaged, sprag G is off, so that H and J turn together and $\omega_6 = \omega_C = \omega_B = \omega_4$; thus 4-5-6 have no relative motion; everything is turning together.

However, a study of this phase of the problem is beyond the scope of this book. Perhaps an ordinary spur or helical pair with an efficiency of about 97% in series with the planetary train is helpful. Sometimes two planetary trains in series are used.

Because of the advantages of planetary trains, there is renewed interest in them. Thus, the current technical journal literature would yield rewarding study for those who must become expert in this area. See references **88** to **92** and search the literature for others.

PROBLEMS

Note to Student. Always specify the sense of an angular velocity unless it is indefinite by virtue of the definition of the train.

Ordinary Gear Trains

671. A double-reduction helical gearbox similar to that in Fig. P132 has tooth numbers as follows: $N_1 = 28$ (attached to motor shaft and out of sight), $N_2 = 96$, $N_3 = 32$, $N_4 = 112$; $n_1 = 2400$ rpm CL. What is n_4?

672. The gear train driving a supercharger on an airplane engine is an ordinary reverted train (axes fixed). Input is at 2800 rpm of 1 ($N_1 = 68$ teeth); gear 1 drives 2 ($N_2 = 20$, $P_d = 7.7498$, $\phi = 22\frac{1}{2}°$). On the same shaft with gear 2 is gear 3 ($N_3 = 105$, $P_d = 14$, $\phi = 21.0878°$), which drives 4 ($N_4 = 56$). The rotor of the supercharger is on the same shaft with 4. What is the angular velocity of the supercharger? Give a sketch of the layout and directions of rotation. (Data courtesy Pratt & Whitney Aircraft.)

673. A motor running at 1500 rpm is connected by a train of gears to a drum shaft as suggested by Fig. P133: 1 = 3-thread worm RH, $N_2 = 90$, $D_3 = 6$ in., 4-pitch pinion, $N_4 = 100$, $N_5 = 15$

(bevel), $N_6 = 35$ (bevel), $N_7 = 30$, $N_8 = 12$; $D = 60$-in. drum. A cable winds on the drum and is used to raise and lower a mine hoist. (a) How long will it take to raise the hoist 800 ft.? (b) If the worm has a RH thread, does a point on the side of the drum nearest you move toward the right or left? Using all gears as given, what single change can be made that would reverse the direction of rotation of D?

Ans. (a) 3.96 min.

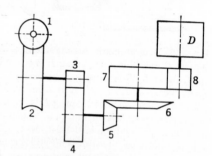

Fig. P133. Problem 673.

Courtesy Westinghouse Electric Corp., Pittsburgh
Fig. P132. Problem 671. Motor with Integral Gear Reducer.

674. In a worm gear box similar to Fig. 153, p. 244, worm A has 4 threads RH, gear B has 32 teeth, worm C has 4 threads RH, and gear D has 36 teeth. If the input is at $n_A = 1800$ rpm, what is n_D? Do A and D turn in the same or opposite direction? If one of the worms were left hand (LH), what would be the relative sense of n_A and n_D?

675. What is the angular velocity of the last gear of the following train if the first gear makes 125 rpm CL?

12 in.
6 in.
20 in._____6 in.
    12 in.
    20 in._____48 T (annular)
        24 T, 12 in.

Ans. 45 rpm CC.

676. In a 3-speed transmission, similar to Fig. 159, the tooth numbers are as follows: $N_1 = 22$ teeth, $N_2 = 32$, $N_3 = 28$, $N_4 = 26$. The shaft with gear 4 connects with the differential whose pinion has 12 teeth and ring gear 58 teeth (Fig. 166). The wheels are 30 in. in diameter. Indicate the train in the manner of the example of § 147. If the engine turns 2000 rpm, what will be the speed of the car?

Ans. 27.3 mph.

677. The same as **676** except that the engine speed is 3600 rpm.

678. Let the number of teeth in the gears of the transmission shown in Fig. 159 be: $N_1 = 19$, $N_2 = 27$, $N_3 = 24$, $N_4 = 26$, $N_5 = 18$, $N_6 = 30$, $N_7 = 18$, $N_8 = 16$. The ratio in the differential is 4.25 and the outside tire diameter is 28 in., and the speed of the car is 12 mph. Find the speed of the engine when the car is (a) in high gear, (b) in intermediate gear, (c) in low gear, (d) in reverse.

679. In a thread-cutting train on a lathe similar to that of Fig. 157(b), $N_1 = N_4 = 24$, $N_2 = N_3 = 21$; threads per inch on the lead screw = 8 RH; all gears have $P_d = 12$. Choose suitable tooth numbers for the outside stud 5 and the screw gear 8 for cutting 8 threads per inch on the work.

680. The same as **679** except that 14 threads per inch are to be cut.

Exact Velocity Ratios

681. A velocity ratio of 2.42 is desired. The maximum number of teeth should be 100, the minimum 12. (a) What is the closest approach if the number of gears is limited to two? (b) What is the closest if a 4-gear train may be used?

682. The same as **681** except that $m_\omega = 2.47$ is desired.

683. The value of the desired velocity ratio is $m_\omega = 2.7183$, the natural logarithmic base to five digits. No gear is to have more than 100 teeth or fewer than 12. Decide upon a suitable train if the number of gears is limited to (a) two, (b) four, (c) six.

684. The same as **683** except $m_\omega = \pi = 3.1416$ is desired.

685. A double reduction of $m_\omega = 22$ is desired, in which the limiting tooth numbers are 12 and 100. Specify the number of teeth in each gear.

686. A reverted spur gear train, Fig. P132, is to be designed for $m_\omega = 15$. If the pitch of all gears is the same, specify the number of teeth in each gear.

687. The same as **686** except that for gears 1 and 2, the pitch is 6, and for 2 and 3, the pitch is 4. What center distance is used?

Epicyclic Gear Trains

688. In a reverted epicyclic gear train similar to that of Fig. P134, $N_1 = 100$, $N_2 = 99$, $N_3 = 100$, $N_4 = 99$. If the sun gear 1 is stationary and the carrier A turns 100 rpm CL, what is $n_4 = n_Q$?

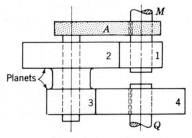

Fig. P134. Problems 688, 689.

What is the train value; the velocity ratio?

Ans. $n_4 = 20.41$ rpm CC.

689. The same as **688** but $N_1 = 1000$, $N_2 = 1001$, $N_3 = 1000$, $N_4 = 1001$.

690. A gear train as in Fig. P135 has gears as follows: $N_1 = 20$, $N_2 = 100$, $N_3 = 40$, $N_4 = 60$, $N_5 = 30$, $N_6 = 70$. The axes for shafts M, R, Q are fixed: $n_M = 500$ rpm CL; $n_R = 50$ rpm CC and drives the sun gear 3. What is the angular velocity of 6?

Ans. 85.75 CC.

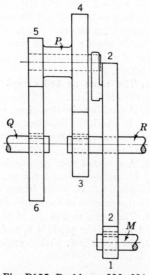

Fig. P135. Problems 690, 691.

691. The same as **690** except that the planetary train is changed as follows: $N_3 = 50$, $N_4 = 45$, $N_5 = 50$, and $N_6 = 45$.

692. In Ferguson's paradox, three outputs are obtained from a single input, say shafts P, D, Q, Fig. P136, to which

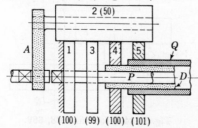

Fig. P136. Problem 692.

gears 3, 4, and 5 are, respectively, attached. If the arm turns 100 rpm CL, what are the rpm and directions of rotations of shafts P, D, and Q? The tooth numbers are shown in parentheses. Gears 1, 3, 4, and 5 mesh with gear 2.

Ans. $-1, 0, +1$.

693. In the train of Fig. P137, the arm turns $n_a = 600$ rpm CL; gear 7 turns $n_7 = 400$ rpm CL. The tooth numbers are shown in parentheses. (a) Determine the angular velocity of shaft P and the velocity ratio n_Q/n_P. (b) If the speed of gear 7 is increased slightly, does the velocity ratio n_Q/n_P increase or decrease? The picture is diagrammatic.

Ans. (a) $m_\omega = 2.88$, (b) increases.

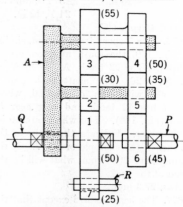

Fig. P137. Problems 693, 694.

694. The same as **693** except that $N_3 = 110$ and $N_4 = 100$.

695. In Fig. P138, the arm A turns 100 rpm CC and gear 1 turns 50 rpm CL.

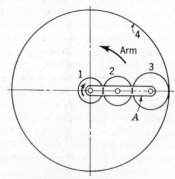

Fig. P138. Problems 695–697.

Tooth numbers are: $N_1 = 16$, $N_2 = 18$, $N_4 = 100$, all standard interchangeable type. What is the angular velocity of 4?

Ans. 76 rpm CC.

696. The same as **695** except that $N_1 = 20$.

697. The same as **695** except that $N_1 = 28$.

698. The tooth numbers of the gears of Fig. P139 are: $N_1 = 200$, $N_2 = 50$, $N_3 = 25$, $N_4 = 45$, $N_5 = 30$. If $n_5 = 50$ rpm CL, what is n_a?

Ans. 4.55 CC.

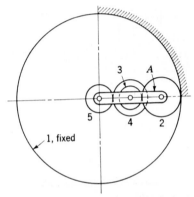

Fig. P139. Problems 698–700.

699. The same as **698** except that $n_a = 30$ rpm CL and n_5 is desired.

Ans. 330 CC.

700. The tooth numbers of the gears of Fig. P139 are: $N_1 = 200$, $N_2 = 20$, $N_3 = 70$, $N_4 = 18$, $N_5 = 72$; $n_5 = 50$ rpm CL. What is n_a?

Ans. 175 CL.

701. Figure P140 shows a Yale and Towne hoist, the operation of which is as follows: The hand-chain wheel A turns gear F. Gears D and E are fastened together. Gear C is a stationary internal gear meshing with D-D. When the hand-chain wheel turns, F drives E-E which, because D must roll around C, revolve around F. The load sheave B rotates at the same speed as the arm to which E-E are attached. This is a reverted epicyclic gear train, F-E-D-C, with C stationary. The diameters of the various wheels and gears are: handwheel $A = 10$ in., load wheel $B = 3\frac{1}{4}$ in.,

gear $D = 1\frac{7}{8}$ in., gear $C = 7$ in., gear $E = 3\frac{5}{8}$ in., gear $F = 1\frac{1}{2}$ in. (a) If the hand chain is moved 100 fpm, how fast does the load move? (b) Use the principle of virtual velocities (zero friction), § 96, and compute the load that can be raised by a 50-lb. pull on the hand chain.

Ans. (a) 3.25 fpm, (b) 1540 lb.

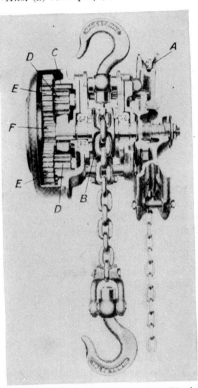

Fig. P140. Problem 701.   Chain Block.

702. A speed reducer between an airplane engine and the propeller is similar to Fig. 165, p. 265. However, there are 10 planet pinions instead of 3. The engine is directly connected to the sun gear 1 and turns at 3800 rpm cruising. The ring gear 3 is "stationary." The propeller turns with the arm; $N_1 = 69$, $N_2 = 23$, and $N_3 = 115$. What are the angular velocity of the propeller and the velocity ratio?

Ans. 1425 rpm. (Data courtesy Pratt and Whitney Aircraft.)

703. The tooth numbers of the gears in a transmission similar to that of Fig. 169 are: $N_1 = 51$, $N_2 = 13$, $N_3 = 24$, $N_4 = 39$, $N_5 = 12$, $N_6 = 63$. Let the ratio in the differential be 3.2 and the speed of the engine be 4200 rpm. The diameter of the tires is 29.5 in. What is the speed (miles per hour) of the car in each gear? See caption to Fig. 169.

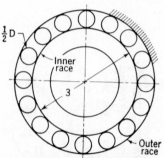

Fig. P141. Problem **704.**

704. Let Fig. P141 represent diagrammatically a ball bearing on an electric motor shaft. The outer ring (and race) is stationary. If the speed of the motor is 1800 rpm, determine the angular velocity of the balls and the linear speed of their center. The balls are in pure rolling.

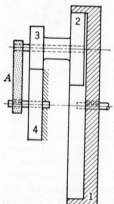

Fig. P142. Problem **705.**

705. A speed reducer between an airplane engine and the propeller is defined diagrammatically in Fig. P142. The

engine crankshaft is directly connected to the internal gear 1; $N_1 = 120$. There are three pairs of planets 2, 3, which turn together; $N_2 = 45$, $N_3 = 27$. The carrier A and the propeller are splined to the same shaft. The sun gear 4 is "stationary"; $N_4 = 48$. If the motor is turning 2500 rpm CL, what is the angular velocity of the propeller?

Ans. 1500 rpm CL.

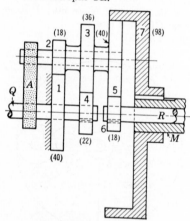

Fig. P143. Problem **706.**

706. In Fig. P143, the shaft Q is keyed to gear 4 and turns 1000 rpm, say CL. What are the angular velocities of shafts R and M, keyed, respectively, to gears 6 and 7? The tooth numbers are in parentheses.

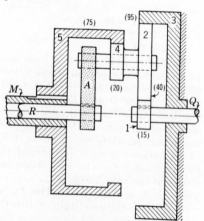

Fig. P144. Problem **707.**

707. In Fig. P144, shaft Q, turning at 1000 rpm, is keyed to gear 1; internal gear 3 is stationary; the arm is attached to shaft R; internal gear 5 drives the hollow shaft M. What are the angular velocities of R and M (sense same as or opposite to Q)?

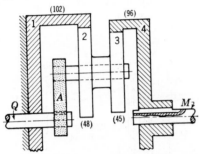

Fig. P145. Problems 708–710.

708. In Fig. P145 is shown diagrammatically an arrangement capable of high velocity ratios, the lower part cut away. The input is via shaft Q, keyed to the arm A. Internal gear 1 is stationary. Output gear 4 is attached to shaft M. Tooth numbers are in parentheses. If $n_Q = 600$ rpm, what are n_M (and sense) and the velocity ratio?

Ans. $m_\omega = 256$.

709. The same as **708** except that the tooth numbers are: $N_1 = 202$, $N_2 = 100$, $N_3 = 99$, and $N_4 = 200$.

Ans. $m_\omega = 10,000$.

710. Choose tooth numbers for a gear train arranged as shown in Fig. P145 with a velocity ratio of closely 300. See reference **91**.

711. The D. O. James Gear Manufacturing Company manufactures a line of epicyclic-gear motor reducers, one of which is represented by Fig. P146. Gear 1 receives power from the motor; the carrier C delivers it. This is a series connection of 3 epicyclic trains, with arms (planet carriers) A, B, and C; A and 4 are integral; so are B and 7. The tooth numbers are: $N_1 = 15$, $N_2 = 36$, $N_3 = 87$ $(P_d = 8)$; $N_4 = 15$, $N_5 = 24$, $N_6 = 63$ $(P_d = 6)$; $N_7 = 15$, $N_8 = 18$, $N_9 = 51$ $(P_d = 4.5)$. Determine the

velocity ratio and the speed of shaft Q if the motor turns 1750 rpm.

Ans. $n_Q = 11.25$ rpm.

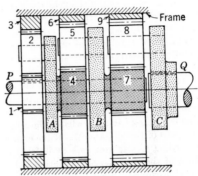

Fig. P146. Problems 711, 712.

712. The same as **711** except that there are 4 trains (add gears 10, 11, 12) and tooth numbers are: $N_1 = 15$, $N_2 = 42$, $N_3 = 99$ $(P_d = 6)$; $N_4 = 15$, $N_5 = 27$, $N_6 = 69$ $(P_d = 4.5)$; $N_7 = 15$, $N_8 = 24$, $N_9 = 63$ $(P_d = 4)$; $N_{10} = 19$, $N_{11} = 23$, $N_{12} = 65$ $(P_d = 4)$.

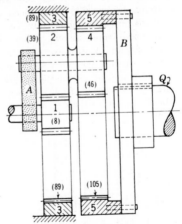

Fig. P147. Problem 713.

713. Winsmith, Inc., produces a line of planetary gear reducers using a train of helical gears like the one shown diagrammatically in Fig. P147. The high-speed pinion is integral with the drive shaft. Internal gear 3 is stationary. The output is via gear 5 and the attached member *B*. Tooth numbers are: $N_1 =$ 8, $N_2 = 39$, $N_3 = 89$, $N_4 = 46$, $N_5 =$ 105. If the gear 1 turns 1750 rpm, what are the speed of Q and the velocity ratio? Compute the velocity ratio to 5 places. *Ans.* $m_\omega = 49{,}652$.

714. In the differential of Fig. 167(a), p. 267, the tooth numbers are as follows: $N_1 = 11$, $N_2 = 54$, $N_3 = N_4 = 16$, $N_p = 11$. The right rear wheel is on the ground, the left one is jacked up, the transmission is in high gear. If the engine is turning 540 rpm, what is the angular speed of the left wheel? If the engine is turning CL as you look at it from in front of the car, in what direction is the wheel turning? Show a sketch tracing the motions and justifying your answer. *Ans.* 148 rpm.

715. Figure P148 indicates part of a train of mechanism used to vary the speed of the driven shaft Q. Power comes in on shaft M, turning gear 1 CL at a constant speed of 350 rpm. Gear 3 acts as the arm carrying the bevel planet gears 2, 2; it rotates about the axis, and it is driven by gear 5. The speed of gear 5 is varied automatically by a variable-speed transmission. What are the turns of Q and its direction of motion, if (a) gear 5 turns 300 rpm CL, (b) gear 5 turns 400 rpm CL? Take the direction of turning as viewed from the right. Numbers in parentheses are tooth numbers.

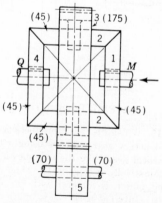

Fig. P148. Problems 715, 716.

716. The same as **715** except that gear 5 turns counterclockwise.

717. The input motion in Fig. P149 is 100 rpm CC (as viewed from right to left) of shaft M. The tooth numbers are as given in parentheses. If gear 5 (and shaft R) are stationary, what is the angular velocity of shaft Q (and gear 4)? In what direction does the arm turn looking down toward it?

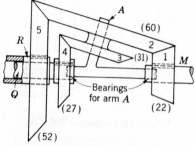

Fig. P149. Problems 717, 718.

718. The same as **717** except that $n_5 = 30$ rpm CL.

719. The input for the train of Fig. P150 is via shaft M (fixed axis), keyed with gears 1 and 2 and turning at 300 rpm. Shaft R is a stationary axle on which both the cluster 3, 4 and gear 8 (attached to the arm A) rotate. The output is via gear 7 keyed to Q. The planets are gears 5 and 6. What are the speed of Q and its sense relative to M? (A study of this basic arrangement is given in reference **88**.)

720–730. These numbers may be used for other problems.

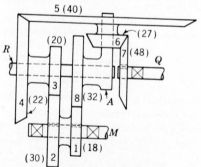

Fig. P150. Problem 719.

10

Flexible Connectors and Friction Wheels

157. Introduction. Flexible connectors, such as flat belts, V belts, roller chains, are useful for transmitting power between shafts whose center distance is too great for gearing; and they are also used as conveyors and hoists. If a certain velocity ratio must be maintained under all load conditions, some form of toothed wheels must be used, because when friction is relied on to transmit power, some slippage is inevitable.

In this chapter, we shall discuss flat belts, V belts, timing belts, roller chains, inverted chains, various friction wheels, and some variable-speed drives. The kinematics of these mechanisms is ordinarily quite simple, but there are several other points worth noting. Moreover, an acquaintance with what is being done is often suggestive of what can be done in a new situation, so that the engineer needs to know about these things.

158. Flat Belts. Flat belts are made of a number of different materials, including leather, fabric, rubber-impregnated fabric, and combinations of leather and synthetics. Since drives dependent on friction would not be used if a precise velocity ratio were important, this ratio is taken as

$$(a) \qquad m_\omega = \frac{\omega_1}{\omega_2} = \frac{n_1}{n_2} \approx \frac{D_2}{D_1} = \frac{r_2}{r_1},$$

where the subscript 1 refers to the driving pulley and 2 to the driven pulley. If the ratio between shafts is unusually large, it may be desirable to bring about the speed change through one or more intermediate shafts, variously known as *countershafts* or *jackshafts*.

With the judicious use of guide pulleys (see *law of belting* below), belts may be arranged to drive any two shafts in space if such drives seem to be necessary (see Figs. 170, 171, and 172); but preponderantly, flat belts connect parallel shafts. If an *open-belt* arrangement is used, Fig.

277

174, the shafts turn in the same direction. When the belt is crossed, Fig. 175, the shafts turn in opposite directions. Because of the consequences of the belt rubbing on itself, a crossed belt should be used only for a good reason.

All belt drives must satisfy the **law of belting,** which says that *the center line of the belt as it approaches a pulley must lie in a plane perpendicular to the axis of that pulley;* otherwise, the belt will run off the pulley. The

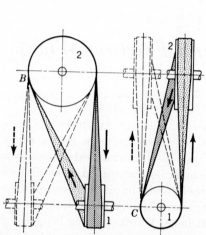

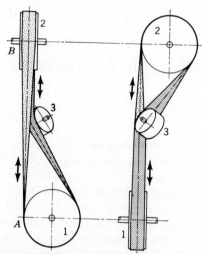

Fig. 170. Quarter-turn Belt. The condition to be met to satisfy the law of belting is that a plane passed through the middle of the face of one pulley perpendicular to its axis must be tangent to the face of the other pulley. For the belt shown in solid lines, the upper pulley 2 must rotate clockwise; otherwise the strand *B*-1 would run off pulley 1 and strand *C*-2 would run off pulley 2. The pulleys can be arranged for either direction of rotation of 2, as suggested by the dotted position, but once installed it cannot be reversed.

Fig. 171. Reversible Quarter-turn Belt. If the main pulleys 1 and 2 are located as defined in Fig. 170 (strand *AB* is perpendicular to both main axes) and if a guide pulley 3 is used in a position such that the belt between it and each of the main shafts satisfies the law of belting, this system can drive in either direction.

belt may *leave* a pulley at some angle other than 90° with the shaft, Fig. 170, without disadvantageous effects. As a practical matter, it is also necessary for the pulleys to be crowned (or else have side flanges) in order for the belt to stay on. The action of a belt on a crowned pulley is suggested by Fig. 173.

In the position shown in Fig. 173(a), the point *M* on the belt has just made contact with the pulley. If the belt does not slip, this point *M* on the belt moves to point *N* on the pulley in a quarter turn, and the belt will have moved toward the center plane of the pulley. When M on the belt has reached point N on the pulley, the belt is still bent, so that the foregoing action repeats itself as long as the belt is not centered. The **height of crown,** which may be of the order of $\frac{1}{8}$ in./ft. of face width, depends upon the width, length, and speed of the belt; the wider and longer the belt and the greater its speed, the smaller the crown required.

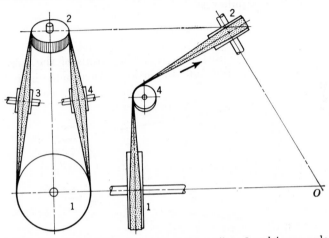

Fig. 172. Belt for Intersecting Shafts.   Two guide pulleys 3 and 4 are needed in order to satisfy the law of belting. In the left view, observe that the axes of the guide pulleys are normal to lines that are tangent to the pulleys 1 and 2. In the right view, notice that the faces of the guide pulleys are tangent to planes perpendicular to the main axes at the mid-section of each pulley.

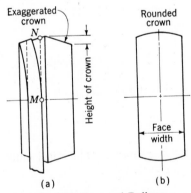

Fig. 173. Crowned Pulleys.

159. Length of Flat Belts.   No matter what the configuration of the pulleys, the length of belt is computed on the basis of the geometry of the layout (or measured with a steel tape). Then in order to provide a reasonable **initial tension,** the unloaded belt is made somewhat shorter (see handbooks and manufacturers' recommendations) and stretched onto the pulleys.   It is convenient to consider the length in parts: for example, (1) the length not in contact with either pulley, plus (2) the length in contact with the larger pulley, plus (3) the length in contact with the smaller pulley.

For an open belt, we see from triangle ABE in Fig. 174 that the length *not* in contact with a pulley is

(**b**) $$2[C^2 - (R - r)^2]^{1/2},$$

in which C is the center distance, R is the radius of the larger pulley, and r is the smaller radius (all dimensions are usually in inches).   If

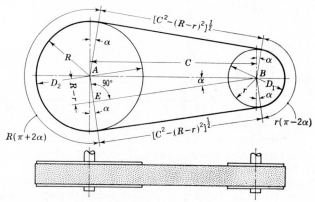

Fig. 174. Open Belt.

$\alpha = \sin^{-1} AE/AB = (R - r)/C$, Fig. 174, the length in contact with the larger pulley is

(**c**) $$R(\pi + 2\alpha) = R\left(\pi + 2\sin^{-1}\frac{R - r}{C}\right).$$

The length in contact with the smaller pulley is

(**d**) $$r(\pi - 2\alpha) = r\left(\pi - 2\sin^{-1}\frac{R - r}{C}\right),$$

in which the angles are in radians.   The total length is the sum of the parts (**b**), (**c**), and (**d**), or

(31) $$L = 2[C^2 - (R - r)^2]^{1/2} + \pi(R + r) + 2(R - r)\sin^{-1}\frac{R - r}{C}.$$

Since pulley sizes are expressed as their diameters, it is convenient to use D's.   Also, the nature of the precision required is such that an approximation is satisfactory.   If α is not too large, we may use α for $\sin \alpha$ with little error.   The term in the brackets can be expanded by the binomial theorem,* with negligible terms discarded.   These operations give (D_2 = larger pulley diameter)

(e) $$L \approx 2C + 1.57(D_2 + D_1) + \frac{(D_2 - D_1)^2}{4C},$$

an equation often found in handbooks.

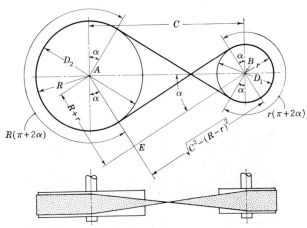

Fig. 175. Crossed Belt.

For the crossed belt, Fig. 175, the length BE not in contact is found as before, triangle AEB; $2[C^2 - (R + r)^2]^{1/2}$.   The length of belt in contact with the larger pulley is

$$R(\pi + 2\alpha) = R\left(\pi + 2\sin^{-1}\frac{R + r}{C}\right).$$

The length in contact with the smaller pulley is

$$r(\pi + 2\alpha) = r\left(\pi + 2\sin^{-1}\frac{R + r}{C}\right).$$

The total length is the sum of the parts, or

(32) $$L = 2[C^2 - (R + r)^2]^{1/2} + (R + r)\left(\pi + 2\sin^{-1}\frac{R + r}{C}\right),$$

in which the angle is in radians.   Observe that the length of an open belt depends upon both the *sum and difference* of the radii (or diameters) and that the length of the crossed belt depends only upon the *sum* of the radii.   The exact open-belt formula can be solved by trial and error

* $(1 + x)^n = 1 + nx + \dfrac{n(n - 1)}{2!}x^2 + \dfrac{n(n - 1)(n - 2)}{3!}x^3 + \cdots$

if one of the unknowns is R or r (D_2 or D_1) as happens in designing step cones, Fig. 176. Each step necessarily accommodates the same belt length L.

Belt drives, flat and V, are quiet in operation. Because of their comparatively large deformation (stretch) under a particular load, they absorb a larger proportion of any shock coming on one shaft than do more rigid bodies and thus transmit less force to the other shaft. Also, any belt with a relatively large unit deformation, particularly flat belts, undergoes a phenomenon known as *creep*. Creep occurs because the *driving* pulley receives a tightly stretched belt and delivers the belt with the slack-side tension in it. Since the force in the belt decreases as it

Courtesy South Bend Lathe Works, South Bend, Ind.

Fig. 176. Step Cone on Lathe. The belt is shifted manually from step to step according to the spindle speed desired. This arrangement is part of the lathe from which the train of Fig. 157 is taken. The back gears may be meshed at will. By their use, the number of available speeds of the spindle for a constant-speed driving pulley is doubled: in this case, four speeds without back gears (4-step cone) and four speeds with back gears. Given the diameters of the steps on this step cone, the diameters of the steps on the mating cone on the drive shaft must match the same belt length. Knowing L and D_1, say, solve for D_2 from equation (e) or (31). There would be an economic value in having the driving and driven cones identical.

passes about the driving pulley, the deformation decreases and the driver delivers a shorter length of belt than it receives. Hence, with less length leaving than approaching, the belt velocity is less than that of a point on the surface of the pulley. There is also some slippage in any friction drive. Slip and creep add, and together are usually called *the* slip. They also add on the driven pulley, where a point on the pulley face moves somewhat slower than the belt. The order of magnitude of the total slip effect would be some 2% to 3% for heavily loaded flat leather belts on iron or steel pulleys.

160. V-belt Drives. V-belt drives are particularly suited for short-center drives. While they nearly monopolize the field for light power transmission, for example, the fan belt on automotive engines, they are also common in transmissions of much greater power.

Because the sides of the belt are sloping and fit into V-shaped grooves, the normal force, and therefore the frictional force for a particular coef-

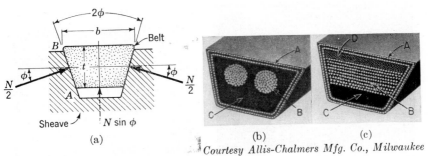

(a)   (b)   (c)

Courtesy Allis-Chalmers Mfg. Co., Milwaukee

Fig. 177. V Belts. A certain tension in the belt produces a radial force $N \sin \phi$ and a total normal force N on the sides of the groove.

ficient of friction, is much greater for a given tension in the belt than if the belt were resting on a flat surface [Fig. 177(a)]. The details of the construction of V belts vary with the manufacturer, but in all cases, there are some *tension members B*, Fig. 177, generally made of cotton, rayon or other synthetic fibers, or occasionally steel wires or cables. All belts have a resilient cushion material C, which is natural or synthetic rubber in most cases; sometimes other synthetic compounds are used for a reason, for example, oil-resistant neoprene. If the oily condition is not too severe, the outer cover only may be neoprene-impregnated. Some sort of "rubberized" fabric cover A, Fig. 177, is common to all V belts.

Courtesy Allis-Chalmers Mfg. Co., Milwaukee

Fig. 178. V-belt Sheave. This sheave has a split, tapered bushing, which is the usual practice for stock sizes. The groove angle 2ϕ is 34°, 36°, or 38°.

The grooved members over which the belts run are called **sheaves**, Fig. 178. A particular size of V belt (**5**) in a particular groove will "ride" at a sort of average diameter called the **pitch diameter**. This is the diameter used in calculating the belt length and belt speed. Manufacturers' catalogs give pitch diameters for standard sheaves and belts. The length of V belt is computed from equation (**e**), in which D_1 and D_2 are the pitch diameters of the sheaves.

Multiple V-belt drives are commonly used on industrial drives, Fig. 179(a). In case one of the belts breaks, the other belts will carry the load without loss of production time until it is convenient to repair. However, a new belt should not be used with old belts, because the new belt, being unstretched and shorter, will take nearly all of the load and, being overloaded, will soon wear out and break. Replace all belts with a

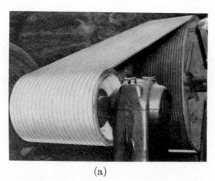

(a)   (b)

Courtesy Gates Rubber Co., Denver

Fig. 179. V-belt Drive.   In (a) is seen a 20-belt drive in a flour mill; in (b) is shown a double-V drive for steam rolls.

matched set.   With the use of guide pulleys, V belts can be used to connect nonparallel shafts.   See the law of belting, § 158.   V-belt drives are also designed with the smaller driving member a grooved sheave and with the larger and more expensive member having a flat face (perhaps

Courtesy United States Rubber,
New York

Fig. 180. Timing Belt.

it is a flywheel); these drives are called V-flat drives.   Double V belts are made so that either side may run in a groove (the belt is approximately hexagonal); this makes it feasible to drive sheaves with both sides of the belt, putting reversed bends in the belts, Fig. 179(b).

161. Timing Belt.   A belt that combines some of the advantages of belts and chains (§ 162) is the so-called timing belt, Fig. 180.   Because of the materials of which it is made, it is quiet; because of the teeth, there is no slippage and little initial tension is needed.   Once the belt is mounted, the angular relationship of the driving and driven members remains the same.   The manufacturer states that belt speeds over 16,000 fpm are feasible.   The tensile load is carried by small steel cables embedded in the neoprene backing; thus the deformation under load is minimized.   Moreover, the steel provides a comparatively good load-carrying capacity.   The teeth are neoprene and are surfaced with a nylon fabric facing, since experience suggests that this surface has good wearing and frictional properties.   The pitch surface is taken along the center plane of the wire cables.   Belts with five different pitches (0.2, $\frac{3}{8}$, $\frac{1}{2}$, $\frac{7}{8}$, and $1\frac{1}{4}$ in.) are available.

162. Roller Chains.   A single roller chain is shown in Fig. 181, which also gives the names of the parts.   Because toothed wheels are used, there is no slippage and the rotating members maintain their original angular relationship—except for a variation of angular velocity ratio while a link is coming into full contact with the sprocket (§ 163).   Chain drives in a general way fit into situations where the shafts are too far apart for gears and yet could be said to be on short centers, also in environments where belting would be inappropriate for various reasons (for example, high temperature) and where friction could not safely be

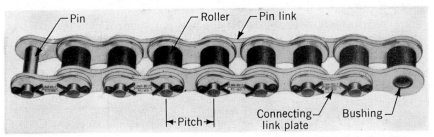

Fig. 181. Roller Chain.   When the chain bends around the sprocket, the bushing and pin move relatively to each other.

relied upon as the driving force.   Chains run on **sprockets** (see manufacturers' catalogs for sprocket details).   The standard sprocket tooth has been so proportioned that the rollers of a new chain ride low in the tooth space.   As the chain stretches in use, mostly because of wear in the joints, its pitch increases and the chain supposedly rides higher in the tooth space, still making contact with most of the teeth.   The length of a chain is measured along its center line.   Multiple chain widths, double, triple, etc., are widely used to provide greater power-transmitting capacity.

163. Chordal Action of Chains.   Chordal action is explained for the 7-tooth sprocket of Fig. 182, in which the effect is exaggerated (as few as

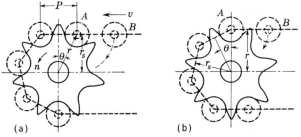

Fig. 182. Chordal Action.   The angle $\theta = 180/N_t$; $\sin \theta = P/2r$, or the effective pitch diameter is $D = 2r = P/\sin \theta$; N_t = number of teeth in the sprocket; P is the pitch of the chain.

7 teeth are unlikely). In Fig. 182(a), the roller A, which has just seated, is seen to be at a vertical distance r_s from the center of the sprocket. This radius r_s is smaller than the radius r of the center of the roller after a rotation of θ, as shown in Fig. 182(b). If we assume that this sprocket turns at a constant speed of n rpm, the speed of the center line of the chain changes from

$$v = 2\pi r(\cos \theta)n = 2\pi r_s n \qquad \text{to} \qquad v = 2\pi rn$$

and back to the lower speed during every cycle of tooth engagement. Since change of velocity means acceleration, there is a periodic force ($F = ma$) conducive to vibration and noise. Evidently also, since the chain changes its radius from r_s to r to r_s in each cycle, there are radial accelerations which compound the difficulty. If the two sprockets are of

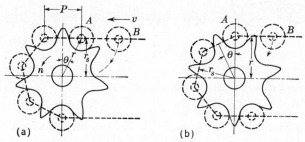

(a)   (b)

Fig. 182. Repeated.

different size, a constant velocity ratio will not be transmitted. These accelerations make it necessary to limit the speed of a chain and the minimum number of teeth to be used on the sprocket. A 21-tooth sprocket may be run satisfactorily at a higher speed than a 16-tooth sprocket. A small chain (less mass) may be run satisfactorily at a higher speed than a large chain. Another factor is the impact with which the chain roller contacts the tooth. Some impact arises because of small differences in the pitch of the teeth and chain. Another impact effect arises as the chain wraps on the sprocket. The fact of this impact can be seen by assuming that the rollers seat on the bottoms of the tooth spaces and imagining the sprocket stationary in Fig. 182(a). Then let the chain AB wrap onto the sprocket with angular speed n CL (same relative motion). It is seen that roller B, rotating about A, strikes the tooth while its center is moving with a speed of $v_B = 2\pi Pn$ ips if P is the chain pitch in inches. This impact not only contributes to limiting the speed but is the source of wear on the teeth and roller. This impact is observed to be a function of the chain pitch (given speed), while the chordal action is a function of the number of teeth. The change of radius is $r - r \cos \theta = r(1 - \cos \theta)$. With a large number of teeth,

$1 - \cos \theta$ approaches unity and the chordal action approaches zero. Actually, the number of teeth on the smaller sprocket will probably be between 16 and 30 and usually an odd number.

164. Inverted-tooth Chains.   Inverted-tooth chains, commonly called silent chains, consist of a series of laminations or links, Fig. 183, which are

(a)   (b)

Courtesy Link-Belt Co., Chicago

Fig. 183. Silent Chain Drive.   The details of the construction of the chain are clearly seen in (b).

assembled by means of pins and bushings to any desired width of chain. The chains of different manufacturers differ in the details of the construction of the joints; Fig. 184 shows another manufacturer's method.

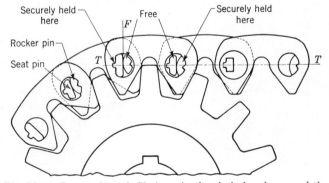

Fig. 184. The Morse Inverted-tooth Chain.   As the chain bends around the sprocket, the rocker pin rolls on the seat pin.

Fig. 185. Morse Hy-Vo Sprocket Tooth.

In the traditional silent chain, the tooth face and the surface of the link in contact are plane surfaces, Fig. 184. Recent developments aim to approach conjugate action between the sprocket tooth and the chain tooth, the purpose being to reduce or eliminate chordal action. The Morse Company makes a curved-profile (involute) tooth to engage a straight-face link (rack), Fig. 185. Thus the contact action is analogous to that between a pinion and rack. With chordal effects reduced, higher chain speeds become possible.

There are many designs of chains suited to various uses: block chains, Fig. 186, for low speed, low power, and low initial cost; a number of chains used for conveyors; and the ordinary link chains for exerting force (hauling, hoists, etc., Fig. P140, p. 273). Space limitations preclude detailed descriptions.

Fig. 186. Block Chain.

165. Friction Drives. Any of the gear forms in which the pitch surfaces roll on one another, principally spur and bevel, can be and are frequently used as friction drives. To transmit power by friction, a normal force is necessary,

(Frictional force) = (coefficient of friction) × (normal force).

Spur friction wheels are shown in Fig. 187; bevel wheels in Fig. 188. As for the corresponding toothed wheels, the velocity ratio is inversely as the radii to the point of contact, slippage neglected.

Fig. 187. Spur Friction Wheels.

Fig. 188. Bevel Friction Wheels.

Bevel wheels take forms analogous to bevel gears, Fig. 189. If cone 1 is the driver in each situation of Fig. 189, the velocity ratio without slippage is $m_\omega = \omega_1/\omega_2 = r_2/r_1$. Since $r_1 = OM \sin \gamma_1$ and $r_2 = OM \sin \gamma_2$, Fig. 189, the velocity ratio of two rolling cones is

$$(f) \qquad m_\omega = \frac{\omega_1}{\omega_2} = \frac{n_1}{n_2} = \frac{r_2}{r_1} = \frac{\sin \gamma_2}{\sin \gamma_1}.$$

The shaft angle Σ is seen to be $\Sigma = \gamma_2 \pm \gamma_1$, where the $+$ or $-$ depends upon the configuration. For external bevels, $\Sigma = \gamma_2 + \gamma_1$; hence, for a

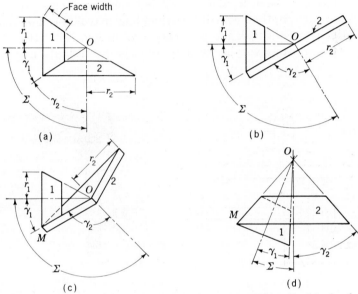

Fig. 189. Bevel Cones. In (a), the shafts are at right angles. In (b), the shaft for **2** has been moved until the angle between shafts is $\Sigma = 90° + \gamma_1$ and the cone has become a flat disk analogous to a crown gear. In (c), the angle Σ between shafts is greater than $90 + \gamma_1$, and the contact is on an internal element of the pitch cone **2**, a situation analogous to an internal bevel gear. However, it is not necessary for Σ to be greater than $90° + \gamma_1$ in order to have internal contact, as shown in (d).

specified shaft angle and velocity ratio, either γ_1 or γ_2 can be eliminated in equation (f). Also, a graphical method of finding the cones for a particular velocity ratio and shaft angle is shown in the Appendix, § 182.

A cone drive is sometimes a useful device for reversing the direction of rotation of the driven shaft. If, in Fig. 188, there were another mating pinion cone (either friction or toothed) on the right-hand side of the big cone at B, the pinion cones could be placed far enough apart that a shift of the pinion shaft P toward the right would engage the pinion cone 1, and a shift of the shaft leftward would engage the other pinion cone, thus reversing the direction of rotation of the big cone.

166. Variable-speed Drives. Variable-speed drives play an important and indispensable role in modern industry. Change gearboxes, such as the automotive transmission, allow a change of speed ratio in steps, but there are myriad situations in which it is necessary or advantageous to be able to change the speed ratio by practically differential amounts at a time, for example, in regulating the speed of conveyor lines, automatic welders, cement kilns, inspection tables, metering systems, printing presses, stokers, and wire-coiling machines and in certain computing machines (for very small forces, the slippage is negligible). Most variable-speed devices have friction drives, but there are some exceptions. The following descriptions constitute a sampling of available mechanisms.

The principle of the disk and wheel is shown in Fig. 190. It is appropriate for transmitting varying velocity ratios between shafts intersecting

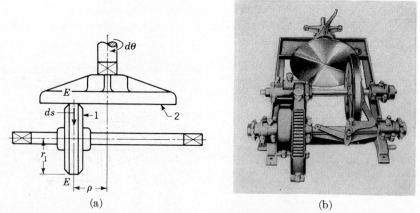

(a)   (b)

Courtesy Rockwood Mfg. Co., Indianapolis

Fig. 190. Plate and Wheel. In (b) is seen an illustration of a large wheel-and-disk transmission. However, this device is more common in small sizes, for example, to provide a speed adjustment for the turn table of a record changer. In the changer application, the arrangement is modified so that the driver is another disk similar to 2, located below the wheel in (a), and the wheel becomes an idler between driving and driven disks.

at right angles, for example, drives on sensitive drills and in calculating machines. Rolling is assumed to occur on the circle EE, so that the velocity ratio, wheel 1 driving, is $m_\omega = \omega_1/\omega_2 = r_2/r_1 = \rho/r_1$; and it is varied by varying ρ (moving the wheel along its axis). If the wheel is moved to the right-hand side of the disk's axis, the disk's direction of rotation is reversed. As a power-transmitting device, the wheel is generally faced with the softer material, such as a synthetic or leather, which results in a coefficient of friction larger than would be obtained in metal-to-metal contact. Moreover, if the disk is stalled by an overload, the softer wheel takes the wear distributed about its circumference; the wearing of a flat spot is less likely. Theoretically, a zero velocity ratio (infinite speed of the disk) is obtainable ($\rho = 0$). However, the practical limit involves the torque necessary to turn the driven disk plus other elements that may move with it. We can say $m_\omega = \rho/r_1 = F\rho/(Fr_1) = T_o/T_i$, where T_o is the output torque and T_i is the input torque for a particular frictional force F. Thus, at some small radius ρ, the output torque will cease to be enough to move the driven mechanism and the machine will no longer function properly because of excessive slippage.

With the disk 2 as a driver, Fig. 190(a), the principle of operation is also used in computers as an integrator. When disk 2 has an angular displacement of $d\theta$, the distance moved by a point at the radius ρ and on the surface of wheel 1 is $ds = \rho\, d\theta$; or $s = \int \rho\, d\theta$. Thus, if the input to disk 2 is regulated to give the proper sequence of values of $\rho\, d\theta$, the linear movement of a point on wheel 1, the output, gives the value of the integral. For this application, frictional resistance to varying ρ is reduced by replacing wheel 1 with two steel balls and adding a rotating shaft contacting the balls opposite the disk; the motion is then from disk through the balls to the shaft; the linear movement of the surface of the shaft is the integral.

V-shaped grooves whose sides are adjustable in a manner to make grooves wider or narrower are a feature that has been used in a number of ways to obtain different speed ratios. Thus, sheaves with movable sides are used with V belts to give speed adjustment as in Fig. 191. Let the upper shaft A drive the lower shaft B (small radius at B, large one at A). When the sensing system decides that a speed change should be made, it causes the shaft C to turn, which decreases the groove width on B and increases it on A, or vice versa, causing the V belt to climb to a larger radius on B and to move to a smaller radius on A, decreasing the speed of B.

Courtesy Allis-Chalmers Mfg. Co., Milwaukee

Fig. 191. Variable-speed V Drive.

These movements are simultaneous, the belt fitting both sheaves at all settings. A speed change of the driven shaft of some 20% is easily obtained.

The Reeves variable-speed transmission, Fig. 192, operates on the same principle, with movable conical disks whose distance apart can be controlled manually or automatically. On each wide side of a rubber belt are bolted hardwood blocks, as seen in Fig. 192, whose ends are beveled and tipped with leather in order to improve the coefficient of friction.

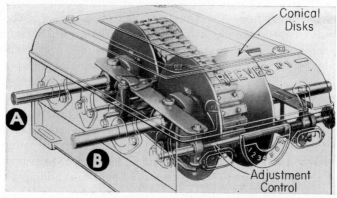

Courtesy Reeves Pulley Co., Columbus, Ind.

Fig. 192. Reeves Variable-speed Transmission. Constant-speed shaft is A. Adjustment in illustration is for maximum speed of driven shaft B.

Courtesy Link-Belt Co., Chicago

Fig. 193. P.I.V. Transmission.

The Link-Belt Company manufactures a similar speed-change mechanism that provides positive drive, which they call the P.I.V. gear, Fig. 193. As seen, the faces of the conical disks have "teeth" cut into them. The disks are mounted with the teeth of one staggered with respect to the other, the projection on one being opposite a valley on the other. The portion of the chain which comes into contact with the disks is made up of laminations, thin, hardened sheets of steel, each one free to slide from side to side. When the chain engages the disks, some of the laminations slide to the right into grooves and others to the left into grooves. For any setting, a particular lamination is in a valley on one disk and on top of a tooth on the other disk.

The epicyclic principle is also used in variable-speed devices. In one case, V belts connect planet sheaves to coaxial sheaves in a reverted train.

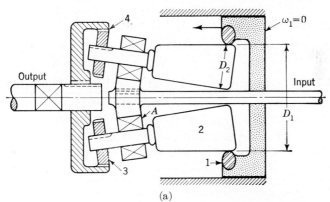

(a)

The sheaves are adjustable after the manner of Fig. 191. The Graham variable-speed transmission, Fig. 194, has an internal ring gear 4 which turns with the output shaft. Planets 3 and tapered rollers 2 are carried by the arm A. The tapered rollers are mounted with the outside element horizontal and they roll on the traction ring 1. Ring 1 is stationary except that it can be moved axially by operating a handle, permitting one to dial the desired speed ratio by changing the effective diameter D_2. As seen, the drive between 1 and 2 is frictional, the normal force being due to centrifugal action. This force is a function of the speed of the arm, the mass of the rollers, and the loca-

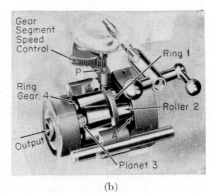

(b)

Courtesy Graham Transmissions, Inc.,
Milwaukee

Fig. 194. Graham Variable-speed Transmission. In (b), the pin P extends into a slot in the control assembly via which the ring 1 is moved rightward or leftward.

tion of their center of gravity. These transmissions are built in several models, but usually in fractional-horsepower sizes, and they can be automatically regulated.

167. Hoisting Tackle. Hoisting tackle or a block and tackle consists of an arrangement of rope (or chain) and pulleys which provides a mechanical advantage. A *mechanical advantage* exists when a relatively small force or torque input to a mechanism results in a larger force or torque output, and its magnitude is the load moved or the resistance overcome *divided* by the force (or torque) applied. If these forces are the result of actual measurement, the mechanical advantage accounts for the loss due to friction. For ideal systems, in the absence of friction, an ideal mechanical advantage M_a is easily computed. From the point of view of motion and displacement, it is defined as

$$(\textit{g}) \qquad M_a = \frac{\text{velocity of point at which force } F \text{ is applied}}{\text{velocity of the load or resistance } W},$$

$$(\textit{h}) \qquad M_a = \frac{\text{displacement of point at which } F \text{ is applied}}{\text{displacement of the load or resistance } W},$$

where F is the input force and W the output. Generally, it is fairly easy to determine the displacements. One might ask and answer this question: *If the load W is moved 1 ft., how far does the applied force F move?* For example, let W in Fig. 195(a) be raised 1 ft. Observe that there will

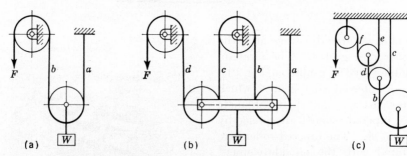

(a)   (b)   (c)

Fig. 195. Block and Tackle Arrangements.

therefore be a foot of slack in both ropes a and b. To take up this slack, F must move 2 ft. The ideal (no friction) value of M_a is then 2. Apply this same form of reasoning to other arrangements of ropes and pulleys.

A differential hoist, Fig. 196, may be made for use with ropes, but customarily a chain is used in formed sheaves (Fig. P140, p. 273) to provide a positive drive. In this mechanism, there is a duplex pulley, M-N, Fig. 196, with two diameters D_1 and D_2. An endless rope or chain passes onto the larger diameter M at c, off at d, onto the smaller diameter N at e, off at f, then around the load pulley and back to c. To find the ideal mechanical advantage, assume that M-N turns once CL. The length of rope a which moves onto M is πD_1; the length of rope b which moves off of N is πD_2. The net amount by which the loop ba is shortened is thus $\pi(D_1 - D_2)$; so the load moves upward a distance of $\pi(D_1 - D_2)/2$.

During the same movement, F is displaced a distance πD_1 (think of it moving in a straight line); hence,

(i)
$$M_a = \frac{\pi D_1}{\pi (D_1 - D_2)/2} = \frac{2D_1}{D_1 - D_2},$$

in which it is seen that a very large mechanical advantage can be obtained by making D_1 and D_2 nearly the same size.

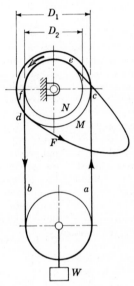

Fig. 196. Differential Hoist.

168. Closure. As the reader has observed, the kinematics of most of the devices of this chapter is rather simple. However, the complete design of some of these items may consume hundreds of man-hours and involve many troublesome problems. Additional information as found in books on design and in specialized works is needed.

PROBLEMS

Belting

731. Taking the steps suggested in the text, show all details of the derivation of equation (*e*), § 159.

732. The velocity ratio between two shafts is to be 12. The step-down is through a countershaft. If the smallest pulley, which is on the driving shaft, is to be 6 in. in diameter, find suitable diameters for the other 3 pulleys.

733. (a) A 16-in. pulley drives a 48-in. pulley, with a center distance of 10 ft., via an open rubber belt. What is the length of belt? (b) The same as (a) except that the belt is crossed.

Ans. (a) 28.6 ft., (b) 29.1 ft.

734. An 8-in. sheave drives a 24-in. sheave with a center distance of 5 ft. (a) What is the required length of V belt? (b) The nearest stock length actually available is 174.5 in. What center distance should be used?

Ans. (a) 14.3 ft.

735. A certain machine shop buys a second-hand lathe on which the lead screw (see Fig. 157, p. 253) is driven by stepped pulleys with three steps. It was found after purchase that the lead screw rotated too fast at the minimum speed for best results in this shop. It was decided to renew the driven stepped pulley to provide for a slower speed. With the belt on the smallest step of the driving pulley, the velocity ratio was 2.26. An increase in the velocity ratio to 3.125 would give a sufficiently slow speed to the lead screw. The following data were taken from the machine: distance between centers of shafts, 25.5 in.; diameters of the steps on the driving cone, 3.75, 5, and 6.218 in., respectively. (a) What should be the diameters of the steps on the new driven cone? (b) In belt design, the smallest arc of contact is often needed. What is the smallest arc of contact in this installation?

Ans. (a) $11\frac{3}{4}$, $10\frac{11}{16}$, $9\frac{21}{32}$ in.

736. The distance between the center lines of the shafts of two stepped pulleys is 5 ft. The driving cone turns at a constant speed of 150 rpm. The driven cone is to turn at a maximum speed of 450 rpm. The steps on the driving cone are 8, 10, and 12 in., respectively. (a) What should be the diameter of the smallest step on the driven cone? (b) What should be the diameter of the other steps if an open belt is used; if a crossed belt is used? (c) What are the rpm of the driven cone with the belt on the two other steps? (d) In belt design, the smallest arc of contact is often needed. What is the smallest arc for an open belt; for a crossed belt?

737. Stepped pulleys with three steps are to be used to vary the velocity ratio between two shafts 100 in. apart. These pulleys are to be made alike. The middle step on each pulley must therefore be the same size. If the diameter of the middle step is 10 in., what must be the diameters of the other steps for a maximum velocity ratio of about 1.5?

Chain Drives

738. A roller-chain drive is used to step up the velocity, the velocity ratio being 0.3. The driving sprocket has 60 teeth and turns 230 rpm. Find the number of teeth in and the rpm of the driven sprocket.

739. What is the ratio of the maximum chain velocity to the minimum chain velocity when the number of teeth in each sprocket is: (a) 6; and (b) 20? Does the variation in either case appear to be enough to cause trouble for medium or high speeds? (c) Let the pitch of the chain be 1 in. and let the sprocket turn at 900 rpm. Compute the average momentary acceleration as the chain undergoes the chordal action.

740. The same as **739** except that the number of teeth is: (a) 7; (b) 25.

Bevel Friction Drives

741. Show that for internal cones, Fig. 189(d), which are to have a velocity ratio of m_ω and axes at an angle Σ,

$$\tan \gamma_1 = \frac{\sin \Sigma}{m_\omega - \cos \Sigma}.$$

742. Show that for external cones, Fig. P151, which are to have a velocity ratio of m_ω and axes at an angle Σ,

$$\tan \gamma_1 = \frac{\sin \Sigma}{\cos \Sigma + m_\omega}.$$

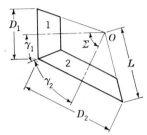

Fig. P151. Problem **742.**

743. Two shafts that intersect at $75°$ are to turn with a velocity ratio of 1.8. Motion is to be transmitted through external rolling cones. (a) Compute the angles γ_1 and γ_2, Fig. 189. (b) Check this solution graphically (see Appendix). (c) Let the diameter of the driving cone be 10 in. and the face width be one-quarter of the length of a cone element. Compute the size of the other cone and the face width.

744. The same as **743** except that $\Sigma = 120°$.

745. The same as **743** except that the shaft angle is $25°$ and the driven cone is an internal cone.

746. A disk 2 is to be driven by a rolling cone 1 with $m_\omega = 2.1$. The cone diameter is 4 in. The angle between the shafts as their axes approach the cone center may be either of two values. What are these angles? What is the largest disk diameter?

747. For two shafts which intersect at right angles, specify a friction device which will provide a velocity ratio of 3 when (a) rolling cones are used, (b) one of the contacting surfaces is a plane.

Variable-speed Transmissions

748. In Fig. 190, the wheel is 4 in. in diameter and the disk is 22 in. in diam-

eter. The center line of the wheel may be moved to within 1 in. of the outside of the disk's center. The wheel turns at 250 rpm. What are the maximum and minimum speeds of the disk?

749. In a Reeves variable-speed transmission, the constant-speed shaft turns at 380 rpm. The minimum speed of the variable-speed shaft is 154 rpm and its maximum speed is 6 times the minimum. What are the maximum and minimum velocity ratios?

750. In a Graham transmission, Fig. 194, $D_1 = 6$ in., $N_3 = 30$ teeth, $N_4 = 90$ teeth. The driving motor turns 1750 CL. What is the output angular velocity when contact between the traction ring and tapered roller is (a) at $D_2 = 2.5$ in., (b) at $D_2 = 1.8$ in.? (c) At what diameter D_2 would the output speed be zero?

Ans. (a) 350 rpm CL, (b) 194.4 rpm CC, (c) 2 in.

751. The same as **750** except that $N_3 = 25$ and $N_4 = 100$.

752–755. The illustrations for these problems are Figs. P152 to P155, inclu-

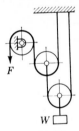

Fig. P152. Problem **752.**

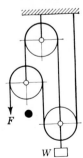

Fig. P153. Problem **753.**

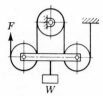

Fig. P154. Problem 754.

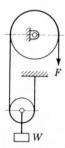

Fig. P155. Problem 755.

sive. For the problem assigned, find the ideal mechanical advantage. If F moves 1 fps, how fast does W move in ips?

756. Figure P156 shows the end view of a differential windlass. The crank against which F is acting is keyed to and

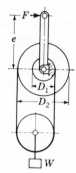

Fig. P156. Problem 756.

turns the shaft on which are sheaves of diameters $D_1 = 8$ in. and $D_2 = 12$ in.; $e = 14$ in. What is the ideal mechanical advantage?

757. The same as **756** except that $D_2 = 9$ in.

758. A force $F = 100$ lb. is applied to a differential hoist, Fig. 196, wherein $D_1 = 10$ in. and $D_2 = 9$ in. What load could be raised if there were no frictional losses?

(There are not 758 problems in this book. See unused numbers at the end of each chapter.)

11

Miscellaneous Devices

169. Introduction. Inasmuch as kinematics relates to all movements, there is no end to devices which might be described. There is the matter of fluid controls, of fluid couplings, and of other devices involving the movement of fluids. This aspect of kinematics may be covered in a study of fluid mechanics. We shall not have the space to mention mechanisms in which there are points that do not move in plane motion (**128**), for example, steering-gear mechanisms for automotive vehicles.

170. Ratchets. Because of its simplicity, the ratchet is one of the most common devices for producing intermittent motion. It consists of a driving **pawl** C (also called a *detent* or *catch*) attached to an oscillating link A, and a **ratchet wheel,** Fig. 197. If there is an undesired tendency for the wheel to turn backward, another pawl D with a fixed center is placed in a position to prevent backward motion. The arm A and the wheel turn about the same axis but are independent.

Ratchets are often used as "feed" devices wherein it is necessary to vary the amount the wheel turns. One method of regulating is to vary the angle of oscillation of the arm, making the pawl feed $1P$, $2P$, or $3P$, etc., where P is the pitch (but see § 171).

The profile of the teeth should be such that the pawl will not slip out of engagement. To assure this attribute, the normal MN to the tooth face at the point of contact M should pass between centers O and Q, Fig. 197(a). The smaller the teeth, the smaller may be the angular displacement of the wheel. However, a point is reached at which the teeth are not sturdy enough to carry the load. The effect of smaller teeth in the holding operation can be obtained from multiple fixed pawls, Fig. 198. By the device of using two or more pawls, D and E, of different lengths side by side, the maximum backward motion can be reduced to an amount equal to the pitch divided by the number of pawls.   Similarly, a fine feed (small displacement of the wheel) may be secured without decreasing the pitch by using multiple *driving* pawls of different lengths.

In order to move the wheel in one direction during each oscillation of the link that carries the driving pawls, an arrangement similar to that in Fig. 199, called a **double-acting ratchet,** may be employed.   Pawls with

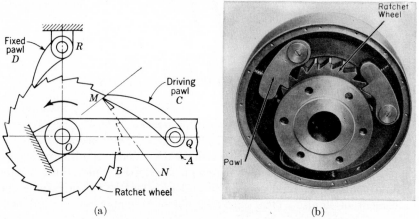

(a)   (b)

Courtesy Foote Bros. Gear & Machine Corp., Chicago

Fig. 197. Ratchet Gearing with Fixed Pawl.

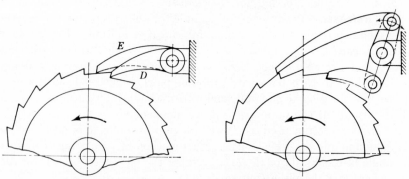

Fig. 198. Double Pawl.   **Fig. 199. Double-acting Ratchet.**

"hooks" at their ends so that they *pull* the wheel around, instead of pushing, are sometimes used.

When it is necessary to have an intermittent drive in either direction from the same pawl, a **reversing pawl,** designed as in Fig. 200, may be serviceable.   The side of the pawl opposite the driving face is curved to facilitate the sliding of the pawl over the teeth on the return stroke. The **jack ratchet** is familiar to most of those who have changed tires on automobiles.

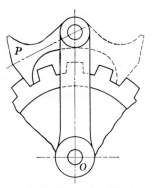

Fig. 200. Reversing Pawl.

171. Masked Ratchet. A common method of governing the amount of feed (turn of the wheel) is by means of a **masked ratchet.** Suppose the pawl moves through the same angle every time, and as seen in Fig. 201, let the angle be measured by 8 teeth on the wheel. The mask M, of slightly larger diameter than the outside diameter of the wheel, is cut away so as to leave a maximum of 8 teeth exposed. The holes 1, 2,

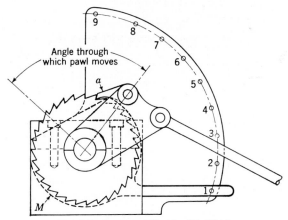

Fig. 201. Ratchet Wheel with Mask.

3, etc., in the fixed plate are so located that if the handle operating the mask is moved to position 2, one tooth is shielded and the wheel turns through an arc subtended by seven teeth; if the handle is moved to position 8, seven teeth are shielded and the wheel turns through an angle subtended by one tooth; etc. The width of the pawl should be equal to the width of the ratchet wheel plus the width of the mask. The mask can be automatically controlled. Masked ratchets are found on inking mechanisms for printing presses, positive lubricating systems, etc.

172. Silent Ratchet. A silent ratchet, so-called because of the absence of the click of the pawl, is one that depends on friction as the driving force. A large normal force is generated by wedging action, as evident in Fig. 202. In Fig. 202(a) and (b), the members R are balls (or rollers). When, for example, A turns CC, the balls R move toward the narrower section until wedging occurs and the internal member drives the external member. The designs of Fig. 202(b) and (c) are ones in which the link B is an oscillating driver with conventional ratchet motion. The design of

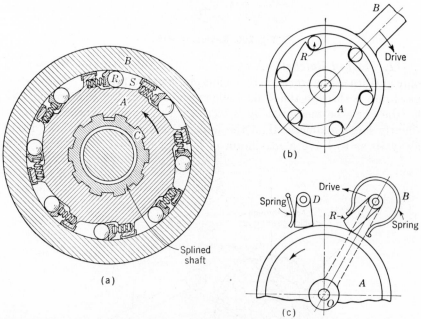

Fig. 202. Silent Ratchets. In (a) and (b), A drives CC or B drives CL; in (c), B drives CC and friction pawl D prevents A from turning CL. Springs are generally applied to keep the members in contact so that wedging action is sure and immediate.

Fig. 202(a) can also be operated in this manner, but this type is utilized as an ***overrunning clutch.*** The typical action of an overrunning clutch is that both members A and B turn in the same direction; but if A is the driver, it may turn slower than B and cease driving, permitting B to turn independently of A except when A is actually driving. Rollers provide greater power transmitting capacity than balls; ***sprags,*** Fig. 203, have more capacity than rollers for the same over-all dimensions of the connection and are therefore favored in heavy-duty industrial applications. See Fig. 168. Each of the foregoing devices is variously used as a brake or stop on inclined conveyors to prevent their backward motion, as

Courtesy Formsprag Co., Van Dyke, Mich.

Fig. 203. Sprag-type Overrunning Clutch. The sprags *R*, which are kept in contact with the members by the pressure spring, wedge tight for one direction of drive and release for the other direction.

feed mechanisms, in indexing devices, and in general wherever a pawl and ratchet could be used.

173. Universal Joints. If two shafts are unintentionally misaligned, as they would be by a slight amount in any event, they may be satisfactorily connected by a flexible coupling (5). When a significant misalignment of intersecting shafts is necessary, a universal joint is the usual answer. In its traditional form, Figs. 204 and 205, it is also called a Hooke's joint

Courtesy Spicer Mfg. Corp., Toledo, Ohio

Fig. 204. Universal Joint, Automotive Drive Shaft.

after Robert Hooke, an obstreperous genius and contemporary of Newton's (it is sometimes called a Cardan joint after the Italian who is credited with being the first to invent it). It consists of a driving yoke 1 fixed to the input shaft, a cross link *C*, and a driven yoke 2, Fig. 205, with turning pairs at *A*, *B*, *M*, and *N*, as can be seen in Fig. 204.

The disadvantage of Hooke's joint is that it does not provide a constant angular velocity ratio between driving and driven shafts except when the acute angle δ between shafts is zero. This can be seen from Fig. 205(b) and (c) which show the true magnitudes of: MN in (b), and AB in (c). In Fig. 205(b), the speed of point M is $v_m = r(\cos \delta)\omega_1$. This vector is perpendicular to the axis of rotation of 2; hence, the angular velocity of 2 is $\omega_2 = v_m/r = \omega_1 \cos \delta$, or

(**a**)
$$m_\omega = \frac{\omega_1}{\omega_2} = \frac{\omega_1}{\omega_1 \cos \delta} = \frac{1}{\cos \delta}.$$

Figure 205(c) shows the joint after a 90° turn, where we note that both M and A are at a distance r from the axis of 1; hence $v_A = r\omega_1$, where the vertical vector e represents this velocity. The illustration shows that this vector is a component of v_m which is normal to the axis of 2.

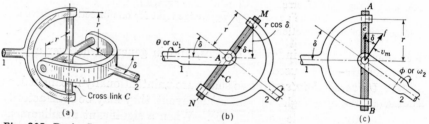

(a)　　　　　　　　　　　　　　(b)　　　　　　　　　　　　(c)

Fig. 205. Basic Components of Hooke's Joint. When 1 has turned $\theta = 90°$ (and 180°, 270°, 360°), the driven member 2 has turned $\phi = 90°$ (and 180°, 270°, 360°); but at other values of θ, θ and ϕ are not the same.

Thus $v_m = r\omega_1/\cos \delta$; therefore $\omega_2 = v_m/r = \omega_1/\cos \delta$ and the velocity ratio is

(**b**)
$$m_\omega = \frac{\omega_1}{\omega_2} = \frac{\omega_1}{\omega_1/\cos \delta} = \cos \delta.$$

The equations (**a**) and (**b**) show that the speed of the driven shaft varies from a minimum of $\omega_1 \cos \delta$ to a maximum of $\omega_1/\cos \delta$ during the 90° turn. During the next 90°, the driven speed returns to its minimum, and the cycle repeats every 180°.

From spherical trigonometry, the relation between the angular displacement input θ, the output displacement ϕ, and the acute shaft angle δ is as follows:

(**c**)
$$\tan \phi = \tan \theta \cos \delta.$$

Differentiation of (**c**) with respect to t for a particular shaft angle δ ($\cos \delta$ = constant) gives the relation between angular velocities:

(**d**)
$$\omega_2 = \frac{\omega_1 \cos \delta}{1 - \sin^2 \delta \sin^2 \theta}.$$

Another differentiation (ω_1 = constant) gives

(e)
$$\alpha_2 = \frac{\omega_1{}^2 \sin^2 \delta \cos \delta \sin 2\theta}{(1 - \sin^2 \delta \sin^2 \theta)^2}.$$

Inspection of the various foregoing equations, especially (a) and (b), readily reveals that the angle δ can soon become large enough that the change in the angular velocity ω_2 is relatively large. At other than the slowest speeds, the corresponding accelerations and the consequent vibrations can be intolerable. Sometimes, a satisfactory solution of this predicament is to use two universal joints with an intermediate member D, Fig. 206, as is found on some automotive drives (one after the transmission, another before the differential). In any event, the intermediate member varies in speed, but its mass and inertia may be small enough that the accelerations cause no trouble. The conditions to be met in order for the second universal joint to compensate the variations introduced by the first joint are [see Fig. 206(a)]: the angle between 2 and D

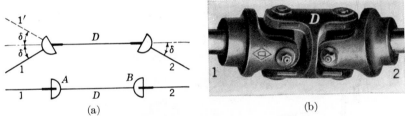

Fig. 206. Compensating Universal Joints.

must be the same as the angle between 1 and D, namely, δ; if the shafts 1 and 2 lie in the same plane, the yokes A and B lie in the same plane as shown; if the shafts are not in the same plane, yoke A is in the plane of 1 and D, yoke B is in the plane of D and 2. It often happens that space is at a premium and that two universal joints are undesirably bulky, no matter how short the intermediate member D is made. In response to such need, at least three universal joints which do maintain a constant velocity ratio have been invented and are in use (**131**). These are the Rzeppa, Bendix-Weiss, and the Tracta. Return to Fig. 205 and imagine a complete rotation with the mind on what the cross link C does. The equal-radius points A, B, M, and N of course remain on the surface of a sphere, but the plane $AMBN$ changes its orientation [sloping at $90 - \delta$ in Fig. 205(b), vertical in (c), etc.]. Thinking of it another way, we see that the driven point M, for example, is at different radii for different orientations of the cross link, because of the "wobbling" action. Hence, we may conclude that if the drive through the intermediate element can be designed so that the forces transmitting the power to the driven shaft always remain in a single plane, there will be no change in radii of the driving points and therefore no change of velocity of the driven shaft.

It can be proved that the plane of these driving contacts, whose trace is
NN in Fig. 207(a), should be perpendicular to the plane of the connected shafts and that it should bisect the obtuse angle between the
shafts, as in the Bendix-Weiss joint shown in Fig. 207. The feature of
this design is the nonconcentric intersecting races or grooves E, F. The
center of curvature of driving groove E is on the center line of shaft 1;

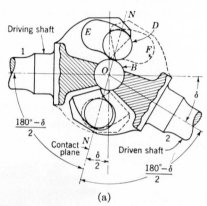

(a)

Courtesy Bendix Aviation Corp., South Bend, Ind.

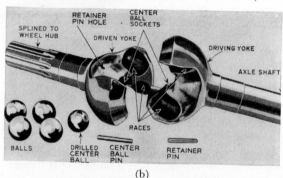

(b)

Courtesy Bendix Aviation Corp., South Bend, Ind.

Fig. 207. Constant-velocity Universal Joint.

the center of curvature of driven groove F is on the center line of shaft 2;
but these centers are separated from each other [in Fig. 207(a), they are
outside of the circle B, which is the center ball—see Fig. 207(b)], so that
no matter what angle shaft 2 makes with axis 1, the arcs representing
these grooves intersect so that the line NN through this intersection and
through O bisects the angle $180 - \delta$. (Try it with your compass.)
Hence each ball in a half race E and a half race F is forced to assume a
position at all times in a bisecting plane and remain there throughout the
revolution, thus maintaining a constant velocity ratio.

Appendix

174. Suggestions on Graphical Solutions. For graphical solutions, use a hard lead (4H or 5H) with a *sharp point*. Sharpen point frequently. Draw lines lightly until solution is complete. Use a softer pencil with a sharp point to go over principal lines, leaving all construction lines light. Use a soft pencil with a sharp point, about H or HB, for standard engineering lettering and other freehand work. Never use script on the drawing. Work as accurately as you can, so that your solution will be more meaningful.

175. Degrees to Radians. For convenience, some conversion equivalents are given in Table 4.

Table 4. Degrees into Radians

Deg.	Rad.	Deg.	Rad.	Deg.	Rad.	Deg.	Rad.	Deg.	Rad.
5	0.08727	30	0.52360	65	1.13446	105	1.83260	145	2.53073
10	0.17453	35	0.61087	70	1.22173	110	1.91986	150	2.61800
$14\frac{1}{2}$	0.25308	40	0.69813	75	1.30900	115	2.00713	155	2.70526
15	0.26180	45	0.78540	80	1.39626	120	2.09440	160	2.79253
$17\frac{1}{2}$	0.30544	50	0.87267	85	1.48353	125	2.18166	165	2.87979
20	0.34907	55	0.95993	90	1.57080	130	2.26893	170	2.96706
$22\frac{1}{2}$	0.39270	57.3	1.00000	95	1.65806	135	2.35620	175	3.05433
25	0.43633	60	1.04720	100	1.74533	140	2.44346	180	3.14159

176. Areas under Curves. Areas under curves are frequently useful in studying curved surfaces, cams, or rolling curves.

A trapezoidal area is equal to one-half the sum of the parallel sides $(a + b)/2$ times the perpendicular distance h between the parallel sides; $A = h(a + b)/2$.

A half segment of a circle, *abd*, Fig. 208, has the area

$$A_1 = \frac{r^2}{4} (2\theta - \sin 2\theta).$$

$$\sin 2\theta = 2 \sin \theta \cos \theta; \qquad Oa = r - bc.$$

The area 2, Fig. 208, is equal to the rectangle *abcd* minus area A_1. An area such as 3, Fig. 208(b), is the half segment *egk* minus the half segment *fgh*.

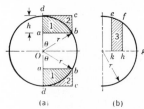
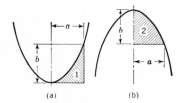

Fig. 208. Circle.　　　　　　　　　**Fig. 209. Parabola.**

Parabolic areas are:

$$A_1 = \frac{ab}{3}, \text{ Fig. 209(a)}; \qquad A_2 = \frac{2ab}{3}, \text{ Fig. 209(b)}.$$

A half segment of an ellipse, Fig. 210, has the area

$$A_1 = \frac{1}{2} \left(xy + ab \sin^{-1} \frac{x}{a} \right).$$

The total area of an ellipse is πab.

177. Dividing a Line into Parts of Any Proportion. Suppose that it is desired to divide the distance AB, Fig. 211, into

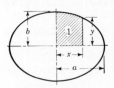

Fig. 210. Ellipse.

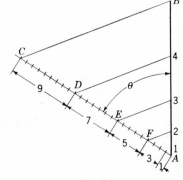

Fig. 211.

parts proportional to $1:3:5:7:9$ (constant acceleration). Draw a line AC from one end of AB and making a convenient angle θ with AB. Lay a scale along line AC, count off the proper number of divisions, and mark. Connect the last mark C with the other end B of line AB and draw $D4$, $E3$, $F2$, etc., parallel to CB. The resulting divisions made on AB are proportional to the divisions which have been made on AC. For best accuracy the length AC should be such that the angles CBA and BCA are, say, between 60° and 120°. This is done by using a scale which results in AC

being little different in length from AB. Make angle θ at least $45°$ so that the parallels $D4$, $E3$, etc., will be more accurately parallel to CB.

178. Basic Racks for Gear-tooth Systems. The basic racks for interchangeable gear teeth, taken from the ASA standard B6.1-1932, are shown in Figs. 212 to 215. The approximate composite tooth is basic

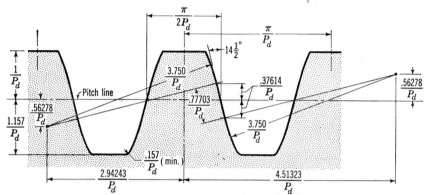

Fig. 212. Approximate Basic Rack, $14\frac{1}{2}°$ Full-depth Composite System. In the exact basic rack, the curved parts of the profile ($r = 3.75/P_d$) are cycloidal curves instead of arcs of circles (epicycloid at tips, hypocycloid at roots).

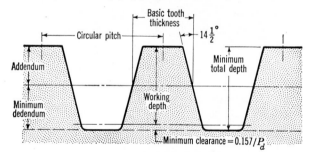

Fig. 213. Basic Rack, $14\frac{1}{2}°$ Full-depth Involute System. See § 118 for values of dimensions shown.

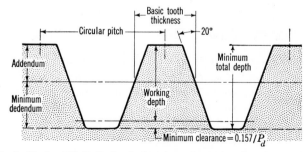

Fig. 214. Basic Rack, $20°$ Full-depth Involute System. See § 118 for values of dimensions shown.

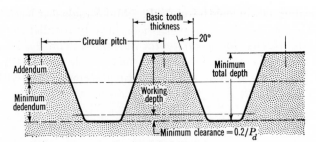

Fig. 215. Basic Rack, 20° Stub-tooth Involute System. See § 118 for values of dimensions shown.

for form cutters. The standard referred to contains additional information, including diameters of gear blanks for long-and-short addendums (see reference **141**).

179. Drawing an Involute. The circle of Fig. 216 represents a base circle. Choose some small angle $aO1$ as a unit and lay out several of them consecutively: $1O2$, $2O3$, etc. Draw tangents to the base circle at 1, 2, 3, etc. Lay off from 1 along its tangent a distance $1b$ equal to the arc $1a$;

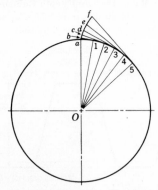

Fig. 216. Drawing an Involute.

lay off from 2 along its tangent a distance $2c$ equal to the arc $2a$; similarly, $3d$ equal to arc $3a$; etc. A curve through the points a, b, c, etc., is an involute. That portion of an involute gear tooth outside the base circle follows this curve.

180. Drawing Gear Teeth. One method of drawing gear-tooth profiles is to develop an involute as described in the previous article, then by trial and error find one or more circular arcs which closely follow that part of the involute curve that is the profile, and use the corresponding radii for drawing teeth. Each gear has its own base circle from which the involute for its teeth must be generated.

Another method of drawing gear teeth is to use a table of radii prepared by someone else, sometimes called the odontograph method. See

Keown and Faires "Mechanism," fourth edition, McGraw-Hill. Also
templates can be made or purchased for the purpose.

For drafting purposes, realistic teeth can be drawn quickly by using the
fact that the radius of curvature at any point of an involute is along the
normal to the curve to the point of tangency of the normal with the base
circle; for example, the radius of curvature at point e, Fig. 216, is $4e$.
A method of utilizing this fact is described in the following procedure for
drawing gear teeth.

   1. Draw the pitch, addendum, and dedendum circles.

   2. Draw the base circle from a computation of its diameter, or by
graphical construction. Graphically: draw a tangent to the pitch circle
at a convenient position, say p, Fig. 217; draw the line of action at the
pressure angle ϕ; draw the base circle tangent to the line of action.

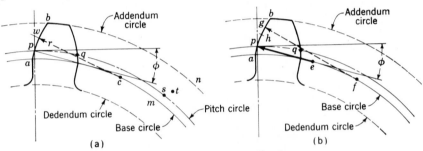

Fig. 217. Drawing Gear Teeth.

   3. Decide upon the position or approximate position where a profile is
desired, as at p (or s). (a) Then estimate by eye a point, such as w (or t),
which is about midway between the ends a and b (or m and n) of the utilized parts of the involute, Fig. 217. Through this point, which may or
may not turn out to be directly on the profile, draw line wc tangent to
the base circle; the radius of curvature of the true involute through w is
$r = wc$. With center at c and radius cp (where p is a point through
which a profile is desired), draw arc ab as an approximation of the involute.
If w is a little closer to a than to b, the profile would probably be better.
(b) If a closer approach to the involute is desired, divide the distance ab
into roughly two equal parts ah and hb, Fig. 217(b). Remember that at
this point no profile is drawn (as at mn). From approximately the middle
positions between a and h and between h and b, draw lines tangent to the
base circle, say, pe and gf. Then ep is a sort of average radius for the
true involute between a and h, and fg is the same between h and b.
With center at e and radius ep, draw an arc through p, where a profile is
desired, about equidistant on each side of p, say to h. Then with center
at f and radius fh, draw arc hb, completing the "involute" part of the
profile. If a still closer approximation of the involute is desired, use three

or more radii. It is understood that the foregoing method is for the purpose only of drawing gear teeth to look reasonably like gear teeth and should not be used in determining any manufacturing dimensions.

4. With the radii as found in (3) and centers always on the base circle, draw as many teeth as desired. The pitch circumference is marked off into distances pq equal to the circular thickness ($P_c/2$) and the profiles are put through these points.

If the dedendum circle is inside of the base circle, as shown, draw that part of the profile inside the base circle as a *radial* line tangent to the involute at the base circle, except of course for the fillets. If the base circle is inside of the dedendum circle, point a, the bottom end of the involute could be taken at the working-depth circle.

181. Drawing Cycloidal Curves. A **cycloid** is the curve generated by a point on the circumference of a (generating) circle rolling on a straight line. Let the circle with radius OT, Fig. 218, be the **generating circle**

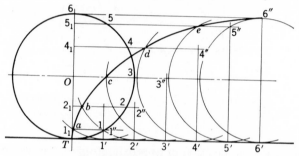

Fig. 218. Drawing a Cycloid.

which rolls on line $T6'$. Divide the semicircumference of the circle into any convenient number of equal parts, say six, and lay off these arcs on the line $T6'$. On the straight line, $T1' = \text{arc } T1$, $1'\text{-}2' = \text{arc } 1\text{-}2$, etc. Through the points $1'$, $2'$, $3'$, etc., erect perpendiculars to $T6'$; and through the points 1, 2, 3, etc., on the circumference of the generating circle, draw lines parallel to $T6'$. These lines parallel to $T6'$ intersect the diameter of the generating circle at points, 1_1, 2_1, O, 4_1, etc., and they intersect the perpendiculars just erected at points $1''$, $2''$, $3''$, etc. Now from $1''$, lay off $1''a = 1\text{-}1_1$; from $2''$, lay off $2''b = 2\text{-}2_1$, etc. A smooth curve through the points a, b, c, etc., is the cycloid as traced by point T on the generating circle.

The **epicycloid** is a curve traced by a point on the circumference of a *generating* circle rolling on the outside of another (directing) circle. In Fig. 219, the radius of the generating circle is OT; of the directing circle, AT. Divide the semicircumference of the generating circle into any convenient number of equal parts, as six, and from T, lay off arcs on the directing circle equal in length to the arcs just obtained on the generating circle; that is $T1' = T1$, $1'\text{-}2' = 1\text{-}2$, etc., Fig. 219. Draw radial lines

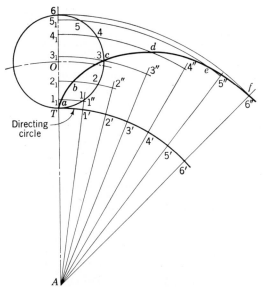

Fig. 219. Drawing an Epicycloid.

from A through $1'$, $2'$, $3'$, etc. With A as a center, draw arcs 1-1_1, 2-2_1, 3-3_1, etc., locating points 1_1, 2_1, 3_1, etc., and points $1''$, $2''$, $3''$, etc. Then from $1''$, lay off on the arc $1''$-1_1 the distance $1''a = 1$-1_1; from $2''$, lay off $2''b = 2$-2_1, etc. A smooth curve through the points a, b, c, etc., thus obtained is the epicycloid as traced by T on the generating circle.

The **hypocycloid** is the curve generated by a point on the circumference of a generating circle rolling on the inside of another (directing) circle. Since the construction of the hypocycloid, shown in Fig. 220, is the same

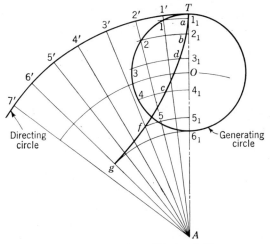

Fig. 220. Drawing a Hypocycloid.

as for the epicycloid, no detailed explanation will be given.   If the diameter of the rolling circle is equal to the *radius* of the directing circle, the resulting hypocycloid is a radial line of the directing circle.

182. Layout of Rolling Cones.   Given shaft OA driving shaft OB and let the velocity ratio $m_\omega = 5/7$.   At *any* point a on OA, erect a perpendicular ab and lay off 7 equal spaces, locating point b.   At *any* point c on OB, erect a perpendicular cd and lay off 5 spaces, each equal to those on ab, and locate point d.   Through b and d, draw lines parallel to OA and OB, respectively, and through the intersection of these lines at C, draw CO.   The line CO is a common element of the two desired cones.   Cones of any practicable diameters may be drawn, as suggested by the dotted outlines.

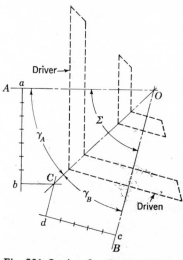

Fig. 221. Laying Out Rolling Cones.

183. Drawing   Teeth   on   Bevel Gears — Tredgold's   Approximation. As stated in § 139, bevel gear teeth must theoretically be generated from a spherical surface, and since such a surface is not developable, it is customary to resort to an approximate method, known as Tredgold's approximation.

Let OB and OA, Fig. 222, be the axes of the pitch cones 1 and 2.   For the moment, imagine the cone 2 to be the *base* cone and imagine it split along the element PO.   Then imagine that this surface is unwrapped from itself (as in unwrapping a string from a cylinder in generating an involute). Note that point P is always the same distance from O and that therefore it moves on the surface of a sphere in generating the tooth form at the large end.   In the approximation, we assume that the generating point P moves in a *plane* perpendicular to the element PO.   If P moves in a plane which is normal to PO, the profile generated is the involute from the circle of radius BP.   The reason for the following procedure should now be evident.

Draw the arc PF of radius BP, which is the back-cone radius.   At the angle of obliquity with a tangent to this arc, draw a pressure line and determine the base circle.   Now draw a tooth profile by methods previously described on the arc PF, using the pitch at the large end of the bevel gear.   In a similar manner, determine a profile for the mating gear. These profiles are the approximately correct forms of the bevel gear teeth at the large end when viewed in the direction of a pitch-cone element (PO).

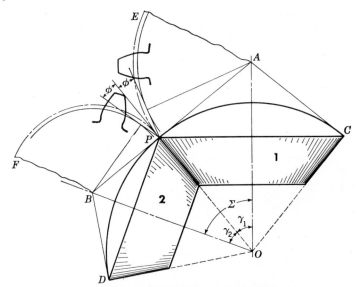

Fig. 222. Tredgold's Approximate Tooth Form.

The method of completing the drawing of the teeth is indicated in Fig. 223. The outline 1-2-3-4-5 at C is obtained for the desired system of gear teeth. The circles on the front view are obtained by projection and the distances 1'-1', 2'-2', etc., are laid off equal to 1-1, 2-2, etc. After the teeth have been drawn on the front view, the side view is obtained by orthographic projection.

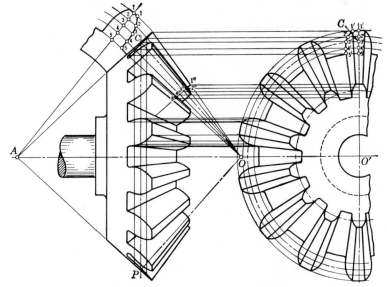

Fig. 223. Drawing Bevel Gear Teeth.

184. Some Useful Mathematical Relations. No attempt has been made to include here all relationships that may be needed in the solution of problems in this book. It is presumed that the reader will recall the correct manner of using these equations.

TRIGONOMETRY

Law of sines: $\dfrac{\sin A}{a} = \dfrac{\sin B}{b} = \dfrac{\sin C}{c}$.

Law of cosines: $a^2 = b^2 + c^2 - 2bc \cos A$.

Function of angles:

$$\sin 2\theta = 2 \sin \theta \cos \theta \qquad \sin^2 \theta + \cos^2 \theta = 1$$

$$\cos 2\theta = \cos^2 \theta - \sin^2 \theta = 2 \cos^2 \theta - 1 = 1 - 2 \sin^2 \theta$$

$$\sin \frac{\theta}{2} = \left(\frac{1 - \cos \theta}{2}\right)^{\frac{1}{2}} \qquad \cos \frac{\theta}{2} = \left(\frac{1 + \cos \theta}{2}\right)^{\frac{1}{2}}$$

$$\sin^2 \theta = \frac{1 - \cos 2\theta}{2} \qquad \cos^2 \theta = \frac{1 + \cos 2\theta}{2}$$

$$\sin (\alpha \pm \beta) = \sin \alpha \cos \beta \pm \cos \alpha \sin \beta$$

$$\cos (\alpha \pm \beta) = \cos \alpha \cos \beta \mp \sin \alpha \sin \beta$$

$$\tan (\alpha \pm \beta) = \frac{\tan \alpha \pm \tan \beta}{1 \mp \tan \alpha \tan \beta}$$

$$\sin (90 \pm \alpha) = + \cos \alpha \qquad \cos (90 \pm \alpha) = \mp \sin \alpha \qquad \tan (90 \pm \alpha) = \mp \cos \alpha$$

$$\sin (180 \pm \alpha) = \mp \sin \alpha \qquad \cos (180 \pm \alpha) = - \cos \alpha \qquad \tan (180 \pm \alpha) = \pm \tan \alpha$$

Function of frequently used angles:

	0°	14½°	15°	20°	30°	45°	60°	75°	90°
sin	0	0.250	0.259	0.342	0.5	0.707	0.866	0.966	1
cos	1	0.968	0.966	0.940	0.866	0.707	0.5	0.259	0
tan	0	0.259	0.268	0.364	0.577	1	1.732	3.73	∞

List of References

This is not a bibliography. It includes books and papers to which acknowledgment to *specific* material could be made, but it also includes works where the help was less tangible. These references should be helpful to those who wish to pursue further study.

1. W. G. Green, "Theory of Machines," Blackie, Glasgow.
2. Faires and Chambers, "Analytic Mechanics," Macmillan.
3. H. A. Rothbart, "Cams," Wiley.
4. A. B. W. Kennedy, "Mechanics of Machinery," Macmillan.
5. V. M. Faires, "Design of Machine Elements," Macmillan.
6. Kloomok and Muffley, Plate Cam Design, *Prod. Eng.*, vol. 26, no. 2, p. 156.
7. J. A. Hrones, An Analysis of the Dynamic Forces in a Cam-driven System, *Trans. ASME*, vol. 70, p. 473.
8. D. B. Mitchell, Tests on Dynamic Response of Cam Follower Systems, *Mech. Eng.* vol. 72, p. 467.
9. D. A. Stoddart, Polydyne Cam Design, *Machine Design*, vol. 25, no. 1, p. 121; no. 2, p. 146; no. 3, p. 149.
10. W. D. Cram, Practical Approaches to Cam Design, *Machine Design*, vol. 28, no. 22, p. 92.
11. H. A. Rothbart, Basic Factors in Cam Design, *Machine Design*, vol. 28, no 21, p. 107.
12. G. F. Kennison, Cycloidal Motion Cams, *Machine Design*, vol. 28, no. 1, p. 141.
13. H. A. Rothbart, Cam Dynamics, *Machine Design*, vol. 28, no. 5, p. 100.
14. D. G. Anderson, Cam Dynamics, *Prod. Eng.*, vol. 24, p. 107.
15. R. C. Johnson, Cam Profiles, *Machine Design*, vol. 28, no. 25, p. 129.
16. R. C. Johnson, Cam Dynamics, *Machine Design*, vol. 29, no. 3, p. 105.
17. W. M. Dudley, New Methods of Valve Cam Design, *Trans. SAE*, vol. 56, p. 19.
18. A. F. Gagne, Jr., Design of High Speed Cams, *Machine Design*, vol. 22, no. 7, p. 108.
19. C. N. Neklutin, Designing Cams, *Machine Design*, vol. 24, no. 6, p. 143.
20. Kloomok and Muffley, Determining the Maximum Pressure Angle, *Prod. Eng.*, vol. 26, no. 5, p. 155; Calculating Minimum Radius of Curvature, *Prod. Eng.*, vol. 26, no. 9, p. 187; Evaluating Dynamic Loads, *Prod. Eng.*, vol. 27, no. 1, p. 178; Design of Cam Followers, *Prod. Eng.*, vol. 27, no. 9, p. 197.

21. T. Weber, Jr., Cam Development by Evolute Analysis, *Machine Design*, vol. 28, no. 2, p. 117.

22. Holowenko and Hall, Cam Curvature, *Machine Design*, vol. 25, no. 11, p. 148.

23. R. T. Hinkle, Cam Pressure Angles, *Machine Design*, vol. 27, no. 7, p. 187.

24. L. S. Linderoth, Jr., Calculating Cam Profiles, *Machine Design*, vol. 23, no. 7, p. 115.

25. A. S. Gutman, Cam Dynamics, *Machine Design*, vol. 23, no. 3, p. 147.

26. H. H. Pan, Cam Curves and Cutter Pitch Curves, *Machine Design*, vol. 29, no. 14, p. 137.

27. F. R. E. Crossley, "Dynamics in Machines," Ronald.

28. J. B. Hartman, "Dynamics of Machinery," McGraw-Hill.

29. A. R. Holowenko, "Dynamics of Machinery," Wiley.

30. A. Svoboda, "Computing Mechanisms and Linkages," McGraw-Hill.

31. Rosenauer and Willis, "Kinematics of Mechanisms," Associated General Publications, Sidney, Australia.

32. J. S. Beggs, "Mechanism," McGraw-Hill.

33. Doughtie and James, "Elements of Mechanism," Wiley.

34. R. T. Hinkle, "Kinematics of Machines," Prentice-Hall.

35. A. H. Church, "Kinematics of Machines," Wiley.

36. J. H. Billings, "Applied Kinematics," 3d ed., Van Nostrand.

37. A. Sloane, "Engineering Kinematics," Macmillan.

38. Mabie and Ocvirk, "Mechanisms and Dynamics of Machinery," Wiley.

39. R. C. Johnson, Geneva Mechanisms, *Machine Design*, vol. 28, no. 6, p. 107.

40. "Transactions of the First Conference on Mechanisms," published by *Machine Design;* also in vol. 25, no. 12, pp. 174–220.

41. "Transactions of the Second Conference on Mechanisms," published by *Machine Design.*

42. "Transactions of the Third Conference on Mechanisms," published by *Machine Design.*

43. F. Seybold, Mathematical Analysis of the Geneva Movement, *Am. Machinist*, vol. 64, no. 20, p. 793.

44. S. Dudnick, Angular Relationship of a Geneva Mechanism, *Machine Design*, vol. 28, no. 24, p. 91.

45. A. H. Candee, Kinematics of Disk Cam and Flat Follower, *Trans. ASME*, vol. 69, p. 709.

46. A. E. R. de Jonge, A Brief Account of Modern Kinematics, *Trans. ASME*, vol. 65, p. 663.

47. F. D. Jones, "Ingenious Mechanisms for Designers and Inventors," vols. I, II, The Industrial Press.

48. H. L. Horton, "Ingenious Mechanisms for Designers and Inventors," vol. III, The Industrial Press.

49. Morse, Ip, and Hinkle, Dynamic Characteristics of Mechanisms, *Machine Design*, vol. 27, no. 6, p. 169.

50. F. D. Furman, "Cams, Elementary and Advanced," Wiley.

51. Ham, Crane, and Rogers, "Mechanics of Machinery," 4th ed., McGraw-Hill.

52. G. J. Talbourdet, personal communication.

53. Devanit and Hartenberg, A Kinematic Notation for Lower-pair Mechanisms Based on Matrices, *Trans. ASME*, vol. 77, p. 215, *J. Appl. Mechanics.*

54. H. G. Conway, Straight-line Linkages, *Machine Design*, vol. 22, no. 1, p. 90.

55. L. F. Welanetz, Graphical versus Computer Techniques, *ASME Paper 56-A-48.*

56. M. Goldberg, New Five-bar and Six-bar Linkages in Three Dimensions, *Trans. ASME*, vol. 65, p. 649.

57. Keator and Crossley, 3-D Mechanisms, *Machine Design*, vol. 27, no. 9, p. 204.

58. Hrones and Nelson, "Analysis of the Four Bar Mechanism," Wiley.

59. Hinkle, Ip, and Frame, Acceleration in Mechanisms, *Trans. ASME*, vol. 77, p. 222, *J. Appl. Mechanics*.

60. Hall and Ault, How Acceleration Analysis Can Be Improved, *Machine Design*, vol. 15, no. 2, p. 100, no. 3, p. 90.

61. Wolford and Hall, Second-acceleration Analyses of Plane Mechanisms, *ASME Paper* 57-A-52.

62. T. P. Goodman, An Indirect Method for Determining Accelerations in Complex Mechanisms, *ASME Paper* 57-A-108.

63. Publications of the American Gear Manufacturers Association (AGMA), available to members.

64. ASA Standard B6.1-1932., Spur Gear Tooth Form.

65. The Fellows Gear Shaper Co., "The Involute Curve and Involute Gearing."

66. Earle Buckingham, "Analytical Mechanics of Gears," McGraw-Hill.

67. D. W. Dudley, "Practical Gear Design," McGraw-Hill.

68. Barber-Colman Co., "Hob Handbook."

69. The Fellows Gear Shaper Co., "The Internal Gear."

70. Michigan Tool Co., "Involutometry and Trigonometry," Denjam & Co.

71. Earle Buckingham, "Spur Gears," McGraw-Hill.

72. A. H. Candee, A Simple Method of Determining the Thickness of Involute Gear Teeth, *Machine Design*, vol. 28, no. 10, p. 90.

73. A. H. Ahlberg, Gear Tooth Action, *Machine Design*, vol. 26, no. 7, p. 121.

74. Hans Jeans, Designing Cam Profiles with Digital Computers, *Machine Design*, vol. 29, no. 22, p. 103.

75. F. W. Kinsman, Designing Nonstandard Spur Gears, *Machine Design*, vol. 27, no. 6, p. 195.

76. E. P. Pollitt, Motion Characteristics of Slider-crank Linkages, *Machine Design*, vol. 30, no. 10, p. 136.

77. Wolford and Hall, Second-acceleration [Jerk] Analyses of Plane Mechanisms, *ASME Paper* 57-A-52.

78. H. E. Merritt, "Gears," Pitman.

79. Oliver Saari, Designing Right-angle Helical Gears, *Machine Design*, vol. 25, no. 7, p. 145, July, 1953.

80. W. A. Tuplin, Designing Crossed Helical Gears, *Machine Design*, vol. 22, no. 8, p. 125, August, 1950.

81. W. D. Nelson, Spiroid Gearing, *ASME Paper* 57-A-162.

82. A. S. Beam, Beveloid Gearing, *Machine Design*, vol. 26, no. 12, p. 220.

83. M. F. Spotts, Enlarged Center Distance System for Improved Gear Tooth Forms, *Prod. Eng.*, vol. 16, p. 339.

84. A. H. Candee, Formulas for Involute Curve Layouts, *Prod. Eng.*, vol. 19, p. 145.

85. Anonymous, Packaged Speed Reducers and Gear Motors, *Machine Design*, vol. 29, p. 122.

86. Earle Buckingham, "Manual of Gear Design," The Industrial Press.

87. C. C. Stutz, "Formulas in Gearing," Brown & Sharpe Mfg. Co.

88. Laughlin, Holowenko, and Hall, Epicyclic Gear Systems, *Machine Design*, vol. 28, no. 24, p. 129.

89. R. N. Abild, Epicyclic Gear Systems, *Machine Design*, vol. 27, no. 8, p. 171.

90. E. I. Radzimovsky, Planetary Gear Drives, *Machine Design*, vol. 28, no. 3, p. 101.

91. W. A. Tuplin, Compound Epicyclic Gear Trains, *Machine Design*, vol. 29, no. 7, p. 101.

92. V. W. Bolie, A New Method for Producing Continuously Adjustable Gear Ratios, *ASME Paper* 57-A-146.

93. W. S. Miller, Speed Reducers and Gear Motors, *Machine Design*, vol. 29, no. 6, p. 121.

94. L. F. Spector, Mechanical Adjustable Speed Drives, *Machine Design*, vol. 27, no. 6, p. 178.

95. R. C. Rodgers, "Adjustable Speed Drives," published by *Machine Design*.

96. A. Benson, Gear-train Ratios, *Machine Design*, vol. 30, no. 19, p. 167.

97. J. W. Edgemond, Jr., Epicyclic Gears for Control Mechanisms, *Prod. Eng.*, vol. 28, no. 2, p. 194.

98. Creech and Randle, Planetary Gear Train Analysis, *Machine Design*, vol. 22, no. 5, p. 101.

99. M. E. Cushman, How Planetary Reducer Aids Compact Design, *Machine Design*, vol. 17, no. 4, p. 135.

100. H. E. Golber, Rollcurve Gears, *Trans. ASME*, vol. 61, p. 223.

101. H. E. Peyrebrune, Application and Design of Noncircular Gears, *Machine Design*, vol. 25, no. 12, p. 173.

102. W. W. Sloane, Utilizing Irregular Gears for Inertia Control, *Machine Design*, vol. 25, no. 12, p. 193.

103. F. W. Hannula, Designing Noncircular Surfaces, *Machine Design*, vol. 23, no. 7, p. 111.

104. P. Grodzinski, Applying Eccentric Gearing, *Machine Design*, vol. 26, no. 7, p. 147.

105. Lockenvitz, Oliphint, Wilde, and Young, Noncircular Cams and Gears, *Machine Design*, vol. 24, no. 5, p. 141.

106. J. H. Billings, Computer Mechanisms, *Machine Design*, vol. 27, no. 3, p. 213. See reference **36**.

107. Pike and Silverberg, Designing Mechanical Computers, *Machine Design*, vol. 24, no. 7, p. 131, no. 8, p. 159.

108. P. T. Nickson, A Simplified Approach to Linkage Design, *Machine Design*, vol. 25, no. 12, p. 196.

109. Pike, Silverberg, and Nickson, Linkage Layout, *Machine Design*, vol. 23, no. 11, p. 105.

110. Hain, K., Drag-link Mechanisms in Synthesis, *Machine Design*, vol. 30, no. 13, p. 104.

111. R. S. Hartenberg, Complex Numbers and Four-bar Linkages, *Machine Design*, vol. 30, no. 6, p. 156.

112. Wolford and Nicklas, Linkage Design Techniques, *Prod. Eng.*, vol. 25, no. 12, p. 133.

113. P. T. Nickson, Linkage Design, *Machine Design*, vol. 25, no. 7, p. 137.

114. Shaffer and Cochin, Synthesis of the Four-bar Mechanism When the Position of Two Members Is Described, *Trans. ASME*, vol. 76, p. 1137.

115. Hartenberg and Denavit, Kinematic Synthesis, *Machine Design*, vol. 28, no. 18, p. 101.

116. W. J. Carter, Kinematic Analysis and Synthesis Using Collineation-axis Equations, *Trans. ASME*, vol. 78, p. 1152.

117. F. Freudenstein, Approximate Synthesis of Four-bar Linkages, *Trans. ASME*, vol. 77, p. 853.

118. F. Freudenstein, On the Maximum and Minimum Velocities and the Accelerations in Four-link Mechanisms, *Trans. ASME*, vol. 78, p. 779.

119. Hall and Tao, Linkage Design—a Note on One Method, *Trans. ASME*, vol. 76, p. 633.

120. J. Boehm, Four-bar Linkages, *Machine Design*, vol. 24, no. 8, p. 118.

121. Devanit and Hartenberg, Systematic Mechanism Design, *Machine Design*, vol. 26, no. 9, p. 167.

122. A. S. Hall, Jr., Alternate Four-bar Mechanisms, *Machine Design*, vol. 30, no. 9, p. 133.

123. F. Freudenstein, An Analytical Approach to the Design of Four-link Mechanisms, *Trans. ASME*, vol. 76, p. 483.

124. Kaplan and Pollick, Kinematic Analysis, *Machine Design*, vol. 26, no. 1, p. 153.

125. P. Grodzinski, Mechanisms for Uniform-velocity Reciprocating Motion, *Machine Design*, vol. 26, no. 2, p. 141.

126. Hain and Marx, How to Replace Gears by Mechanisms (Linkages), *ASME Paper* 58-SA-33.

127. G. Talbourdet, Mathematical Solution of 4-bar Linkages, *Machine Design*, vol. 13, nos. 5–7.

128. F. R. E. Crossley, 3-D Mechanisms, *Machine Design*, vol. 27, no. 8, p. 175.

129. A. H. Rzeppa, Universal Joint Drives, *Machine Design*, vol. 25, no. 4, p. 162.

130. H. H. Mabie, Constant Velocity Joints, *Machine Design*, vol. 20, no. 5.

131. F. Freudenstein, Four-bar Function Generators, *Machine Design*, vol. 30, no. 24, p. 119.

132. F. Freudenstein, Structural Error Analysis in Plane Kinematics Synthesis, *ASME Paper* 58-SA-12.

133. S. J. Tracy, Jr., "How to Solve Problems in Engineering Kinematics," published by author.

134. Earle Buckingham, "Gear Ratio Tables for 4-, 6-, and 8-gear Combinations," Industrial Press.

135. Hartenberg and Denavit, Cognate Linkages, *Machine Design*, vol. 31, no. 8, p. 149.

136. G. W. Michalec, Analog Computing Mechanisms, *Machine Design*, vol. 31, no. 6, p. 157.

137. Kaplan and Korth, Cyclic Three-gear Drives, *Machine Design*, vol. 31, no. 6, p. 185.

138. E. H. Schmidt, Cycloidal-crank Mechanisms, *Machine Design*, vol. 31, no. 7, p. 111.

139. D. C. Allais, Mirror Image Cams, *Machine Design*, vol. 31, no. 2, p. 136.

140. C. W. Allen, Point-position Reduction, *Machine Design*, vol. 31, no. 2, p. 141.

141. V. M. Faires, "Kinematics," McGraw-Hill.

Index

Date Due